POST-INDUSTRIAL SOCIETY

Edited by

A. Bruce Boenau
Katsuyuki Niiro

Gettysburg College
Senior Scholars' Seminar 1982-83

UNIVERSITY
PRESS OF
AMERICA

LANHAM • NEW YORK • LONDON

Copyright © 1983 by

University Press of America,™ Inc.

4720 Boston Way
Lanham, MD 20706

3 Henrietta Street
London WC2E 8LU England

Co-Published by arrangement with
Gettysburg College.

Library of Congress Cataloging in Publication Data

Senior Scholars Seminar (1982-83 : Gettysburg College)
 Post-industrial society.

 Bibliography: p.
  1. Technology—Social aspects—Congresses.
2. Industry—Social aspects—Congresses. I. Boenau,
A. Bruce. II. Niiro, Katsuyuki.
T14.5.S46   1983   303.4'83   83-19668
ISBN 0-8191-3613-1 (alk. paper)

HN
49
.V64
B48
1991

TABLE OF CONTENTS

iii

PREFACE

In 1974 the Gettysburg College faculty established an interdisciplinary seminar for outstanding seniors specially selected by the faculty Committee on Interdepartmental Studies. Each year, under the direction of two faculty members, the Senior Scholars' Seminar investigates a topic bearing on the future of humankind.

In 1982-83 the topic for this seminar was "Post-Industrial Society." Although the catchwords of a day are often no more than momentary fads, Harvard sociologist Daniel Bell's term for contemporary society in the United States, parts of Western Europe, and Japan seems to have endured. His book, The Coming of Post-Industrial Society (1973), was a seminal work which gave rise to countless studies around the world. Nonetheless, as a term, "post-industrial society" is imprecise, is variously used, and survives because people cannot agree on what should take its place. It refers to a new society emerging out of and superseding an industrial society, but does not specify its nature. Bell himself carefully avoided adopting a single label to denote the essence of this new society.

Some scholars point to technological developments as the key to understanding the changes in advanced industrial societies. Computers, television, and communications satellites - to mention but a few of the dazzling innovations transforming our existence - have both accelerated and expanded the processing and transmission of information. New lines of enterprise utilizing the new technologies have risen to the fore, displacing the earlier primacy of agriculture and, later, manufacturing. Some writers emphasize the growth of the service sector, which now accounts for a majority of the work force in some developed countries. Bell, acknowledging the importance of the technological changes and the rise of the service sector, stresses the central importance of information as a strategic resource in the post-industrial era. Others point to affluence as both the precondition for, and the predominant characteristic of, the new society. Finally, some authors emphasize the possibility of striving for "post-materialistic" goals - like the quality of life or individual self-actualization - in a society which has answered such primary questions of existence as how to

v

provide for safety and satisfy material needs.

One can spot ample room for controversy in such attempts to isolate a single characterizing feature of post-industrial society. That society may be all of these: a high technology society, a service economy, an information society, an affluent society, and more. Yet, one ought not to prefer an easy eclecticism to more rigorous attempts to achieve a systematic conception of the structural components of post-industrial society. When that is understood, perhaps a new term will take the place of what now can only point to, but not fully conceptualize, a social transition.

The Senior Scholars read several books and articles in common, attended lectures and films, met in discussion with prominent resource persons and with each other, and explored an individual topic in depth. It is our pleasure now to present the results of their efforts, a series of papers on major topics relating to post-industrial society.

We are sincerely grateful to a number of people whose efforts meant much to the success of the seminar:

Visiting resource persons: Daniel Bell, Harvard University; Roger Benjamin, University of Minnesota; Ronald Inglehart, University of Michigan; Thomas Pepper, Hudson Institute; John Wicklein, Corporation for Public Broadcasting.

Members of the Gettysburg College faculty and staff: Kim S. Breighner, David T. Hedrick, Janet Hertzbach, Anna Jane Moyer, Samuel A. Mudd, Ralph A. Sorensen, and William P. Wilson.

And we wish to acknowledge our admiration and affection for the students who comprised this seminar. It was a special pleasure for us to work with such talented and likable people.

May, 1983   A. BRUCE BOENAU
                                             KATSUYUKI NIIRO

THE EXPANDING FIELD OF ROBOTICS AND

ITS EFFECT ON EMPLOYMENT IN THE UNITED STATES

Denice Andrews

As technology continues to hurtle mankind into the latter half of the twentieth century, the complexities of automation are becoming manifest in the ever-widening field of robotics. The use of robots in industries, particularly in repetitious assembly line production, has become economically feasible due to the increasing costs of human labor. However, this technological advance also entails many problems. The most obvious one concerns the subsequent employment and adjustment of those persons displaced by robots. This question and others raised in connection with the utilization of robots are becoming increasingly important. Industries in the United States are turning to the employment of robots in an attempt to meet the competitive demands of the international market of the 1980's. The advantages and disadvantages of robots will be dealt with following a general analysis of the current situation in the United States' robot industry. The effects of robotics on U.S. labor, unions, and the government will also be examined to aid in the development of a few techniques to help ease the growing pains surrounding the aggrandizement of robotics.

The Origins of Robotics

The word "Robot" comes from a Czech word meaning "Worker". It was first introduced in the play "R.U.R." written by Karl Capek in 1921. R.U.R. stands for Rossum's Universal Robots which were artificial men created by Rossum to work for mankind. Although originally designed to be laborers, society's leaders discover that the robots make excellent soldiers. Eventually, the robots rebel against being used to fight wars by destroying mankind and taking over the world.[1] This plot reflects the "Frankenstein motif" used to create the terrifying robots who always desire to harm humanity.

Isaac Asimov, a science fiction writer, invented the word "Robotics" to describe the technology used to construct

1

his fictional robots. He also developed "the three laws of Robotics" which have become standard in many science fiction robot stories:

1. A robot may not injure a human being, or, through inaction, allow a human being to come to harm.
2. A robot must obey the orders given it by human beings except where such orders conflict with the first law.
3. A robot must protect its own existence as long as such protection does not conflict with the first or second law.

These laws were established to dispel the negative aura which usually surrounds man's concept of robots. Many of today's unemployed may argue that Asimov's laws are being broken as robots advance into American industries. Robotic developments allow greater numbers of humans to be replaced, destroying human dignity, income, and skills. Is this not injurious to mankind? Asimov wrote many short stories dealing with a future world of robots, constantly addressing the conflict between human and mechanical, artificial beings. The robots flourished on the outer planets and represented a technological salvation for the earthmen headed for certain destruction on an overloaded planet. In today's society a more detailed study of America's robots is necessary before judgment can be passed on these technological wonders.

Robots for Today and Tomorrow

In the real world of the 1980's the interpretation of robotics is slightly different from Asimov's conceptions. The Robot Institute of America defines a robot as a "Reprogrammable multifunctional manipulator designed to move material, parts, tools, or specialized devices, through variable programmable motions for a variety of tasks."[2] The key word is reprogrammable. This concept makes the robot more practical since it can be adapted to a variety of functions as the demands of society alter the styles or materials used in production. This versatility separates the other automated machinery from the superiority of robots.

There are currently only about 22,500 robots in the world. Japan, which utilizes almost 14,000 of the mechanical beings, is the leader in this technological development.[3] The United States has roughly four thousand robots and three-quarters of them are employed in the auto industry.

Substantial growth in the production of robots is expected throughout the nineteen eighties and nineties. Robot sales are forecasted to grow 30 to 35 percent per year and the American market alone is projected to be worth $2 billion by 1990. The fastest growing sector of the robot market is expected to continue to be the assembly line robots, which could account for as much as one-half of the total sales by 1990.[4]

The auto industry has primarily used robots for welding and painting purposes. These jobs are dangerous and thus far laborers have not regretted losing them to the mechanical workers. In fact one ex-painter for General Electric admits feeling bitter but then actually grateful to the robot replacement for possibly saving his life.[5]

However, robots are starting to invade more territory. Will labor resistance continue to be low when robots perform more desirable jobs? A recent study conducted by Carnegie-Mellon University concluded that robots could execute four million of the nine million existing factory jobs by 2000. One million of these jobs, i.e. welding and painting, could be done by simple robots and three million more jobs would require the services of "seeing and feeling" robots. These more complex robots would be capable of performing jobs in assembly, packaging and inspecting.[6] Once these smarter robots are perfected, 65 to 75 percent of the entire factory labor force could eventually be replaced.[7] However, such an occurrence is closer to the twenty-first century than to the present time frame.

Today the least expensive computer controlled industrial robot with five servoed degrees of freedom, one parallel jaw gripper and a crude sense of touch and vision costs $60,000 and is incapable of lifting even 1/10 its own weight.[8] So the robot is far from perfected. Yet as the microchip continues to become less expensive and rapid technological advances continue to occur, an affordable, "smarter" robot may soon be available.

Currently, many robots are "taught" the necessary mechanics of their job by first being led through the motions by a human in order to set the program in their

3

computer brain. Then this program is replayed and the robot can perform the movements on its own. Another method used to program the robot's actions is through a Cartesian Coordinate System in a three dimensional field (i.e. X,Y,Z axes). The points of the motion are programmed into the computer which controls the functions of the robot from point to point.

Although these methods are serviceable, the prime impediment to widespread industrial use of robots remains their lack of sensory perceptions which would allow them to adjust to unstructured work conditions. One of the first uses of industrial robots was to unload diecasting machines. No vision sensors are necessary for this job because the part to be grasped is always in the exact same position. This job was tedious as well as physically demanding, due to the weight of the parts and the fumes emitted by the machines.

Automobile spot welding has become one of the largest uses of industrial robots. In fact, over 50 percent of the cars built today are spot welded by robots. Auto manufacturers continue to invest in robotics, striving to implement them in a greater variety of tasks.

Therefore, robot technicians are always trying to develop a cheaper and smarter model. Unimation, America's largest robot manufacturer, has recently developed a new robot called "PUMA", which stands for "Programmable Universal Machine for Assembly". It is more intelligent than some of their earlier models and contains a cheap microprocessor brain which gives the robot a rudimentary sense of touch. This robot costs only $35,000 and is promptly being purchased by auto manufacturers.[9]

Escalating labor costs coupled with the decreasing costs of computer technology have ignited a growing interest in robots. Within the last few years their popularity has become quite noticeable. The Robot Institute of America, formed in 1974, holds an annual conference to demonstrate the new models of robots for the coming year. This past year 27,000 people attended the Robots VI Conference compared to only 6,000 participants in 1981.[10] This increase is understandable, because only recently has the high cost of wages and benefits made the expense of robot investments economical. This type of capital investment

4

also requires a number of years for profits to be realized. Unimation, responsible for 41 percent of U.S. robot production, manufactured robots for 14 years before showing a profit. Other companies are also just recently reaping a profit from their robots.[11] Projections are not unreasonable that more than 1/3 of assembly work will be roboticized by 1990 in comparison to the 5 to 10 percent of this type of work being done by robots now.

Competition from Japan may be one of the major reasons behind the burgeoning U.S. interest in robotics. The U.S. auto industry just happens to be one of the biggest investors in robots, and it is certainly experiencing sales difficulties from Japanese imports. Its current interest in robots may be due to a desire to become more competitive with Japan's lower labor costs or a result of the easy adaptability of assembly line production runs to robot installations.

At a Chrysler assembly plant in Delaware, one electrician oversees thirty robot welders. Where thirty men once welded sixty cars per hour, robots can now do the job more accurately and turn out one hundred cars per hour.[12]

General Motors has also experienced increased productivity which stemmed from roboticization. It reports that one robot replaces 1.7 workers in an assembly plant and 2.7 workers in a manufacturing plant.[13] By 1983, GM plans to invest $200 million in its most massive high technology automation program to date. General Motors currently employs 150 robots which are used mainly to assemble the bodies of their new X car compacts. The new robots will be able to do much more. Over the next three years 800 robots will be installed to weld the bodies of new cars. GM is in the process of modeling its plant after the Italian auto company, Comau. Using the Italian system, called Robogate, a computerized clamping system welds the car together in two or three steps. It used to take human welders forty to fifty steps to accomplish this process. This system will cut production costs by 70 percent and reduce body rattles and vibrations. By 1990, General Motors will have invested $1 billion to install 13,000 more robots. These robots include 5000 PUMA robots which will complete most of their automobile assembly work, 4000 more robots will load and unload machines and another 1500 robots will paint the cars.[14]

5

The auto industry is not the only one investing in robotics. Rumors have been circulating that General Electric plans to install robots which will displace 15,000 workers and GE confirms that it is initiating a sweeping automation program.[15] The installation of robots is moving at a rapid pace. In the beginning of 1979, the company used only 2 robots. But by the end of the year, GE had installed 26 robots and the company would like to have 1000 robots spread throughout its 8 plants by the late 1980's. This increase will lead to the eventual replacement of one-half of General Electric's assembly workers.[16] GE has improved the quality of its products as well as increased productivity and, consequently, expects to begin realizing a profit from this investment in technology in two years. So far the union has not objected, because the jobs the robots have taken were considered undesirable. The workers have simply been retrained for other jobs.

However, robots are now beginning to enter some of the light assembly line jobs where 40 percent of General Electric's blue collar workers are employed.[17] GE may not be able to retrain all of these workers. Many older laborers may be incapable of learning new technological skills or there just may not be enough jobs for human laborers to perform. These technological advancements may cause considerable discord with the union and slow General Electric's projected schedule. Depending on the cost of settling with the union, it may take longer than two years to pay back the investment.

Nevertheless, the American market for robots is starting to flourish. The American robot market is also becoming very competitive. Unimation is having trouble filling all its orders so competitors are rapidly entering the market. Westinghouse, GE, and Cincinnati Milacron are giving Unimation some opposition. However, the toughest competition has come from IBM. Unlike General Electric and Westinghouse, which are dependent on technology from abroad, IBM has an edge in the research and development area since it has been making its own robots for over a decade. Over the past six months, IBM has quietly supplied fifteen American firms with its robots. It is currently testing an assembly line robot which is superior to the other models in three ways: 1) It can do more delicate and precise work; 2) Its computer brain is larger and contains a more intelligent

6

software system; and 3) It can monitor as well as correct its own mistakes.[18] However, IBM's robot is still a member of the current generation of robots.

Vision is considered to be the key to the next generation of robots. This development would allow the work environment to be less structured and less precisely arranged. Instead of having to position the piece exactly where the robot's arm has been programmed to pick it up, the robot could "see" and pick the correct piece out of a bin, for example.

Several vision systems are currently being tested. The British Science Research Council (BSRC) has created a robot which utilizes a parallel array processor composed of many microprocessors. However, this mechanism costs $140,000 and ideally requires a system 100 times more powerful to make it effective in its operations.[19] As the price of silicon chips decreases, the BSRC hopes to reduce costs and make this type of robot economically available to company investors. Another vision system, the SRI International, includes a television camera, a computer interface, a microcomputer, and a software package in its mechanical hardware. Is it complicated? Yes, and so research continues as roboticists strive to create a robot which can perform all the amazing activities humans do so easily everyday.

A tactile sense would also greatly increase the potential application of robots. They would be able to sense when a gripper did not operate correctly and call for a human supervisor to repair it. Researchers at MIT have developed artificial skin for robots which is composed of sheets of rubber laced with wire. Electric currents pulsating through the wires in the top layer pass the current down to lower layers as pressure is applied. A microprocessor records the voltage of the different sheets of rubber forming an image of the object the robot is "touching".

Microphones are being adapted to robotic uses so a robot can respond to simple voice commands by converting sound waves into numerical sequences which are then compared to number sequences stored in its computer memory. Many experiments are being conducted in this area on computer language systems as technicians attempt to develop more "friendly" software and dispel some of the human fear of machines.

Thinking robots may still be a product of the distant future. But, meanwhile, robots will continue to become practical in other areas. Joseph Engelberger, president of Unimation, Inc., predicts that robots will be commonplace in the home by the late 1980's.[20] The robot will have to know the floor plan, how to recognize dirt when it vacuums, how the furniture is arranged and what to dust. These tasks require a rather sophisticated system. Therefore, some experts feel that robots will not become household items until the 1990's. By that time the cost of these "Metal Maids" would only be $4000 to $8000 (in 1980 dollars).[21]

However, just last Christmas, Neiman-Marcus advertised a $15,000 robot that was programmed to walk the dog, take out the garbage, water the plants and sweep the floors. This past December one magazine carried an article advertising a "Built-it-Yourself Robot" kit. The completed robot, manufactured by Heath company, can serve drinks, act as an alarm clock, prowl around for intruders and even threaten to call the police if an intruder is discovered. This robot, named "Hero I", also has been programmed with thirty-three phrases built into its memory. "I do not do windows" is just one of its snappy comments. The drawbacks to this little wonder include the fact that it can not lift more than one pound at a time, nor can it climb stairs or recharge its own batteries. "Hero I" can also be hooked up to a computer in order to program it to do more elaborate tasks, such as following orders from a cassette. Heath is also offering a fully constructed "Hero I" for $2495. At this price a complementary 1200 page Robot Education Course is included.[22] "Hero I" is still a very rudimentary type of robot and the price is exorbitant for the type of services performed. But prices are continuing to fall and advances are continuing to occur, making a domestic robot more and more desirable.

Commercially, robots would be ideal for nuclear power plant jobs as well as underwater exploration, mining and construction. The implementation of robots in nuclear power plants would allow the breeder reactor to be permanently sealed, eliminating the threats of sabotage and theft of nuclear materials. Robots could work in radioactive areas to perform emergency repairs without endangering the lives

8

of human laborers.[23]

Robots could easily perform underwater mining and construction jobs. The contraction or expansion of air filled chambers would control the robots' bouyancy allowing them to search the ocean floor for minerals or to operate drilling and mining equipment.

Robots could also be invaluable tools in space exploration and even space colonization. Robots will not need elaborate life support systems and one-way trips would be possible. Space robots could transmit their experiences back to earth for human interpretation if they were equipped with the proper type of sensory perceptions.

In the field of medicine a roboticized arm has been developed which functions quite similarly to a human appendage. Pairs of these robot arms have been mounted on tables and linked to carts which shuttle from the patients' rooms to a storage cart, responding to simple vocal commands.

Many offices have roboticized mail carts which follow a metal strip throughout the building, stopping at each office to deliver the day's mail. The carts are very unflexible as they beep along the way to warn people out of their paths. Unimation Inc. is even working on a robot capable of plucking chickens and, in Australia, roboticists are designing a robot that can shear sheep.

One of the most amazing feats robots have performed so far is taking place in Japan. Fujitsu Fanuc Ltd. opened a robot plant last year which is capable of reproducing robots. About one hundred new robots are "born" each month. Although human laborers are still needed for the final assembly, this operation entails the use of only one hundred workers to keep the plant operating around the clock. This is only one-fifth the number of workers used in a normal plant of a comparable size.[24]

Robotics is definitely spreading throughout America, displacing human laborers. The following section looks at the advantages of robot labor to American society.

The Benefits of a Roboticized Society

The use of robots in factories allows producers to decrease the labor costs associated with the manufacturing of their products. The savings accrued by the use of labor-saving robots can be passed on to the consumer in the form of lower prices. Initially, the construction of completely roboticized factories will be extremely expensive. However, after the first one has been completed, the rest will be easier to reproduce and costs will fall. Also, since robots require no heat, air conditioning, lunchroom or lavatories, overhead costs could be considerably reduced. The chairman of General Motors, Roger B. Smith, recently stated that for each dollar an hour rise in wages, one thousand more robots become economically pratical.[25]

By using robots U.S. companies are able to make their products more competitive abroad and keep work in the United States. For example, the lower labor costs have brought television assembly back into the United States. There is also a homeward drift in the assembly of microchips since machines are able to check the chips more accurately.[26]

Examples of greater productivity are available at plants throughout the world. In Longbridge, England, robots have increased BL's capacity to produce cars by 342,000 a year. Utilizing 70 percent fewer human laborers it was able to reduce production time to only one shift a day. The workers alone were able to produce sixteen cars per person in a year, but with the aid of robots manufactured twenty-three cars per person per year.[27]

Nissan, which is Japan's second largest car manufacturer, currently operates the most advanced robot assembly plant in the world. At its Zama factory 96 percent of the welding is done by robots. Sixty-seven cars were produced per man per year and Nissan expected to increase this figure to eighty-one by the end of 1981. Production takes place at seventy-four automated assembly stations and twelve manned ones. Each of the manned stations contains two workers so only twenty-four people are employed at its Zama plant.[28]

In the United States, Yamazaki Machinery Works Ltd. has recently opened a new $15 million "manless" factory which is referred to as a flexible manufacturing factory. This plant in Kentucky employs only six people and twenty robots to

10

produce sixty machine tools a month. Yamazaki will hire one
hundred people to assemble the parts, but once the "smarter"
robots are on the market their jobs may also become
roboticized.[29]

When robots perform the task, more accuracy is possible
and flawless, high quality products are manufactured every
time. Often the robot can achieve greater precision simply
because the job is physically dangerous for human laborers.
As previously mentioned, robots can work in areas such as
nuclear power plants and mining shafts where human health
and safety standards are questionable factors. They can
also get into tight spots for welding purposes and do an
extremely accurate job.

For instance, when arc welding, a robot can hold the
torch on a particular spot 90 percent of the time compared
to a human accuracy level of only 30 percent. Even though
the robot can work no faster than a human welder, a robot
can produce three times more output per shift than its human
counterpart, because it does not need to take any breaks.
If this robot works for approximately 1/3 the wages of a
human arc welder, the total productivity gain is 900 percent
per shift![30]

Therefore, the initial cost of the robot is justified
in terms of net productivity gain. Although it currently
may cost $70,000 to install the robot and pay for its
accessories, it will be labor saving and will pay for itself
in only 2 years. The robot operating time is 97 to 98
percent while human workers lag by about 9 percent due to
lunch breaks, coffee breaks, and sick days.[31]

The use of robots can be advantageous to the laborer,
the consumer, and the factory owner. There exists an
overall trend toward a total upgrading of skill level in
blue collar and some while collar work. This phenomenon
will be reinforced and pushed along by the introduction of
more robots into the workplace. Alex Mair, a vice president
for the technical staffs of GM corporation, predicts that by
the turn of the century as many as 50 percent of the workers
will be skilled tradesmen compared to the current 16
percent.[32] This can have a beneficial effect on the
well-being of society since many studies have found a
positive correlation between the level of job grade and
mental health.

On behalf of the consumer, not only will lower labor costs and higher productivity lower the cost of products, but there will be a greater choice in factory-made products as robots make it more profitable to utilize small-batch production methods.

This will benefit both the consumer and the producer. The higher quality of the product will tend to increase the quantity demanded at any given price. For the producer this requires less spent on inspection and lowers the costs generated by rejected, poor quality items.

Increased productivity, lower labor costs, and quality products are some of the benefits of a roboticized society. However, as with all technological advances, there exist negative aspects and these problematical areas will be discussed in the next section.

The Costs of Roboticizing America

There are a number of problems associated with the widespread introduction of robots in the U.S. economy. Many people fear the dehumanization of the workplace will affect the mental psyche of the laborers. The human laborer becomes isolated among machines and may lose the ability or even the desire to communicate effectively with his human neighbors. One seems to be trading the repetition of assembly line work for new stresses created by isolation and constant monitoring of machines. Will the human laborer assigned to watch the robots find himself unable to deal with human society? As skills are lost to robotic performance, what will these laborers do?

Although jobs will be created to repair, program and monitor the robots they will not be as numerous as the ones which are lost. There is some dispute as to how severe this gap will be. A general consensus predicts that a higher level of skill will be required for most of the jobs remaining after robots are installed. David Gay, an analyst for General Motors, feels that GM needs more capital and less personnel to compete with foreign firms. More than 200,000 people would be replaced, those least capable of moving to the next skill level. Moreover, these people are accustomed to high pay scales which are unavailable elsewhere.[33]

Michigan will be the hardest hit and is taking steps now to try to prevent its whole economy from collapsing. Michigan is already the largest robot manufacturer in the United States. However, this state is facing the loss of 100,000 jobs by 1985, and only 50,000 jobs are expected to be created in the robot manufacturing industry in the next decade. In order to entice more high technology firms into the state, unemployment benefits have been cut by one-quarter to create tax credits for research and development. A fund of $600 million has also been set aside to provide loans for development. The state government is establishing an education-training institute which plans to conduct $200 million worth of robotic experimentation by 1990. The state educational system is also trying to aid the jobless laborers. The University of Michigan is offering programs to retrain the displaced workers and give them skills for today's changing world.[34]

A recent House Education and Labor Committee report anticipates that in contrast to past economic recessions, "hundreds of thousands" of skilled workers will remain unemployed during the recovery because of structural changes in the American economy.[35] The skills most likely to be needed in the next few decades will probably be those which are relevant in the service sector of the economy. Data Resources, Inc., predicts a massive flow of labor into the service sector. During the next decade, the service industry is projected to grow by 7.5 million jobs. Unfortunately, the unskilled workers will find themselves in the lower strata of the service sector. At a recent Congressional hearing, Roger Vaughan predicted that the unskilled workers will be condemned to minimum wage jobs in the "consumer-service" sector (i.e. custodian or janitor).[36] Even this type of occupation may be performed by robots in the near future. Employment in services, especially health care jobs, has been increasing more rapidly than in any of the manufacturing sectors. The demand for engineers and computer operators also outstrips the current supply. Therefore, a restructuring of our labor training programs is imperative in order to prepare laborers for different types of jobs in a roboticized society. Skills lost to robots must be replaced with new skills which are more relevant to our post-industrial society.

To accomplish this task, new methods of handling the

13

labor surplus will be required to deal with unemployment caused by robot installations. In the past, companies have mainly relied on attrition and early retirements to handle the excess labor created by the introduction of robots into their workplaces. However, this method no longer works fast enough and retraining seems to be the best solution proposed so far. "I'm convinced that retraining will be a major need for American Industry over the next ten years," states Robert Sohl, director of Xerox's International Center for training and Management Development.[37]

A case study done on the printers of New York City's three largest newspapers, three years after the union reluctantly allowed automation to move into the pressroom, has investigated the happiness of the printers following the implementation of new technologies. The terms of the contract assured the printers lifetime employment. The management used many incentives to attempt to push the printers into an early retirement. They offered a bonus to the printers who would retire earlier than their sixtieth birthday and gradually increased the sum as fewer and fewer workers were needed to run the presses. Many of the printers faced a dilemma - should they stay on the job and be bored, or retire and face boredom at home?

In the next few years this problem is going to become even more rampant as greater numbers of robots join the work force. This increase will be slightly offset by a decrease in laborers entering the labor market. Since the children of the baby boom have joined the labor force during the seventies, the number of persons entering the labor market will fall by approximately 5.5 million people in the next 20 years.[38] This decline will alleviate some of the pressure on the labor market. However, during 1981, the number of robots in the labor force was expected to increase 30 percent. Two-thirds to three-quarters of the workers displaced will be less than forty-five years old and most of them will have to go through some sort of retraining program because their particular job skill has been rendered useless by robot performance.

Many of the larger companies in the United States are providing this re-education for their employees. For example, both Ford and GM have agreed to set up a retraining school for 107,000 of their UAW workers.[39] AT&T spends $1.1 billion annually to train 500,000 of its employees and

70 percent of these are employees who must be retrained for new technologies.[40] Just this past year, Unimation trained more than 1500 displaced workers at 440 locations for robot repair and robot programming jobs. The courses typically required thirty-six hours of classes.[41]

Job grades are becoming more centralized as lower skills are eliminated, and skills formerly needed in the higher levels are often reduced. Greater centralization tends to result in a total downgrading of skill requirements. For this reason, many people feel the whole education system in America must change. James O'Toole, a professor at USC, predicts: "Those who will succeed in the work force will be those who have learned to learn."[42] The unthinking jobs will be done by robots. Human laborers will be needed to monitor, repair and program the robots. Therefore, jobs will be created, but the unemployed will probably not be able to fulfill the necessary requirements. This discrepancy is already apparent whenever one glances through the want-ads - as President Reagan has suggested - and still sees double-digit unemployment statistics printed on the front page of the newspaper. The net outcome will be better pay for better jobs following a very painful process to achieve this plateau.

Therefore, with the utilization of robots comes a restructuring of skills required for working. There will be less demand for manual dexterity, physical strength and traditional craftsmanship. Instead, more emphasis is going to be placed on formal knowledge, adaptation and perceptual attitudes. It may be better to leave job training to the businesses, and allow schools to concentrate on teaching the fundamentals (i.e. reading, writing and arithmetic). Unskilled workers will be relegated to babysitting the machines. But the semi-skilled worker is going to be affected the most. Skills acquired on the assembly line are usually of a non-transferable value. This problem was illustrated by the closing of a Mack Truck company in Plainfield, New Jersey. It took the laborers six months to a year to find other jobs, and most of these were at wages considerably lower than those to which they were accustomed.[43]

The problem of increased leisure time also implies the necessity for re-education. The American citizen must learn how to deal with leisure. The Protestant work ethic has

15

been thoroughly indoctrinated into our society. What do we do when there is not enough work to go around? Should Americans band together and stop robotics' progress in order to maintain the American "right" to a job, no matter how repetitious, boring and even physically dangerous that job may be?

As more attention focuses on leisure activities, more jobs may be created. In 1960, 36 percent of America's household spending was for services. By 1978 this figure had increased to 43 percent. As one becomes more affluent one turns to more leisure activities and entertainment. Since prices of services have always increased faster than prices in the manufacturing sector, higher incomes could be expected in these potential occupations.[44]

One may notice a positive correlation between robotization and the present structure of the education system when considering the rising level of skill requirement. More time spent on learning will increase the average age one begins working, and if there are not enough jobs to go around, retirement age should continue to decline. Therefore, our children should be able to look forward to enjoying a greater proportion of their lives not working. Education is also needed in order to help people adjust to retirement. How to deal with leisure time is a fundamental problem facing everyone as robots continue to take over jobs and force people to adapt to increasingly larger chunks of leisure time.

Humans will also have to adapt during their working time. How much satisfaction does one experience working at a job with robots for partners? The isolation from other humans has already been mentioned as a detrimental effect on outside personal relationships. However, a company in Sweden has experienced a marked improvement in worker attitude after the installation of robots. The robot maintenance man proved to be the most loyal and faithful laborer, missing fewer days of work than those working on the assembly line. Before the introduction of robots, group production had been tried in an attempt to make the work less boring and give the worker a greater sense of satisfaction in seeing the "whole" product constructed rather than just a portion of it. This plan did not have much of an effect, at least not in comparison to the introduction of robots.[45] Japanese workers also become attached to their robots and have started to name them after their favorite movie stars. Oftentimes pride in the laborer's abilities is transferred to the machine's cost and its capabilities.

Although many workers find their new jobs physically easier yet mentally challenging and interesting, many others find themselves bored and less mentally stimulated. Those older workers who are unable or unwilling to learn new skills will find themselves pushed into the unemployment lines. The loss of one's job is a traumatic experience, tearing up the lives of the unemployed worker and his family. The lost sense of worth often leads to deep depressions which can cause serious mental illnesses.

In areas experiencing high levels of unemployment instances of alcoholism, child abuse, wife beating, suicide and mental illness are rising. In one study conducted in Wisconsin, counties which experienced a sizable increase in unemployment from 1979 to 1981 also witnessed a 69 percent rise in abuse cases compared to a 12 percent increase in counties with lower unemployment growth rates.[46] Although joblessness alone is not considered to be the sole cause behind these mental abnormalities, many studies over the last decade have discovered a definite link between unemployment and some types of violence and sicknesses.

Some people point to the decreased need for physical exertion as a potential source of suppressed tension. Mental fatigue requires longer periods of recuperation than does physical fatigue. The shorter working day may be necessary in order to preserve one's mental well-being. Investigations by W.A. Faunce have discovered that in an automated setting workers prefer shorter working days to increased wages.[47] This reduction may decrease some of the resistance to the implementation of robots in industries as people come to value their leisure time more than income advancements.

If greater emphasis is placed on improvements in technology for greater productivity, labor will tend to be less exploited and the need for unions may decline. Perhaps this is the answer to the question of why union membership in the United States has remained constant for the past five years. The role of labor unions in a roboticized economy is the topic of the following section.

The Future Role of Labor Unions

American labor union membership currently represents less than 21 percent of the entire labor force. This percentage is the smallest share of the work force since World War Two.[48]

To return to the case study of the printers for a moment, the effects of automation on union membership were clearly illustrated in the decade following the capitulation to modern technological practices. Their union was powerful enough to keep automation from entering the New York City pressroom for ten years. The union finally acquiesced after negotiating very good terms for their members. The printers were guaranteed a job for the remainder of their working days. However, by the time their eleven year contract runs out, in 1985, there will only be one-quarter as many printers working as when the contract was signed.[49]

This decline in working membership is creating a massive problem for stocking the pension fund as most of the workers are retiring without many laborers entering the profession. The power of the printer's union has been greatly debilitated and it will probably have to merge with other craft unions if it wants to retain any sort of functional service.

The unions can not resist the robots too much longer. Jobs will be lost to automated competitors if technological advances continue to be locked out of industries in the United States. At the moment those industries which are currently investing most heavily in robotics experienced a considerable decline in employment ten years ago. For example, from 1967 to 1977, employment in primary metal and transportation equipment industries decreased substantially.[50] Both of these industries affect the auto and steel companies which have fallen behind foreign competition and are presently investing heavily in robotics. The most the unions can hope to achieve is a good settlement for their members to help ease the trauma of the transition to robotics.

New technologies will no doubt be the central issue in collective bargaining during the eighties. In fact, in 1980, AT&T agreed to set up twenty-five union-management committees to help smooth over the transition to new equipment. The United Auto Workers already has made an agreement which restricts management from shifting jobs to management through the incorporation of new technologies. It still expects union membership to drop to only 800,000 from 1,000,000 by 1990, even after optimistically allowing for a 1.8 percent increase in domestic auto sales.[51] Just this past June, GE signed a contract with 13 of its labor unions agreeing to pay up to $1800 of tuition costs for retraining each worker displaced by robots.[52]

Japan is also starting to hear worried outcries from its unions. Consequently, the government has instigated a two-year study of automation and employment conditions to examine the effects that utilization of robots is having on their employment figures. Japanese guaranteed lifetime employment with their companies have never complained before about the use of robots. In the past, workers whose jobs were taken over by automation were simply transferred to another part of the factory. Now, moving workers to other factories often requires them to relocate in other towns and give up their homes and friends. Japan's current rate of unemployment is only about 2.3 percent but, with the increasing implementation of robots in their factories, many business leaders are beginning to worry about a potential rise in this figure.

At a Sanyo television assembly plant, one hundred workers were replaced by robots. However, most of them were women so Sanyo simply reduced the number by attrition as the women got married.[53] So far neither the unions, government, or the industries have made many efforts to retrain threatened workers. The Japanese usually assume that an industry will take care of its laborers but, in a saturated market, help may be needed from elsewhere.

The power of the unions to make and win demands is certainly being sapped by robotics progress. The strike will no longer be an effective tool of negotiation, as employers - backed by their robots - become capable of operating with few laborers for longer periods of time. New means will have to be designed for productive negotiating.

The merging of unions is one answer to decreasing membership. Another solution is to try to organize the

19

white collar workers who have previously resisted
unionization. However, union leaders are facing an uphill
battle in this respect, because many industries are moving
to the south and southwest where opposition to unions has
historically been rather blatant. Are the unions an
integral part of labor organization under new technologies,
or do they simple represent a system no longer needed
today?

There has been a noticeable shift among workers'
priorities from concern over salaries to concern over the
workplace. Management also recognizes this demand and the
need for unions appears to be waning. Government aid might
be able to provide the worker with enough help and replace
the increasingly limited functions of the labor union. The
next section deals with the role of the government in the
U.S. post-industrial society as robotic developments allow
robots to enter more industries.

The Future Role of American Government

Because Japan has already integrated more robots into
their society, a comparison of the two nations' methods will
be used to develop a few national programs which would aid
the displaced worker and help ease the integration of robots
into the American system.

The Japanese government has played a major role in the
phenomenal growth of Japan's industries since World War Two.
For example, the Japanese government helped create a robot
market so manufacturers could produce robots in volume. It
also set up a leasing system so the robots could be used and
tested by small companies. In this manner, the Japanese
government assumed the risk that private capital is either
unable or simply unwilling to accept. It does not subsidize
the operations per se, but rather the risk involved in the
formation of the company.

At a Congressional hearing concerning the outlook for
the U.S. auto industry, Donald Ephlin, director of the UAW's
Ford department, recently stated that American management
and unions are as good, if not better than, their Japanese
counterparts. However, the United States government does
not support its industries as well as Japan's government
aids its companies.[54] Because the return on an investment
in robots takes at least two years to realize, many American

management teams have stalled their implementation by looking only for short-run profits. Government aid to off set some of the risk would allow management to plan long-run projects.

Some Japanese companies such as Hitachi and Kawasaki are experiencing difficulties selling their arc welding and loading machine robots to mass production factories in Japan. Some complain that the ability to make robots is ahead of the industries' capacity to utilize them. Also many companies prefer to make their own robots, allowing them to cut costs and tailor the robots to their particular needs.[55]

Nevertheless, the National Science Foundation and the National Bureau of Standards have begun to emulate the Japanese system by investing close to $1.5 billion on research and development.[56] Invention of new technologies has always been the United States' forte. Japan has prospered by simply innovating present technologies and sacrificing complexity for speed. Relying on simple conveyor belts and existing technology, Japan is rapidly automating its factories.

The West seems to be proceeding at a slower pace, using a gradual approach to develop more sophisticated, easily adaptable machines. This strategy often requires a more advanced technology which has not been perfected yet. Software programs must be designed to deal with larger, more complex systems which will make the robots more intelligent, more like the human labor they are expected to replace. Government incentives should be offered which would aid the industries' R & D investments in this complex type of robot.

TRW Choate favors government incentives in the form of tax credits to businesses.[57] The government could also involve itself in the training and retraining of laborers to make them better suited for the changing economy. By paying perhaps 25 percent of a company's retraining costs the government could work with industries to channel displaced laborers into the proper fields.

The present administration is not moving too swiftly toward this objective. In 1981, the Labor Department received only $4.1 billion for its fiscal budget to cover retraining costs while private industry expended $30 billion.[58] The 1983 budget allocations severely cut the funds available for retraining purposes. Money for programs which would replace CETA and vocational education systems has been reduced by almost 50 percent. Congress has introduced bills which would "start discussions" on training for high technology jobs. Herbert Striner, a professor in the College of Business at American University, calculates that retraining 1 to 2 percent of the U.S. work force would cost about $6 billion in tuition per year, which could partly be off set by reduced payouts in unemployment benefits.[59] Another way the Federal government could help smooth over the transition to robotics is by providing data to the unemployed concerning available jobs and their locations throughout the United States. This information would aid the unemployed in their search for suitable occupations.

At the base of the unemployed worker's panic is apprehension over income. Our current method of income distribution is based on wages per hour worked, rather than on the amount of output produced. This structural flaw in the distribution of income is at the root of the fear of unemployment and has been great enough to delay, if not totally prevent, technological advances aimed at increasing productivity. Therefore, the system of income distribution will have to be altered if the utilization of robotics is to increase the standard of living for society.

Some possible solutions to this problem of wealth distribution have been proposed. One plan entails letting the displaced worker buy the robot and lease it to the employer so that the worker would still receive an income even though he is no longer working himself.[60] One must answer questions concerning the initial cost of the robot and who will pay for repairs before this method becomes very practical.

Lewis O. Kelso has proposed that the employee be given "Employee Stock Ownership Plans".[61] In this manner the employee would own a part of the company. However, this method closely binds the economic conditions of the laborer to the stability and profitability of the company.

One could simply rely on the Federal Reserve System to use money to build the robots and then pay everyone

22

dividends on the profits gained from the investments. James
Albus, an executive with the National Bureau of Standards,
specifically calls for the establishment of a $10 million
fund to start this venture. This sum would be increased by
a factor of three every year until the net rate of
investment equals the private rate of investment. This
equilibrium could be attained in 25 years, and the annual
public dividend would equal $8000 per citizen.[62] To
combat the inflation which would result from this massive
increase in the money supply, Albus proposes a mandatory
type of savings bond.[63] Since it takes a few years for
profits to be realized on investments in robotics, these
savings bonds, in the form of a surcharge on income tax,
would keep the inflationary pressure down.

Data gathered over the last twenty years clearly shows
a strong, positive correlation between productivity growth
and capital investment. In order for the United States to
break out of its current recessionary slump, its growth rate
of GNP must rise above negative figures. This increase can
be accomplished through capital investments, particularly in
robotics.

The United States has always led the world in
technological developments. However, up until now, these
technologies have been sold abroad and not implemented at
home. Now the U.S. is suffering the consequences of not
modernizing as it finds itself without a comparative
advantage in large, manufacturing industries, such as the
steel and auto industries. Although unemployment is
currently at double digit figures, a retreat to government
subsidization of inefficient production methods will not
solve the American productivity problem. Although the
present level of robot development hinders their
implementation in many industries, the projected
possibilities are beneficial and should be pursued.
Astounding developments using the microchip encourage an
optimistic viewpoint when considering a future world
containing robots.

Conclusion

Fear of change has long stood in the path of progress,
but has permitted societal problems associated with the
change to be dealt with before they could become widespread.
We now face a change which affects the breathing mechanism

of our society. Through the centuries man has toiled for money to pay for his existence. Work represents a method by which human worth is ascribed. Now we are faced with a conflict between the rights of workers to jobs they have held and the values of efficiency and industrial democracy which the United States has championed for over 200 years.

Progress of robotics in the American industrial setting has been delayed by their degree of development, their cost, and most steadfastly by human laborers. The unions have bitterly fought to retain their members' jobs. But because of the current level of foreign competition, U.S. companies must modernize or fail. They can no longer ignore the increased productivity and high quality products available throuh the utilization of robots.

In order to survive in post-industrial society one must adapt the present social system to a modified environment. Bernard Chern, the program director for computer engineering at the National Science Foundation, has observed that the first industrial revolution involved the transfer of physical skills from man to machine and that now "the second industrial revolution involves the transfer of intelligence from man to machine."[64] In order to accomplish this transition with the least amount of trauma, society will have to make changes primarily in the educational system and in the manner in which income is distributed in the United States.

Robotics are exploding on the scene and the United States will find itself shell-shocked and shattered if it does not allow robots to aid its faltering productivity. Acceptance and adjustment to robots in American industries will lead to a better existence in the "Post-Industrial" United States.

ENDNOTES

[1]Isaac Asimov, The Rest of the Robots, (Garden City: Doubleday & Co., 1964), p. xii.

[2]"The Robot Revolution", Editorial Research Reports, May 14, 1982, p. 348.

[3]Ibid., p. 355.

[4]"IBM: Robots Next?", Economist, January 30, 1982, p. 64.

[5]Richard K. Vedder, "Robotics and the Economy", (Washington, D.C.: Joint Economic Committee Congress of the United States, 1982), p. 16.

[6]Tim Miller, "The Coming Job Crunch", National Journal, March 15, 1982, p. 867.

[7]"Robots Join the Labor Force", Business Week, June 9, 1980, p. 63.

[8]James Albus, Brains, Behavior and Robotics, (Petersborough, N.H.: Byte Publications, Inc., 1981), p. 235.

[9]"Robots: Paying the Price for Innovation", Economist, June 7, 1980, p. 79.

[10]"The Robot Revolution", p. 347.

[11]Ibid., p. 347.

[12]Allen Boraiko, "The Chip", National Geographic, November 1982, p. 450.

[13]Jerry Main, "Work won't be the Same Again", Fortune, June 28, 1982, p. 60.

[14]"GM's Ambitious Plans to Employ Robots", Business Week, March 16, 1981, p. 31.

[15]Leopold Froehlich, "Robots to the Rescue?", Datamation, January 1981, p. 96.

[16]"How Robots are already Cutting Costs for GE", _Business Week_, June 9, 1980, p. 68.

[17]Ibid., p. 68.

[18]"IBM: Robots Next?", p. 64.

[19]"Teaching Robots How to See", _Economist_, July 12, 1980, p. 83.

[20]"Artificial Intelligence: Computers that Think Like Human Experts", _Business Week_, July 6, 1981, p. 51.

[21]James Albus, p. 321.

[22]Gurney Williams III, "Build-it-Yourself Robot", _Omni_, January 1983, p. 32.

[23]James Albus, p. 321.

[24]"The Robot Revolution", p. 354.

[25]Richard K. Vedder, p. 13.

[26]"Where Will the Jobs Come From?", _Economist_, January 3, 1981, p. 55.

[27]"The Longbridge Robots Will March over the Transport Union", _Economist_, April 19, 1980, p. 49.

[28]"Car Automation: Robots Change the Rules", _Economist_, April 19, 1982, p. 94.

[29]"Robots get a Warm Welcome in Kentucky", _Business Week_, April 19, 1982, pp. 48, 49.

[30]James Albus, p. 252.

[31]Leopold Froehlich, p. 94.

[32]Tim Miller, p. 867.

[33]U.S., Congress, House, Committee on Public Works and Transportation. _Outlook for the Auto Industry and its Impact on Employment, Industries, and Communities Dependent_

Upon It, Hearing before a subcommittee on Economic Development, 97th Cong., 1st sess., 1981, p. 185.

[34]Tim Miller, p. 868.

[35]"Retraining Displaced Workers: Too Little, Too Late?", Business Week, July 19, 1982, p. 178.

[36]U.S., Congress, House, Committee on Public Works and Transportation. Projected Changes in the Economy, Population, Labor Market, and Work Force, and their Implications for Economic Development Policy, Hearing before a subcommittee on Economic Development, 97th Cong., 1st sess., 1981, p. 41.

[37]"Retraining Displaced Workers: Too Little, Too Late?", p. 183.

[38]"The Speedup in Automation", Business Week, August 3, 1981, p. 63.

[39]Tim Miller, p. 867.

[40]"The Speedup in Automation", p. 66.

[41]"Retraining Displaced Workers: Too Little, Too Late?", p. 181.

[42]Jerry Main, p. 64.

[43]Otto Eckstein, "Perspectives on Employment under Technical Change", Employment Problems of Automation and Advanced Technology, ed. Jack Stieber, (New York: St. Martin's Press, Inc., 1966), pp. 96, 97.

[44]"Where Will the Jobs Come From?", p. 57.

[45]"Love That Robot", Economist, November 10, 1979, p. 114.

[46]The Washington Post, 23 January 1983, section A.

[47]Otto Neuloh, "A New Definition of Work and Leisure Under Advanced Technology", Employment Problems of Automation and Advanced Technology, ed. Jack Stieber, (New York: St. Martin's Press, Inc., 1966), p. 208.

27

[48]"How the Work Force is being Transformed", US News & World Report, April 26, 1982, p. 52.

[49]Theresa Rogers and Nathalie Friedman, Printers Face Automation, (Lexington: D.C. Heath & Co., 1980), p. 103.

[50]Richard K. Vedder, p. 26.

[51]"The Speedup in Automation", p. 62.

[52]"Retraining Displaced Workers: Too Little, Too Late?", p. 185.

[53]"Japan's Robot Invasion Begins to Worry Labor", Business Week, March 29, 1982, p. 46.

[54]Outlook for the Auto Industry and its Impact on Employment, Industries, and Communities Dependent Upon It, (U.S. Congressional Hearing), p. 123.

[55]"Japan: At Last a Subsidized Robot", Economist, December 22, 1979, p. 57.

[56]"The Speedup in Automation", p. 64.

[57]"Retraining Displaced Workers: Too Little, Too Late?", p. 185.

[58]Tim Miller, p. 868.

[59]"Retraining Displaced Workers: Too Little, Too Late?", p. 181.

[60]James Albus, p. 330.

[61]Ibid., p. 331.

[62]Ibid., pp. 331, 332.

[63]Ibid., p. 334.

[64]"Robots Join the Labor Force", p. 73.

BIBLIOGRAPHY

Albus, James. Brains, Behavior and Robotics. Petersborough: Byte Publications, Inc., 1981.

"Artificial Intelligence: Computers that Think Like Human Experts", Business Week, (July 6, 1981), 50-51.

Asimov, Isaac. The Rest of the Robots. Garden City: Doubleday & Co., 1964.

Boraiko, Allen. "The Chip". National Geographic, (November 1982), 421-456.

"Car Automation: Robots Change the Rules", Economist, (April 19, 1980), 93-94.

Froehlich, Leopold. "Robots to the Rescue?", Datamation, (January 1981), 84-96.

"GM's Ambitious Plans to Employ Robots", Business Week, (March 16, 1981), 31.

"How Robots are already Cutting Costs for GE", Business Week, (June 9, 1980), 68.

"How the Work Force is being Transformed", US News & World Report, (April 26, 1982), 51-53.

"IBM: Robots Next?", Economist, (January 30, 1982), 64.

"Japan: At Last the Subsidized Robot", Economist, (December 22, 1979), 57.

"Japan: Robots Get a Warm Welcome in Kentucky", Business Week, (April 19, 1982), 48-49.

"Japan's Robot Invasion Begins to Worry Labor", Business Week, (March 29, 1982), 46-47.

"Love That Robot", Economist, (November 10, 1979), 114.

Main, Jerry. "Work won't be the Same Again", Fortune, (June 28, 1982), 58-65.

Miller, Tim. "The Coming Job Crunch", National Journal, (March 15, 1982), 865-869.

Rezler, Julius. Automation and Industrial Labor. New York: Random House, Inc., 1969.

Riche, Richard. "Impact of New Electronic Technology", Monthly Labor Review, (March 1982), 37-39.

"Robots Join the Labor Force", Business Week, (June 9, 1980), 62-73.

"Robots: Paying the Price for Innovation", Economist, (June 7, 1980), 78-79.

Rogers, Theresa and Nathalie Friedman. Printers Face Automation. Lexington: D.C. Heath & Co., 1980.

Stieber, Jack, ed. Employment Problems of Automation and Advanced Technology. New York: St. Martin's Press, Inc., 1966.

"Teaching Robots How To See", Economist, (July 12, 1980), 83.

"The Longbridge Robots will March over the Transport Union", Economist, (April 19, 1980), 49.

"The Speedup in Automation", Business Week, (August 3, 1981), 58-67.

The Washington Post. 23 January 1983, section A.

Vedder, Richard K. "Robotics and the Economy", (A Staff Study for the Joint Economic Committee Congress of the United States), Washington, D.C. 1982.

U.S. Congress. House. Committee on Public Works and Transportation. Projected Changes in the Economy, Population, Labor Market, and Work Force, and their Implications For Economic Development Policy. Hearing before the Subcommittee on Economic Development, 97th Cong., 1st sess., 1981.

U.S. Congress. House. Committee on Public Works and Transportation. Outlook for the Auto Industry and its

Impact on Employment, Industries and Communities Dependent upon It. Hearing before the Subcommittee on Economic Development, 97th Cong., 1st sess., 1981.

"Where Will the Jobs Come From?", Economist, (January 3, 1981), 45-62.

Williams, Gurney. "Build-it-Yourself Robot", Omni, (January 1983), 32.

ELECTRONIC FUNDS TRANSFER AND ITS

IMPLICATIONS FOR MONETARY POLICY

David W. Scott

A careful inquiry into economic history would indicate that the evolution of exchange mechanisms has been inseparable from the evolution of the economy itself. Advances in economic society have consistently brought about advances in the exchange mechanism of that society so that the latter would facilitate rather than restrain economic activity. Thus, it "is not surprising that as economic society progressed from the primitive-agrarian state, to the commercial-industrial state, and now to the post-industrial state that the exchange mechanism required to accommodate such a society has itself progressed."[1]

The culmination of progress in this area seems to point to electronic funds transfer systems, or EFTS. An effective analysis of EFTS might best address itself to five major objectives: a general definition of EFTS; the basic elements of EFTS; the categories of EFTS; the intermediate effects of EFTS; and, most important, a discussion of the final implications of EFTS for monetary policy. Such an inquiry would indicate that EFTS is merely one step in the continuous evolution of the banking and the monetary control systems. Therefore, EFTS should not be regarded with terror, but with a sense of optimism toward a future electronic banking system wherein traditional monetary control techniques, with appropriate minor modifications, can be more effective than they have ever been.

The development of EFTS is the result of the convergence of such varied forces as the already high and increasing costs of an exchange mechanism based on paper transfers, the easy availability of the required technology, and the combined pressures for change from both private and public institutions. Despite these inherent forces, many would have the public believe that EFTS constitutes a revolution in the financial sphere, causing the foundation of existing institutions to crumble and necessitating the establishment of a new financial order. They call for radical reform of government policies regarding regulation of privacy and security aspects of EFTS as well as a total rethinking of monetary theory. Although such very important topics as privacy and security of EFTS are not included within the scope of this analysis, it must be granted that the potential for problems in such areas is fairly great. Whereas regulations regarding privacy and security of EFTS are drastically needed and, in many cases, already in effect, an investigation into the monetary policy aspects of EFTS would seem to indicate a need for something less than radical change in that realm:

> One of the greatest dangers inherent in discussion of electronic funds transfer is the temptation to treat this new technology as an event in banking so unique that it will cause a fundamental change in the banking business and require a unique treatment of rules and regulations to bring it to function without wreaking damage.[2]

While electronic funds transfer is almost unanimously regarded as important and desirable, it is not exactly revolutionary. Many who are familiar with its concepts and impact view it as evolutionary. This, however, does not imply that an investigation into the potential future impacts of EFT is unnecessary. "If we go down the road to EFTS one step at a time, without ever looking ahead to see where we want to go, we are almost certain to end up someplace we don't want to be."[3] Because EFT is the most far-reaching change the nation's payments system has ever experienced, society must look ahead and prepare for the effects of EFTS, but may do so without fear that it will radically change the financial system and the way in which the Federal Reserve carries out monetary policy.

Definition of EFTS

Electronic funds transfer is "a payments system in which the processing and communications necessary to effect economic exchange, and the processing and communications necessary for the production and distribution of services incidental or related to economic exchange, are dependent wholly or in large part on the use of electronics."[4] Alternately, it is "a computer-controlled accounting system in which all payors [sic] and payees have an account in which will be recorded all transfers of credit balances."[5] It replaces the paper and metal of the traditional payments system with electronic impulses, but still performs the same function, the transfer of value. As EFT develops further, financial institutions will have connection via computer with supermarkets as well as major retail outlets of all kinds. Interconnection will exist among financial institutions, principal employers, insurance companies, brokers, and many households.

In a future "moneyless society," there would be no need for cash, checks, or credit cards. Since the universal existence of a cashless/checkless society seems so futuristic, it has become more common to label the EFTS society as merely a less cash/less check society. As the check did not replace cash, so will EFT not totally replace either cash or checks.[6] The history of the development of EFTS indicates, however, that EFTS will become increasingly important in the financial sphere in the near future. Since the installation of the first automated teller machine at Valdosta, Georgia in 1970, the banking system has installed more than 26,000 of these units.[7] However, only two billion transactions involved electronic payment media in 1981. Estimates by the Bank Administration Institute and Business Week indicate that this figure will be well over 16 billion transactions in 1986, an 800 percent increase in only five years.[8] Thus, EFTS, having arrived on the financial scene fairly quickly, will likely remain intact and become an even more important element of the banking system in the future.

Elements of EFTS

An electronic funds transfer system must have several basic elements. The most important of these are a unique personal identifier, an on-line banking system and numerous on-line retail merchants, as well as many other institutional relationships.[9]

Under an EFTS each individual will have an identification card which will serve as the input to terminal devices which are connected to a master computer switching system. The Personal Identification Project Subcommittee of the American Bankers' Association Committee on Payments Systems has recommended that the personal identification number be the individual's Social Security number plus an additional digit for ensuring identification and proper transmission of data.[10]

The second basic element is the on-line banking system. Each bank in the system must set up either on- or off-premises computer capabilities. The nucleus of this on-line banking system is a regional computer center.

The third basic element is on-line retail merchants. The key link between retailers and the regional computer center is a remote terminal located at the point of sale in the retail establishment. This terminal is on-line to the regional computer center which switches the transmission of data to the cardholder's bank.

There must also be an integrated national network of computer systems to effect the inter-regional transfers of funds, which include on-line communication between regional computer centers, direct lines between private institutions, and the computer wire transfer service of the Federal Reserve System, the FRCS-80. Between 1979 and 1982, there were over one hundred of these banking networks established. Perhaps the largest network, set up in April 1982, includes twenty-six banks from across the country, including three of the nation's largest, California's Bank of America, New York's Chase Manhattan, and Chicago's Continental Illinois National Bank and Trust Company.

Categories of EFTS

Electronic funds transfer systems can be grouped into three broad categories according to the types of operations available: automated teller machines, point-of-sale terminals, and automated clearing houses, known respectively as A.T.M.'s, P.O.S. terminals, and A.C.H.'s. There are also several minor operations available which cannot be grouped within the three broad categories.

Teller machines are machines through which an individual may conduct various routine banking services. Much of the recent development in EFTS has been in terms of these teller machines. They may be located either on a bank's premises or elsewhere. They may be either manned or automatic, and range greatly in complexity, from simple communications terminals to more complicated automated teller machines. The automated terminals can perform services such as receiving deposits, dispensing funds from savings or checking accounts, transferring funds between accounts, making credit card advances, and receiving payments.[11] Customers typically gain access to the teller machine by first inserting a plastic card with a magnetic strip in which account information is encoded, and then entering their personal identification number via a computer keyboard. Automated teller machines will be used in an estimated 6.5 billion transactions in 1986.[12]

Point-of-sale systems, which will be involved in 5 billion transactions in 1986, allow customers to transfer funds from their accounts to merchants in order to make purchases. Such systems can be used for check authorizations and credit card transactions as well as for "debit card" transactions. The "debit card" transactions immediately transfer funds from the customer's account to the merchant's account.

The third major category of EFTS is the automated clearing house, which is similar to the conventional clearing house, in the sense that it represents a system for interbank clearing of debits and credits.[13] The major difference between the two is that the debit and credit items in an automated clearing house are in the form of electronic impulses, rather than paper items in a traditional clearing house. Automated clearing houses, which will be involved in over 2 billion transactions in 1986, are especially well-equipped to handle recurring payment items, such as payroll, social security, or pension payments. Thus, payers can authorize their banks to pay a specified amount to a payee on a specified date.

Several other relatively minor EFTS components are pay-by-phone arrangements, automatic fund transfers, and check truncation. Pay-by-phone and automatic transfer arrangements, whose combined use will exceed 2.5 billion transactions in 1986, serve as a convenience to the bank customer. Through pay-by-phone arrangements, the bank customer uses his or her touch-tone telephone to authorize certain payments. Automatic fund transfers are arranged to transfer funds from personal savings accounts to checking accounts whenever a checking account balance falls below a level previously agreed upon by the bank and the customer. Check truncation, on the other hand, primarily benefits the banks as it lowers the expense of clearing checks by reducing the number of physical handlings per check. Checks are written as usual, but at some point after they enter the banking system, the data on them are captured electronically. The checks are held at the point of interception while the data are transmitted through the financial system via electronic means.[14]

Intermediate Effects of EFTS

Electronic funds transfer systems make it possible for financial institutions to offer a wider variety of services and greater flexibility in managing funds. These innovations have implications for the way individuals and businesses manage their funds as well as for the way our financial institutions operate. Since monetary policy operates through changes in an economy's money supply, the changes induced by EFT in the public's behavior toward money will affect the development and enactment of monetary policy.

A short digression may be in order here to briefly discuss the goals and tools of monetary policy, so that the effects of EFT on such policies can be fully understood. Federal Reserve monetary policy seeks to control the supply of money and credit, thereby regulating the interest rate in order to effect a change in aggregate demand. The ultimate goals of monetary policy include full employment, a stable price level, sustainable economic growth, and an equilibrium in the balance of payments with other countries. The Federal Reserve (the Fed) uses several traditional tools of monetary policy to pursue these goals.

The general tools of monetary policy include required reserve ratios, discount rates, and open market operations. The Fed can adjust the percentage of a bank's deposits which must be held in reserve with the Fed. If the Fed increases the required reserve ratio, banks must hold more money with the Fed, thus reducing the amount available for loans. The Fed's control over the discount rate, basically the interest rate charged on money borrowed from the Fed by institutions subject to federal reserve requirements, has an obvious effect on the availability of money and credit. An increase in the discount rate tends to reduce the amount of bank borrowing, which in turn lowers bank reserves. The most important general credit control weapon of the Fed is open market operations. The Fed has the authority to buy and sell United States securities. When the Federal Open Market Committee (F.O.M.C.) purchases government securities on the open market, it is injecting money into the economy, loosening credit conditions. On the other hand, an F.O.M.C. sale of securities causes a flow of money from the hands of the public into the Fed, tightening credit conditions.

The Fed's selective credit controls, including margin requirements and moral suasion, are important weapons, but by no means as important as the general controls. Margin requirements limit the amount of credit which banks, brokers, and dealers can lend to securities dealers. In addition, there are margin requirements which set a minimum percentage for an investor's equity, thus setting a limit on borrowing for purchases of stocks and convertible bonds. Moral suasion, or "open mouth policy," is a minor selective control. The Fed uses moral suasion when it tells a bank that it feels a particular policy move is not in the public interest. Although this tool is just a guideline, it becomes a more potent weapon when the bank is hoping to borrow from the Fed, or when there is a threat by the Fed of a serious credit investigation!

The question brought about by EFTS is whether it will compromise the effectiveness of these monetary policy tools. The National Commission on Electronic Fund Transfers issued an interim report in February 1977, in which it analyzed the key issues associated with the expansion of EFT. The Commission found that there was "no reason to believe that EFT will, by itself, be a major factor compromising the effectiveness of monetary policy."[15] However, the NCEFT "leans toward the view that some changes in strategy and criteria for monetary policy in an EFT environment may make the Federal Reserve more effective."[16] The final report of the Commission, issued in October 1977, states that "indications are that, with the pace of EFT development to date, the Federal Reserve, other Federal regulatory agencies, the Congress and the States will have ample time to adjust monetary policies and regulations to accommodate EFT."[17] Thus, the evolutionary process is reasonably slow and consistent, requiring no drastic action, only deliberate inquiry into the impact of EFTS. The monetary policy issues affected by EFTS are the appropriate theoretical definition of "money" for monetary policy purposes, the level and variability of the public's demand for money, and the choice between levels and rates of growth of monetary aggregates and interest rates as intermediate targets of monetary policy.

There are several visible consequences of EFTS which indirectly bring about changes in such monetary policy considerations. These will be known here as intermediate effects of EFTS because it is through them that the broader changes are effected. First, there results a substantial reduction in both the cost and time required to complete most payment transfers. Second, increased use of credit transfers and on-line transactions brings a reduction of float, which is that portion of a bank's total deposits which represents items in the process of collection. Thus, float arises from mail delivery, time required for paper processing at depository institutions, and the inability of the Federal Reserve to meet its time schedule for credits and debits. Third, competition among and between banks and other depository institutions increases because of the decreasing importance of geographic market boundaries, broadened product lines, increased ease of account shifting, and the greater ability of non-bank financial institutions to offer third party payments systems. Fourth, there is a large volume of card-related debit and credit transactions. Fifth, direct electronic deposit of payrolls, social security payments, and other income items becomes more prevalent and these transactions themselves occur with greater frequency. Sixth, the allocation of funds by consumers and businesses is increasingly based on interest rate differentials among accounts and instruments offered by depository institutions and non-depository financial institutions. This emphasis on interest rates occurs as a consequence of the reduced cost and time required to shift asset holdings into forms which can be used for payments.[18]

Another very important intermediate effect of EFT on monetary policy is that EFT makes financial information more easily accessible by increasing the ability to accumulate, store, and retrieve large amounts of information at very low cost.[19] In an EFTS, information regarding transactions is entered, recorded, and stored in a form which can be easily and cheaply retrieved. These reports, when combined with the new transfer capabilities, can become cash management systems which can significantly reduce working cash balances, especially valuable to large corporations.[20] A corporation's financial officer can easily receive detailed information on the firm's demand deposits, wherever they may be located. When the treasurer knows the amount of excess funds available, he or she can move them by wire transfer to invest them in interest-bearing assets. Thus, the firm is earning a return on funds which would otherwise have been idle, as well as reducing its holdings of non-interest-bearing demand deposits.

A related advantage is the intermediate effect that this increased availability of financial data has on government monetary policy and decision making. The major institutional constraint on monetary policy is the series of time lags inherent in its implementation. The Commission on Money and Credit found that "general monetary controls since World War II have required from six to nine months to produce a change in the direction of ease, and a further six months for their maximum effect."[21] The initial lag, known as the "recognition lag," represents the span of time between the moment when there is a need for a change in policy and the moment the monetary authority recognizes this need. Next, the "administrative lag" is a brief lag between the moment of recognition of the need and the moment the monetary authority takes the required action. This is followed by the "intermediate lag" which exists between the time action is taken and the time the banking system begins to be faced with changes in interest rates and credit conditions. After the changes in the supply of money and credit have been effected, there is a "decision lag" which exists until there is any change in spending decisions. The final "production lag" is the span of time between the change in spending decisions and the moment the ultimate effects are felt on income and employment. The expansion of EFT should serve to shorten considerably the "recognition lag" because of the wide availability of various financial data. Because of this shortening of the "recognition lag," the entire process can occur in much less time. Thus, EFT can bring about immense improvement in the timing and efficacy of existing stabilization measures.

Final Effects of EFTS on Monetary Policy

The various operations of EFT work through what have here been called intermediate effects in order to bring about changes in the full-scale monetary policy issues. These full-scale issues include alterations in the income velocity of money, the appropriate definition of money, interest rates, and reserve requirements. Several important objectives of monetary policy are sustainable growth in output, a high level of employment, price stability, and a balance in the transactions with other countries. The monetary aggregates, or money supply measures, are related to these policy objectives through a concept known as velocity. The income velocity of a monetary aggregate is the ratio of nominal gross national product to the particular monetary aggregate. In other words, it measures the average number of times in a given period that each dollar of a particular monetary aggregate is spent for currently produced goods and services. Given expectations of the rate of change in velocity, the Federal Reserve regularly estimates ranges for the growth rates of the monetary aggregates which seem to be consistent with the desired or expected behavior of output, prices, unemployment, and the balance of payments.

The potential impact of innovations in EFT in general on monetary policy and specifically on monetary velocity comes mainly through their effect on the ways in which the public manages its money. It has traditionally been difficult for economists to agree on a unique definition of money as a medium of exchange in the United States. For this reason, there are several different definitions of money, including M_1, M_2, M_3, M_4, M_5, and L, each having relevance to a different view of monetary theory. M_1 includes currency, coin, demand deposits, and Negotiable Order of Withdrawal (N.O.W. accounts) deposits. M_2 includes all the elements of M_1 plus such "near-monies" as savings deposits, small denomination time deposits, overnight and short-term repurchase agreements, and money market mutual funds which deal with the public. M_3 includes all the elements of M_2 plus large denomination certificates of deposit, and money market mutual funds which deal with the institutional investor.

One of the reasons individuals hold money is because it reduces the transactions cost related to the purchases of goods and services. However, by holding money balances, individuals give up a potential interest income, thereby incurring an opportunity cost in the amount of the foregone interest. Because currency, coin, and regular checking accounts do not earn interest, people tend to hold less money as interest rates rise. However, as long as there are costs incurred by transferring funds from interest earning money substitutes to money accounts (whether these costs are explicit financial ones or costs in terms of time or convenience), individuals will choose to hold some of their funds in the form of money balances. Individuals try to hold a minimum amount of money which balances the reduction in transfer costs against the interest income foregone.[22]

When viewing the economy as a whole, one can observe that EFT reduces this minimum money balance in several ways.[23] First, because EFT allows a faster clearing of money balances, a given money supply can support a greater number of transactions. It follows that, with EFT, for any given level of national income, a smaller money supply is needed than without EFT. Second, individuals can more easily convert their interest-earning assets to money, because EFT reduces the transfer cost of such shifting between accounts. The result is the demand for a smaller amount of money than was necessary before the development of EFT. These observations are components of the concept of monetary velocity.

Monetary velocity arises from the fact that, in our present economy, money receipts in the form of income do not occur with the same frequency and in the same patterns as monetary expenditures. There is a definite time lag between payments and receipts, and between receipts and expenditures.[24] The concept of monetary velocity is used to measure how quickly the monetary unit is used in purchase and sale transactions during a given time period. This leads to a statement of something which was touched on earlier: "The higher the frequency of payments and receipts, the smaller the average balance required to finance a given level of economic activity."[25] Monetary velocity is thus the result of decisions made by both the consumer sector and the business sector regarding the amount of money balances they wish to hold relative to their disbursements.

The income velocity of money M_1, the ratio of national income, GNP, to the money supply, M_1, has increased in the period since the late 1940's by an average rate of 3 to 3 1/2 percent per year. The Federal Reserve must keep this in mind when setting its targeted rate of growth in M_1. The widespread introduction of EFT will, as discussed earlier, accelerate the increase in the income velocity of money, as well as make it more variable. This would necessitate a downward adjustment in the targets for the rate of growth of M_1 in order to avoid inflationary effects. "What good are

42

open market operations if the turnover rate (velocity) of reserve balances and of transactions accounts cannot be contained?"[26]

The reduction in transfer costs which EFT enables tends to make the amount of money balances people wish to hold more sensitive to changes in interest rates. This increased responsiveness follows from the greater ease with which money holders can shift their funds between money and near-money forms. This would then increase the variability in the relationship between the monetary aggregates and GNP, making changes in the income velocity of money more frequent and perhaps more difficult to predict. Another policy issue is closely involved here, that of defining an appropriate monetary aggregate to use in monetary policy formulation and evaluation. Monetary aggregates, especially such a narrowly defined one as M_1, are compromised in their effectiveness as indicators or targets of Federal Reserve monetary policy.

In an EFT environment, many assets attain a greater degree of "moniness;" that is, they become closer in nature to money. This results from the availability of virtually costless and instantaneous transfer of accounts within depository institutions, between depository financial institutions of the same kind and of different kinds, and, even more important, from the liabilities of nondeposit institutions to the liabilities of depository institutions. Individual and business assets in a mutual equity or debt fund become much closer in nature to money. This is also true of government notes and bonds, privately issued stocks and bonds, commercial paper, and cash values in reserve life insurance policies. Even normal accounts and notes payable become more liquid, often seen by their holders as means of making EFT payments.[27]

In light of the existence of so many types of near-monies, the problem for the Federal Reserve becomes finding an appropriate monetary aggregate which the Fed can control. This dilemma of finding the proper definition of money for monetary policy purposes existed before the advent of EFT, but has certainly been complicated by its development. A redefinition of money for monetary policy purposes seems in order. Many experts suggest that the Fed could overcome many of the problems caused by funds being shifted from one type of account to another by setting its target growth rates in terms of a more broadly defined

43

monetary aggregate such as M_2 or M_3.[28] Others are concerned that a stable relationship between the money supply and level of economic activity cannot be maintained unless the central bank is able to control the net settlement arrangements of transfers under an EFTS.[29] Such concerns may lead to a need for a new monetary aggregate, thus discarding M_2 and M_3 as inadequate measures for monetary policy considerations.

A further aspect of the monetary aggregate problem arises in the granting of lines of credit to consumers which many see as a by-product of EFT.[30] Such credit, granted by both banks and non-banking financial institutions, has the potential to become a major destabilizing force in the economy. The widespread use of overdraft credit may reduce the demand for current account deposits, because it will no longer be necessary to hold excess balances as a contingency for unexpected demands. Installment credit extended via consumer overdraft facilities under EFT should therefore be under the control of the monetary authorities.[31] Such lines of credit should thus be included in the monetary aggregate used to indicate and target economic stabilization policies. Perhaps, as other financial institutions begin to grant overdraft credit and assume other traditional bank functions, they too should be controlled by the Federal Reserve. A possible form of control is for the Fed to pay interest on the balances kept with it from any financial institution, not just commercial banks. If there were too much public liquidity and over-extension of credit, the Fed would post a higher interest rate on deposits held with it, drawing these funds away from all sorts of money markets. Lowered Federal Reserve rates would conversely decrease deposits, thereby increasing liquidity and credit available to the public.[32]

Electronic funds transfer systems also have implications for policy maker's views of those interest rates which are significant for policy formulation and analysis. The policy makers have three tasks in using interest rates as instruments of monetary policy. First, they must determine which real world interest rates most directly affect the spending and saving behavior of each economic sector. Second, they must identify policy instruments which are related to these important interest rates. Third and last, they must manipulate those instruments to try to bring the interest rates to the target

levels.[33] EFTS will complicate all of these tasks,
especially with regard to retail financial markets. EFT is
increasing the number of institutional sources and methods
of borrowing and lending. Each new arrangement results in a
new interest rate to monitor. This leads to finding an
instrument linked to the new interest rate and with which
the interest rate can be influenced.

Another monetary policy instrument, that of reserve
requirements, is also affected by the development of
electronic funds transfer. EFTS increases the degree by
which savings and loan associations and other non-bank
financial intermediaries can attract deposits away from
commercial banks. Initially, this greater percentage of
deposits in savings and loan associations resulted in a
decrease in the Federal Reserve's control over the money
supply. This resulted from the Fed's inability to set
reserve requirements for savings and loan associations
because they were not Federal Reserve members. A portion of
the Depository Institution Deregulation and Monetary Control
Act of 1980 addressed this problem by requiring all
financial institutions which offer transactions accounts to
hold reserves with the Federal Reserve. Thus, the Fed now
has greater control over transactions balances because a
larger portion of the reserve base is directly subject to
reserve requirements.

Conclusion

It should be clear that electronic funds transfer
systems have many significant implications for monetary
policy which work through the intermediate effects to bear
on the income velocity of money, the appropriate definition
of the money supply, effective interest rates, and reserve
requirements. Effects on these issues mirror a change in
the specific intermediate relationships between control
parameters in the economy. However, once these new
relationships are understood and implemented, "the _modus
operandi_ of monetary policy and the channels of monetary
influence on real sector activity will be much the same as
they are today."[34] Accordingly, "the traditional tools of
monetary policy--open market operations, discount loans from
the Federal Reserve, and reserve requirements--will continue
to operate in much the same manner."[35]

Other viewpoints are even more optimistic:

The efficacy of monetary policy directed toward economic stabilization will be greatly improved under a fully implemented electronic funds-transfer system. The time lags that presently restrain monetary policy from being a much needed tool for the construction of flexible, quick-acting stabilization measures may be largely reduced if not eliminated. Because the national system of regional computer centers will be able to provide instaneous [sic] data on nearly every vital segment of the economy, 'monetary cybernetics' will permit the development of short-run forecasting and review-and-control techniques not available at the present time.[36]

A valid interpretation of the past developments and the potential future impacts of electronic funds transfer reflects the need for definite monetary policy changes. However, as this analysis has illustrated, it hardly portends havoc for the economy by bringing with it the need for completely new and different monetary control techniques. Rather, one may view electronic funds transfer as a tool which, with a working understanding of its implications, can assist the monetary authorities, specifically the Federal Reserve Board, in making future monetary policy more efficient and easier to effect.[37]

ENDNOTES

[1]The theme for the first paragraph and the quotation are from Mark G. Bender, EFTS/Electronic Funds Transfer Systems: Elements and Impact. (Port Washington, NY: Kennikat Press Corp., 1975), p. 3.

[2]William F. Baxter, Paul H. Cootner, and Kenneth E. Scott, Retail Banking in the Electronic Age: The Law and Economics of Electronic Funds Transfer. (Montclair, NJ: Allanheld, Osmun and Co. Publishers, Inc., 1977), p. 3.

[3]Virginia Knauer, Statement before the Texas Credit Union League, Houston, Texas, 22 April 1976, cited by Baxter, Cootner, and Scott, Retail Banking, p. 3.

[4]EFT and the Public Interest: A Report of the National Commission on Electronic Fund Transfers, by William B. Widnall, Chairman. (Washington, DC: Government Printing Office, 23 February 1977), p. 1, n. 1.

[5]Dennis W. Richardson, Electric Money: Evolution of an Electronic Funds-Transfer System. (Cambridge: The M.I.T. Press, 1970), p. 95.

[6]Almarin Phillips, "CMC, Heller, Hunt, FIA, FRA, and FINE: The Neglected Aspect of Financial Reform," Journal of Money, Credit and Banking 9 (November 1977): 638.

[7]"Electronic Banking: Networks for Retail Banking: Making Money From Transactions," Business Week, 18 January 1982, p. 71.

[8]Ibid., p. 74.

[9]Richardson, Electric Money, p. 99.

[10]Ibid., p. 100.

[11]William C. Niblack, "Development of Electronic Funds Transfer Systems," in Current Perspectives in Banking, 2nd. ed., ed. Thomas Havrilesky and John Boorman. (Arlington Heights, IL: AHM Publishing Corp., 1980), p. 246.

[12]All projections of 1986 figures used in this section are from "Electronic Banking: Networks for Retail Banking: Making Money from Transactions," Business Week, 18 January 1982, Chart, p. 74.

[13]Niblack, "Development of Electronic Funds Transfer Systems," p. 247.

[14]Mary G. Grandstaff and Charles J. Smaistrla, "The Payments Mechanism: A Primer on Electronic Funds Transfer," Business Review of the Federal Reserve Bank of Dallas, September 1976, p. 11.

[15]Charles J. Smaistrla, "The Payments Mechanism: Electronic Funds Transfer and Monetary Policy," Review of the Federal Reserve Bank of Dallas, August 1977, p. 6.

[16]EFT and the Public Interest, p. xviii.

[17]EFT in the United States: Policy Recommendations and the Public Interest: Final Report of the National Commission on Electronic Fund Transfers, by William B. Widnall, Chairman. (Washington, DC: Government Printing Office, 28 October 1977), p. 197.

[18]Ibid., p. 198.

[19]EFT and the Public Interest, p. 90.

[20]Smaistrla, "The Payments Mechanism: Electronic Funds Transfer and Monetary Policy," p. 8.

[21]Richardson, Electric Money, p. 136.

[22]Smaistrla, "The Payments Mechanism: Electronic Funds Transfer and Monetary Policy," p. 7.

[23]Ibid., p. 7-8.

[24]Richardson, Electric Money, p. 127.

[25]Ibid.

[26]Phillips, "CMC, Heller, Hunt, FIA, FRA, and FINE: The Neglected Aspect of Financial Reform," p. 641.

[27]Ibid., p. 638.

[28]Smaistrla, "The Payments Mechanism: Electronic Funds Transfer and Monetary Policy," p. 9.

[29]Marjorie Greene, "Will Technology Undermine Today's Monetary Control Techniques?" The Banker 131 (August 1981): 30.

[30]Ibid., p. 31.

[31]Mark J. Flannery and Dwight M. Jaffee, The Economic Implications of an Electronic Monetary Transfer System. (Lexington, MA: D. C. Heath and Company, 1973), p. 173.

[32]Greene, "Will Technology Undermine Today's Monetary Control Techniques?" p. 31-32.

[33]Edward J. Kane, "EFT and Monetary Policy," in Current Perspectives in Banking, 2nd. ed., ed. Thomas Havrilesky and John Boorman. (Arlington Heights, IL: AHM Publishing Corp., 1980), p. 282.

[34]EFT and the Public Interest, p. 91.

[35]Flannery and Jaffee, The Economic Implications of an Electronic Monetary Transfer System, p. 198.

[36]Richardson, Electric Money, p. 145.

[37]Greene, "Will Technology Undermine Today's Monetary Control Techniques?" p. 32.

BIBLIOGRAPHY

Allison, Theodore E. Statement before the Sub-committee on Government Information and Individual Rights of the Committee on Government Operations, U.S. House of Representatives, 22 October 1981, cited in Federal Reserve Bulletin 67 (November 1981): 828-32.

Baxter, William F., Cootner, Paul H., and Scott, Kenneth E. Retail Banking in the Electronic Age: The Law and Economics of Electronic Funds Transfer. Montclair, N.J.: Allanheld, Osmun and Co. Publishers, Inc., 1977.

Bender, Mark G. EFTS/Electronic Funds Transfer Systems: Elements and Impact. Port Washington, N.Y.: Kennikat Press Corp., 1975.

Benton, John B. "Electronic Funds Transfer: Pitfalls and Payoffs." Harvard Business Review 55 (July/August 1977): 16-17+.

Candilis, Wray O., ed. The Future of Commercial Banking. New York: Praeger Publishers, Inc., 1975.

de Juvigny, Francois Leonard, and Wadsworth, John E. New Approaches in Monetary Policy. Alphen aan den Rijn, The Netherlands: Sijthoff and Noordhoff International Publishers B.V., 1979.

Dingle, James F. "The Public Policy Implications of EFTS." Journal of Bank Research 7 (Spring 1976): 30-36.

"Electronic Banking: Networks for Retail Banking: Making Money from Transactions." Business Week, 18 January 1982, pp. 70-80.

EFT and the Public Interest: A Report of the National Commission on Electronic Fund Transfers. By William B. Widnall, Chairman. Washington, D.C.: Government Printing Office, 23 February 1977.

EFT in the United States: Policy Recommendations and the Public Interest: Final Report of the National Commission on Electronic Fund Transfers. By William B.

Widnall, Chairman. Washington, D.C.: Government Printing Office, 28 October 1977.

Flannery, Mark J., and Jaffee, Dwight M. The Economic Implications of an Electronic Monetary Transfer System. Lexington, MA: D. C. Heath and Company, 1973.

Grandstaff, Mary G., and Smaistrla, Charles J. "The Payments Mechanism: A Primer on Electronic Funds Transfer." Business Review of the Federal Reserve Bank of Dallas, September 1976, pp. 7-14.

Greene, Marjorie. "Will Technology Undermine Today's Monetary Control Techniques?" The Banker 131 (August 1981): 29-32.

Kane, Edward J. "EFT and Monetary Policy." In Current Perspectives in Banking. 2nd. ed., pp. 277-290. Edited by Thomas Havrilesky and John Boorman. Arlington Heights, IL: AHM Publishing Corp., 1980.

Knight, Robert E. "The Changing Payments Mechanism: Electronic Funds Transfer Arrangements." Monthly Review of the Federal Reserve Bank of Kansas City, July/August 1974, pp. 10-20.

Lieberman, Charles. "A Note on the Impact of Electronic Funds Transfers on the Effectiveness of Monetary Policy." Economic Inquiry 17 (October 1979): 613-17.

McHenry, Wendell, Jr. "EFT in the 1980's." The Banker's Magazine 164 (May/June 1981): 30-33.

Mason, John. "Innovation in the Money System: EFTS and Economic Welfare." The Quarterly Review of Economics and Business 17 (Winter 1977): 43-55.

Niblack, William C. "Development of Electronic Funds Transfer Systems." In Current Perspectives in Banking. 2nd. ed., pp. 244-58. Edited by Thomas Havrilesky and John Boorman. Arlington Heights, IL: AHM Publishing Corp., 1980.

Phillips, Almarin. "CMC, Heller, Hunt, FIA, FRA, and FINE: The Neglected Aspect of Financial Reform." Journal of Money, Credit and Banking 9 (November 1977): 636-41.

Richards, Al. "AFT: A New Monetary Problem." Dun's Review 112 (October 1978): 114-15+.

Richardson, Dennis W. Electric Money: Evolution of an Electronic Funds-Transfer System. Cambridge: The M.I.T. Press, 1970.

Smaistrla, Charles J. "The Payments Mechanism: Current Issues in Electronic Funds Transfer." Review of the Federal Reserve Bank of Dallas, February 1977, pp. 1-7.

_____. "The Payments Mechanism: Electronic Funds Transfer and Monetary Policy." Review of the Federal Reserve Bank of Dallas, August 1977, pp. 6-12.

Winningham, Scott. "Automatic Transfers and Monetary Policy." Economic Review of the Federal Reserve Bank of Kansas City, November 1978, pp. 18-27.

FAMILY FARMING IN TODAY'S POST-INDUSTRIAL

SOCIETY: A RICHES TO RAGS STORY

Kelly A. Kroll

The genesis of the so-called "post-industrial" society causes unique characteristics that differ from the previous dominant agricultural and industrial ages. The purpose of this paper is to examine how the influence of this new post-industrial age has negatively affected the family farm[1] instead of creating an environment in which the family farmers experience high productivity and substantial profits. The arrival of the post-industrial society has brought with it many dangers and obstacles to the family farmer which have evolved over the years. The goal of the family farmer is to benefit from the intense advancements in technology and scientific research, along with the increasing availabilities in educational opportunities, information services, and communication possibilities that are present today. The advent of the post-industrial society implies a 'surplus' of opportunities in various aspects of the social, economic, and environmental scenes, but the family farmer has yet to witness the rewards of such a bounteous society. I hope to show in this paper, the several trends that have unravelled through time, leading us to the explanation of our current situation.

The "first wave", as Alvin Toffler would define it, involved the period in which farming was the chief occupation of the population.[2] During this era, approximately ninety percent of the United States population was employed as either a farmer or a farmer's helper. Labor was the source of power and agriculture was the primary sector where the laborers proved their belief in the Protestant Work Ethic. Within agriculture, the primary unit has historically been the family farm. Most farmers of this time period raised crops for their own family's subsistence instead of for profits in the marketplace. The equipment and tools that were used would today be considered primitive; however, the farming instruments accomplished what they were designed to do-- produce the food needed for family survival.

The "second wave" as Toffler described it, arrived in the early twentieth century, evolving as the 'industrial age'.[3] Capital replaced labor as the source of power; yet even with the use of machinery, long hours of hard work remained for farmers. The industrial age brought with it mechanization, and mechanization caused the displacement of many farm laborers, along with the effects that follow such an occurrence. "Mechanization of U.S. agriculture has been considered a major factor of social, economic, and environmental change in rural areas."[4] Thus, the industrial sector became the dominating influence in the society's labor market. Men took their wives and children from the fields and migrated to the cities where there were factories. Robert Higgs states in "The Ups and Downs of the Farmer":

> Farmers generally led quite isolated lives. The loneliness of such a life must have cut deeper as the number of urban alternatives grew and became more accessible. One hundred and sixty acres was a small world, and many had less.[5]

Rural population diminished and urban population expanded; a change in farming capabilities was taking place simultaneously while the productivity level of farmers increased. One source of this outpouring of farm products was an expansion of measurable inputs.[6] These inputs included:

> 1) improved land on farms 2) investors augmented the stock of other forms of material capital, increasing the stock of farm machinery and equipment 3) the farm labor force grew substantially in absolute terms (though it became a progressively smaller part of the economy's total work force) 4) as the ratio of capital of all kinds (material, human, intellectual) labor increased, output per man-hour rose.[7]

During this mechanization period, approximately forty-eight percent of the population was employed in agriculture. A declining trend in the farming profession was unfolding; as the capital costs of farm production increased, the numbers and characteristics of farms changed markedly.[8] Farms were decreasing in total number and the need to continue farming was diminishing because of high costs and the ability of fewer farmers to produce great quantities of food.

Another trend was floating through the 'waves'--the beginning of the fight for parity[9] by farmers. "The American farmer receives less than his 'fair share' of the national product."[10] The prices paid by the farmers began to be far greater than the prices they received and many financially costly actions occurred. The railroads were charging unjust prices for the transportation of agricultural products, and speculators and land monopolists engrossed the best of public lands.[11] Farmers were forced to borrow money and the Government began to play a role by controlling interest rates and prices, thus, affecting the farmer's income level. This relationship between the farmer and the Government resulted in the second trend that was to continue to affect the farmer's income level in every fiscal year. In order to find the farmer's total income in any year, one must add the sales receipts and production's implicit value, an amount equal to the appreciation of farm capital stock. Even when the farmer's income is correctly computed, it is still lower on the average than that of nonfarm income earners. "Along with unbalanced incomes compared to the rest of the economy, high costs and falling prices (due to growing surpluses) caused the arrival of heavy indebtedness in the farming industry"[12] -- a trend that was also to have a great impact during the "third wave".

Toffler explains the "third wave" as a surge of new ideals and values within the society which will be affected by the rise in technology, education, and informational exchanges.[13] This new wave will involve the shifting of emphasis from the industrial employee to the service-oriented worker. The trend in the labor force is now beginning to show an increase in the numbers of doctors, lawyers, scientists, beauticians, nutritionists, health spa owners, restaurant managers, and computer programmers, operators, and consultants. Where does the family farmer fit into this service sector 'picture'? What has happened to the traditional American family farmer? The number of farmers has fallen to an all-time low, levelling off between three and four percent of the population.

Declining number of farmers. Larger farms. Higher productivity and higher costs. Government intervention. Farmers receiving less than parity. All of these factual trends are causing the disappearance of the family farmer. Will these trends continue to filter into the future 'waves'?

Before we attempt to glimpse into the <u>future</u>, we should look at the <u>current</u> struggles of the American family farmer. How do the trends mentioned above relate to the situation today in regard to farmers on large farms, and in particular, to farmers on the smaller family farms? The answer begins to unfold in the definitions and the facts of the family farm.

The Small Family Farm's Dilemma: The Rival Larger Farms

The larger commercial and commercial-plus farms include approximately a half million families, the small commercial and small family farms include about 1.2 million families.[14] There are also another one million families living in rural areas, most working in off-farm jobs and farming in their spare time, which still meets the current definition of a farm[15] although most have little economic dependence on agriculture. In turn, there are the farmers who have great economic dependence on agriculture, but must consider outside sources for financial aid. "Farmers and their wives have increasingly turned to off-farm jobs to supplement their farm income. Farm families earn more than half their living off the farm."[16]

The current income situation of family farms is in trouble compared to the current income of larger farms. Today, a significant number of very large farms control an increasing proportion of farm resources, sell a growing part of farm output, and enjoy high family incomes.[17] The concentration of resources and output on these large farms has been increasing rapidly in the past few years. It is the prospect of large farms increasingly beseiging control of agricultural resources that underlies much of the uneasiness among economically-weak family farms. The family farmers, therefore, are able to awaken to the implication of the trends toward increasing farm size.

Most had already watched their affluent (commercial size) farm neighbors slowly and gradually take control of farms 'down the road', add it to their farm operation, and begin to look for more farmland to rent or buy. These expansion-minded neighbors were on the upward 'tread mill' that allowed them to take greater advantage of government farm programs, tax laws that allowed write-off of interest costs as a business deduction, investment tax credit programs, preferential taxation of capital gains, and cash-basis accounting.[18]

Most family farms were not in an income position that allowed them to take advantage of these special tax programs, although the hope that they might someday find these programs useful, has led them to support their continuation.

The 1.2 million farm families face a rather bleak future because of their below average incomes, their lack of 'expansion-power', and their inability to compete with their economically stronger rivals. This problem leads us to ask the key questions, "Why should the family farmer survive in the United States?" and "Why cannot the large farmer be the sole provider of food?"

Family farmers are crucial to agriculture because they have the flexibility that is needed to produce crops intensively, while eliminating the necessity to haul the commodities long distances. The larger commercial farmers, on the other hand, specialize in one or two products, along with having to ship the farm output to far away destinations. The smaller plots that make up the family farm adapt quicker to changes in the marketplace; in this case, flexibility creates efficiency.

Thus, we arrive at a definition of the farm problem facing the society in the 1980's. In overall terms, it is to decide which structure of agriculture the nation wishes to have by the year 2000.[19] Or, in the terms of Land Grant University Economists, it is "Who will control U.S. agriculture?"[20] In financial terms, the farm dilemma is the 1.2 million small, family farms that are losing the battle to control farm production. The majority of these farms have below average family incomes, both in terms of farm incomes and nonfarm incomes.[21]

In more positive terms, the family farm problem is how to increase the income levels of these families to enable them to compete with financially stronger farm interests for various additional farm resources (land, machinery,

livestock, etc...).   Obviously, if the $3.0 billion of
government payments that went to all farmers in 1978 had
gone to these family farms, their incomes would have been
substantially better (about $3000 each).[22]   Instead this
group received about one-third of all farm program payments
(about $1000 per farm) which was insufficient to equalize
their economic bargaining power.[23] Unless some additional
assistance is provided to these family farms, most will
eventually cease their farming operations.

     If the assistance for these family farms comes via farm
prices-- parity prices is the farmer's goal--the improvement
in their incomes will be small in comparison to the big
gains of the larger farms, and the family farms' competitive
position will be worsened instead of improved.   Land and
other asset prices will escalate to the point where small
family farms still will confront economic weakness in the
farm asset market.[24]   If the society is seriously
concerned about encouraging a family-farm structure, federal
farm assistance disbursed appropriately could be one means
of achieving that goal. "Current policies give too much to
those with little need and too little to those with much
need."[25]

Government Policies: Adjustments to Encourage   Family Farming

     For the better part of a century, the federal
government has been a persistent and significant force in
the development of American agriculture.   Its intervention
reflects the fact that the agricultural industry is
intricately tied to the social, economic, and environmental
concerns of · our   society.[26]   Understandably,   this
intervention has spurred much controversy, both over its
legitimacy and magnitude.

     Although the federal government has played an active
role in agriculture since the end of the Civil War, the
nature of this involvement has varied widely.   Between the
end of the Civil War and the advent of World War I,
government programs emphasized self-help measures that were
intended to equalize a growing disparity in living
conditions between rural and urbanizing sectors of
society.[27] Emergence of the land-grant system, with its
emphasis on agricultural research and extension programs,
analysis of agricultural marketing systems, and

implementation of national agricultural surveys are a few examples of programs in the late 19th and early 20th centuries designed to equalize conditions between rural and urban America.

After World War I, relationships between the farm population and the federal government changed.[28] Focusing on commercial farmers, the United States Department of Agriculture (USDA) took a more aggressive posture in its relationship to farmers and began to encourage changes in farming practices. During this period, a myriad of commodity-control and price-support programs were enacted and implemented. Through its constituent agencies, the USDA sought to foster voluntary local participation in such programs as price supports, conservation, and rural development while maintaining its traditional emphasis on research and education.[29]

During the mid-1960's and 1970's, a new tier of government regulatory programs was instituted in agriculture.[30] Regulatory programs concerning pesticides and their application, feed additives, and solid waste disposal, among others, were enacted and considerable discussion has ensued as to the need for stronger regulatory programs in such areas as soil conservation and land use.

These newer governmental programs differ from more traditional ventures in several key respects. The differences may be vital to their acceptance by farm operators. First, many of the newer programs mandate behavioral compliance, in contrast to the philosophy of voluntary participation that prevailed earlier.[31] By mandating participation, the new programs are clearly antithetical to the traditional value that farmers have placed on private property rights and on autonomous decision-making.

Second, many of the newer farm programs have been inspired more out of concern for the public good (especially by issues of public health and safety, consumer rights, and environmental quality) than by a desire to strengthen the economic position of farmers (as through increased farm income and parity prices).[32] The Environmental Quality Movement has especially served to focus public attention on the contribution of modern farming practices to environmental disruption. Numerous environmental ills have been traced to these practices, including soil erosion, water pollution, loss of timber and wetlands, destruction of wildlife habitat, overgrazing of grasslands, and the mining of underground water supplies. Ironically, many of these environmental ills have been created and/or intensified by governmental farm programs that encouraged intensive planting. Thus, remedial environmental programs frequently are at cross-purposes with other governmental efforts to enhance farm income and to increase food production.[33]

A third difference is that the newer farm programs have sometimes necessitated activities and financial investments that are contrary to the perceived self-interests of farmers.[34] Programs designed to reduce or control applications of fertilizers, pesticides, and herbicides to satisfy environmental concerns can threaten crop yields. Some farmers may perceive that their compliance with these programs will result in reduced farm income. Even if lower incomes are a possibility for farmers because of this type of legislation, they are concerned with environmental protectionism. After all, what will happen to the farmer if there is no land?

The most influential problem with newer farm programs has been evolving gradually since World War II.[35] The trend that (family) farmers receive inadequate monetary solutions for their financial crises, suggests the need for certain policy strategies for future federal farm programs.

Preceding any discussion of changes, certain national requirements must be met:

> First, any policy should not disrupt the agricultural industry's capacity to provide adequate supplies of food fiber for the Nation and it should leave adequate amounts for exports. Second, any new policy directives should not replace the guidance provided to farmers by the market on what, how, and how much to produce of food and fiber products. Third, any new programs should consider the cyclical nature of certain types of agricultural production and attempt to moderate that instability and where that is impossible, provide farmers with protection against adverse impacts. Fourth, any new programs should treat different agricultural

groups--specifically crop and livestock
producers--in ways that do not impose unnecessary
burdens on food consumers or tax payers to the
benefit of food producers where the end result is
higher farm asset values, higher future
production costs, and inflationary pressures in
the remainder of the economy that ultimately cause
a circular increase in prices paid for by
agricultural producers.[36]

If we follow the recommendations for the new programs, we
should be able to assist the family farmer in an equitable
way.

Farm programs should be restructured to reduce or
alleviate entirely the assistance that is provided to the
already economically powerful (large) farmer. By providing
these units with the largest share of assistance under
current programs, the Government is less able to help the
below-average income farm families because of its stringent
budget decisions.[37] The budget considerations might well
have led to tighter restrictions on payments to financially
strong large farms and instead, provided family farms with
greater monetary support.

A second program could have as a goal the attainment of
higher levels of target prices or income supplement payments
would raise the incomes of family-size farms to the national
average family income.[38] The increase in income would
encourage family farms to expand; this would be beneficial
to society since official data show family farms to have the
lowest cost per unit of output in comparison to other types
and sizes of farms.[39] The growth of family farms might
cause more stability--economically and socially.

A third program could suggest that a limitation be
placed on the amount of credit any farm operation is
eligible to receive from the Commodity Credit Corporation
(CCC).[40] The restriction might be coupled with a low rate
of interest for this limited amount of credit from CCC.[41]
In this way, farmers would be protected against falling
market prices on some minimum amount of production at a
lower cost than at present. Above the minimum amount, the
CCC could still provide credit for stored commodities if
this were deemed desirable, but interest rates would be
competitive. The importance of the interest rate implies

that productionin excess of the CCC limit is subject to the risks of the market, thus, disengaging prices from support prices, a step that could have useful effects on expanding farm exports during periods of surplus production.[42]

The cyclical nature of agriculture makes it a prime candidate for another program--a program allowing taxes to be "smoothed out" over years.[43] For example, income averaging is useful in a year of rapid increases in farm receipts. However, farmers also suffer years in which there are sharp decreases in income. These reductions can impose a serious income burden on farm families especially. One policy possibility would be to substitute the current "loss recovery provisions" that are in the Internal Revenue Code for large corporations, and to apply them to the farmers' situation during low price years.[44] While the details for applying this concept to farms would need far more research, the possibilities seem relevant to the cyclical income situation of agriculture.

A fifth policy change relates to the programs created by the Farmers Home Administration (FmHA). If the FmHA reduced sharply the interest rates for the first $50,000 of credit it provides and raised interest rates to commercial levels on larger figures, family farmers would be able to benefit from the FmHA instead of only the larger farms benefiting.[45]

Finally, another way to raise farm incomes would be for the Government to expand the direct farmer-to-consumer market program established in 1976.[46] Many smaller, family farms might be able to increase their incomes substantially by marketing directly to consumers through publicly-founded "farmer markets". This program has received meager resources and it has not received the administrative push necessary to launch a useful and viable marketing vehicle. Its potential would seem to exceed noticeably its present capacity, in light of rising retail food prices.[47]

This list of program modifications is not comprehensive. It implies, however, a substantial restructuring of the role of the Federal Government in providing economic assistance to farmers. The assistance so far has been, to an influential degree, skewed in favor of large, specialized, commercial farms, especially since

1970.[48]   The payments distributions in the most   recent
past do not appear to have changed that trend.

The trends lead to many questions for future farm
policy: Can there be a combination of income supplement and
disaster payments programs? What should the minimum level of
income be for farm families before assistance is rendered in
time of need?   Competition suggests that only efficient
farms will survive.   Our society today must be able to
provide and support an economy that will enable the family
farmer to exist along with the larger, commercial farmer.
The politicians must be adequate mediators and coordinators
between the family farmers and the   marketplace.   The
policies chosen to affect the "farming world" depend upon
the decisions (therefore, the values) of our country's
official leaders. One government leader, Congressman Robert
Krueger, has a philosophy that emphasizes his idea of the
government's role in agriculture: space 1;on subpara 5,5 I
believe that flexible parity legislation is vital to prevent
thousands of family farmers from being forced out of farming
by bankruptcy.   It is the least expensive   and least
inflationary approach, and would have been a step away from
centralized, federal planning of agriculture.   This
legislation would have helped return decision-making to each
individual farm, thereby allowing farmers to write their own
farm programs. Most important, the legislation would have
granted higher target prices to producers, up to 100 percent
parity, depending upon the amount of acreage each farmer
chose to set aside from production. I intend to continue to
impress upon members of Congress the necessity for such
legislation, which could herald the beginning of a new
American Agriculture policy assuring the farmer a fair
return for his product and greater independence from
government interference. There can be no doubt that new
policy is necessary. The dedicated farmers who lobbied the
nation's capitol, both for themselves and for the thousands
who toiled at home to produce food for a hungry world, have
proven this ...[49]

Congressman Robert Krueger is in a minority. What does this
imply about the values of the majority of the government
leaders concerning farming?[50]

There exists another minority in the debate over the
government's role in agriculture, and this is the farmers
who make up the American Agriculture Movement. The American

Agriculture Movement was created to preserve the family farm system.[51] The American Agriculture Movement is not another farm organization. There are no memberships, dues, secretaries, or presidents. They are a group of individual farmers, ranchers, and agribusinessmen, unified together in order to achieve the fair price of 100% parity for all agricultural products.

The American Agriculture Movement has no political voting power, but they still have the right to voice their opinions. They are tired of government dominance, speculation, manipulation, and big money influence in the farmers' marketplace.

The Government and the farmers of America need to work together; it is going to take a joint effort in order to revive the livelihood of the family farm. The educational quality of man in the post-industrial society, along with high-technological equipment, multitude of information available, and intense communication services, should facilitate the evolution of an adequate relationship between the Government and the farmer. The post-industrial society must be the catalyst for rebuilding the family farm. Not only will the birth of new government policies affect the family farm, but so too will technological improvements and discoveries.

Technology to the Rescue?: Advancements in Agricultural Research

There are many experiments taking place in research labs investigating the possibilities of new farming techniques. The American agricultural research network--which includes various branches of USDA, State agricultural extension agencies, private and state universities and American industry--is now engaged in a multibillion-dollar effort to come up with solutions to many agricultural problems that exist today.[52] The problems include shrinking agricultural acreage, increasing population, decreasing productivity gains in most crops, heavy dependence on petroleum-based pesticides and fertilizers, and genetic vulnerability.

The agricultural research is focusing on offshoots of age-old plant breeding techniques. In conventional breeding, scientists essentially take characteristics from two different varieties of the same species and inbreed them to form a third type containing the best features of the two. The newest experimental plant breeding technology, genetic engineering, could someday allow biologists to design actual new genetic material rather than just manipulate genetic material already present in crops. Since this is done on the molecular level, it offers an entirely new concept in genetic engineering. Raymond C. Valentine, who heads a fifteen-member genetic engineering research team at the University of California, states: "A gene from corn could be theoretically introduced into soybean or potato; a microbial gene could be introduced into a plant. We now have the ability to move any gene from any organism. Any particular strain that we are interested in could be moved from one organism to another."[53]

Most scientists foresee few imminent breakthroughs with genetic engineering and plant breeding. The research taking place now centers on learning exactly what is involved in this extremely complex process. There may be no practical applications in agriculture for decades.[54] G.W. Schaeffer, chief of USDA's Cell Culture and Nitrogen Fixation Laboratory believes that: space 1;on subpara 5,5 The short-term contribution of genetic engineering will be to help our understanding of the plants we are working with. When you are talking about a gene being inserted into a plant cell and being expressed, the problems are ... great. It is very difficult to design an experiment at the cell level that will be meaningful in the agronomic sense and in the setting of the farmer's field. You really can't make that quantum jump in one step. It has to evolve.[55]

Schaeffer also said that there may be "dramatic findings in specific areas" in the near future, but warned that "there is a lot what I call 'biology' to be worked out before a recipient plant can tolerate external DNA and express it ... There are very large problems, and it will take time to ferret them out."[56]

How much time? "I think we are twenty to fifty years away from being able to manipulate genes to our advantage in an agricultural way—maybe longer that that,"[57] said Quentin Jones of the USDA.

Jones explained some of the problems blocking successful application of genetic engineering technology to

plant breeding: space 1;on subpara 5,5 When you look at the
enormous complexity of all the things that go into what we
call environment--temperature, light, moisture, heat, other
biological systems other than plants that we are trying to
grow--for us to get to the point where we can say, 'What we
need is a gene that controls this reaction by this plant'
and then design it by putting together DNA from wherever our
knowledge tells us it should come from, this I think .. most
of us probably won't live to see it.[58]

While most applications of genetic engineering in
agriculture probably will not be in widespread use for
decades, there is a new type of crop production already in
use--and another on the drawing board -- that allows the
growth of crop in media other than conventional farm soil.
Hydroponics is a technique in which crops are grown not in
soil, but in water fed with chemical nutrients in what is
called a "controlled environment"--in other words, indoors.
One indication that hydroponics has a promising future is
that some large corporations have entered the field in
recent years.[59]

Whittaker Corporation, a Los Angeles-based
conglomerate; Control Data Corporation, the Minneapolis
computer and financial services company; and General Mills,
the food producer also located in Minneapolis, all own
so-called "lettuce factories".[60] These computer-run
indoor operations have the advantage of controlling all the
environmental factors involved in lettuce growing, including
the exact chemical composition of fertilizers and other
nutrients, temperature, humidity, and light. These
automated facilities produce as much as one hundred times
more lettuce than conventional outdoor agriculture.[61]
Whittaker's Agri-Systems Division's lettuce factory in
Somis, California, is the first such structure to grow
plants solely with natural light, thus saving the enormous
expense of artificial light. The 1.8 acre, plastic
greenhouse-type facility also produces cucumbers and
tomatoes. Harvesting and packaging is done mechanically
inside the factory, saving additional labor costs.[62]

Plant scientists also hope to develop crops that can
survive in salty soil. For example, heavy irrigation has
left significant salt deposits in the San Joachin Valley,
the heart of California's agribusiness.[63] Researchers at
the Bodega Marine Laboratory at the University of California

are experimenting with special types of barley, wheat, and tomatoes that are being irrigated with seawater (this technique is called aquaculture).[64] In the future, it may even become possible to use seawater to produce crops along sandy coasts and coastal deserts. However, it is also an important goal to improve the salt tolerance of crops for conventional irrigation agriculture on salt-affected soils.

Technological advancements are crucial to the future of American farming. The improvements in crop plants and productivity levels will benefit most farmers, specifically the large, commercial farmers. But what about the small, family farmer? Can he, too, reap the rewards from the advancements in agricultural research? The answer is maybe--if he can attain the sufficient funds and the necessary education to acquire and maintain the high levels of technology. If the current situation continues, the farming family will not have an "equal opportunity" to invest in the technological improvements. This problem could lead to the family farm's vanishing from the earth. The possibility of farming technology causing the elimination of the family farm forces one to ask, "Is technology always good for everyone involved in the profession being affected?" Some of the pessimistic effects of technological advancements and its implications can be seen in The Cornucopia Project.[65] The Cornucopia Project is one strong effort to eliminate the negative aspects of family farming in the 1980's.

Riches to Rags Part I: The Land

"The United States will lose twenty-six square miles of its land today. It will lose another twenty-six square miles tomorrow, and every day this year. But not to a foreign power. We are giving up our land to the ravenous demands of the food system.[66] In 1980 and 1981, three million acres of farmlands were lost because of erosion. Five billion tons of topsoil are displaced annually--enough to cover all five boroughs of New York City with a 13 1/2-foot deep layer of soil.[67] While this is going on, each year, another three million acres of agricultural land are lost to development: new homes, factories, and other structures. We should dedicate the latter three million acres to the requirements of "progress".

Why is it that, in the acres still being farmed, we are

eating up more land than food? Today's food producers are "burning up" the soil with chemicals in order to maintain high yields. When the farmers overuse chemicals (in fertilizers and pesticides), the soil becomes more vulnerable to wind and water erosion. Also, the ways of commerce are now overruling the wisdom of crop rotation. Finally, modern tractors do not permit the strip cropping and contour plowing that are most effective for preventing erosion because the heavy equipment compacts the soil.

These problems are man-made blights. These are symptoms of a food system that is dangerously out of touch with our new world of limited resources; a food system that is headed for disaster, unless we do something about it. The impending disaster is not easy to believe because of our bumper harvests almost every year and our grain-clogged warehouses.[68]

The danger signals are inescapable--we have already lost half our land and gained soaring prices because the food system needs more and more expensive energy in order to operate.

America can produce food at less cost to the land and to the consumer. The nation's citizens do not have to pay for food today with fertile soil they will need tomorrow. Such practices as proper plowing, crop rotation, and the building up of organic matter and nutrients in the soil will permit intensive use of our land for food production, while at the same time preventing erosion and eliminating the need for chemical fertilizers and pesticides.[69] Planning for these changes will take imagination and effort if we are to reverse our "riches to rags" story to the sought after "rags to riches" reality. "Good land is the only foundation for low-cost food production. If our land continues to disappear, any hope we have of avoiding super-luxury prices for food will disappear with it."[70]

In the analysis of agricultural research possibilities and the effects it could have on the family farmer because of financial costs, educational requirements, and removal of his healthy land, one realizes that all farmers must act responsibly if they choose to use high-technological equipment and chemicals. Also, commercial farmers should encourage the development of new discoveries that are definitely safe for the soil and share the benefits with the

family farmer (the large farms could pick up the tab too!).
We should also pay more attention to the agricultural
research system's priorities and methods of action since it
is becoming increasingly clear that the attitude "technology
for technology's sake" is no longer affordable or wise. We
need to develop a system approach to public agricultural
research, concentrating on the development of new
technologies consciously designed to contribute
simultaneously to the solution of many problems spanning
economics, conservation, and efficiency.

Today the land is in danger of being harmed. The
family farmer is also in a 'shaky' position; he is torn
between the real, depressing facts of today's American
farmer and the genuine advantages of living directly with
the land.

Riches to Rags Part II: The Farmer

> I can be my own boss and work without the noontime
> whistle or the impersonal thud of a card sliding
> out of a time clock ... I wouldn't swap the long
> hours for a 9 to 5 desk job or trade in my years
> for a three-piece suit.[71]

>> (Family Wheat Farmer; Loveland,
>> Colorado)

> For months, I've been trying to figure out what I
> did wrong. You get angry. You can cuss Reagan,
> you can cuss the Secretary of Agriculture, but
> there's no one (person) to blame.[72]

>> (Family Cotton Farmer; Floydada,
>> Texas)

> I wouldn't trade my life with anyone. I want to
> remain a farmer for the rest of my life. It's a
> challenge and there is a new adventure every
> day.[73]

>> (Family Corn Farmer; Johnstown,
>> Colorado)

> If it goes on another couple of years this way, I
> may be out of business. That'd be the end of over

a century of my family on this farm.[74]

(Family Soybean Farmer; Cropey,
Illinois)

These are a few examples of the diversity of strong feelings of the post-industrial family farmer. Unfortunately the list of "rags" are far greater than the list of "riches". Will there ever come a time again when the riches will be greater than the rags? And if the switch does not take place in the near future, how long will the traditional, proud spirit of the farmer remain in his innate culture? What does the future hold for the family farmer considering all the odds he is up against?

Conclusion: Family Farms and the Future

A report issued by the Subcommittee on Family Farms of the House Committee on Agriculture in 1956 described the conditions that threatened the family farm and the structure of the agricultural sector at that time. space 1;on subpara 5,5 America cannot afford to allow the substitution of a hired-hand industrial-type of agriculture, for the independent family farmer on the land. The Nation's policies must be shaped to perpetuate the family as the dominant operating unit in agriculture.[75]

Technological change has been the driving force behind the unprecedented growth in the size of the farm in the past three decades.[76] But what is the significance of this technological trend and the other trends discussed in this paper for the future of American agriculture? If present trends continue, very large farms will control an increasing proportion of farm resources and farm output. The family farm shows evidence of continuing to be the dominant operating unit in agriculture, but these family farms will be larger and less independent than today's family farms. Smaller farmer operators will continue to feel the uneasiness as they watch their more affluent neighbors taking control of farms and adding to their operations. Many of these smaller farm families face a gloomy future, with below-average incomes, lack of capital to expand, and inability to compete with their larger 'rivals'.

A January, 1983, television commercial stated that

70

high productivity in agriculture depends on the continued existence of the family farmer; the U.S. is concerned about the family farmer's struggle and must work towards getting rid of the obstacles they face. This commercial demonstrates the relevance and the seriousness of the family farmer's situation in today's post-industrial society. The people of the United States must care about the family farmer or else his existence (lifestyle, job, etc...) is in jeopardy.

Today, we have a controlled economy where the society has neglected to treat equitably the producer. We must work consciously to see that the family farmer's 'fair share' is accomplished. The family farmer has been deprived of his just income because of embargos and other controls by the Government. As soon as the farmers are allowed to produce and sell the amounts that are needed for consumption for all -- both home and abroad -- then, we will witness a farming industry on par with the rest of the economy and a family farmer who will not be in trouble, but one who is a healthy member of the post-industrial society.

ENDNOTES

[1]Family farms are defined as those not operated by hired managers and which use less than 1.5 man-years of hired labor. These are 90% of all farms and account for 60% of the cash receipts. Larger than family farms are defined as not operated by hired managers, but using more than 1.5 man-years of hired labor. These are 8% of all farms and produce about 20% of farm receipts. Industrial farms have differentiated capital ownership, management, and labor. These corporate farms account for 20% of farm output. A farm, in general, is a place with agricultural sales of $1000 or more annually.

[2]Alvin Toffler, The Third Wave (New York: William Morrow and Co., Inc., 1980), section on the "first wave".

[3]Ibid., section on the "second wave".

[4]G.M. Berardi, "Socio-Economic Consequences of Agricultural Mechanization in the United States: Needed Redirections for Mechanization," Rural Sociology, 46 (Fall 1981), 483.

[5]William Sweet, "The Plains States: World's Breadbasket," Editorial Research Report, 1 (1980), 368.

[6]Robert Higgs, The Transformation of American Economy,1865-1914: An Essay in Interpretation (New York: John Wiley & Sons, 1971), p. 95.

[7]Ibid., p. 96.

[8]G.M. Berardi, p. 486.

[9]Parity is simply a 'relationship'. All it means is that the price the farmer gets for what he sells is fair in relation to the price he pays for what he buys.

[10]Robert Higgs, p. 99.

[11]Ibid., p. 96.

[12]Ibid., p. 101.

[13]Alvin Toffler, section on the "third wave".

[14]Farm Income and Farm Structure in the United States, U.S. Congress. House. Committee on Banking, Finance, and Urban Affairs (Washington, D.C.: CRS, 1979). p. 24.

[15]See footnote #1 for the definition of a 'farm'.

[16]Farm Structure: A Historical Perspective on Changes in Number and Size of Farms, U.S. Congress. Senate. Committee on Agriculture, Nutrition, and Forestry (Washington, D.C.: CRS, 1980), p. 30.

[17]Farm Income and Farm Structure in the U.S., p. 25.

[18]For a description of the original 'treadmill' concept as it applied to farmers on their way out of agriculture, see Willard W. Cochrane, Farm Prices: Myth and Reality, chapter 5. "The Agricultural Treadmill." University of Minnesota Press, 1958.

[19]Farm Income and Farm Structure in the U.S., p. 27.

[20]Ibid., p. 28.

[21]Ibid., p. 28.

[22]Ibid., p. 29.

[23]Ibid., p. 30.

[24]Ibid., p. 33.

[25]Ibid., p. 34.

[26]Eric O. Hoiberg and Gordon L. Bultena, "Farm Operator Attitudes Toward Governmental Involvement in Agriculture," Rural Sociology 46 (Fall 1981), 381.

[27]Ibid., p. 381.

[28]Ibid., p. 381.

[29]Ibid., p. 382.

[30]Ibid., p. 382.

[31]Ibid., p. 382.

[32]Ibid., p. 382.

[33]Ibid., p. 382.

[34]Ibid., p. 382.

[35]Farm Income and Farm Structure in the U.S., p. 35.

[36]Ibid., p. 36.

[37]Ibid., p. 37.

[38]Ibid., p. 37.

[39]Ibid., p. 38.

[40]Ibid., p. 39.

[41]Ibid., p. 40.

[42]Ibid., p. 40.

[43]Ibid., p. 41.

[44]Ibid., p. 42.

[45]Ibid., p. 42.

[46]Ibid., p. 42.

[47]Ibid., p. 43.

[48]Ibid., p. 43.

[49]Congressman Robert Krueger, "Krueger Reports from Washington," Fredericksburg Standard, April 26, 1978.

[50]Ibid., see list following the article.

[51]Mr. S. Edward Murphy's American Agriculture Survival Kit, p. 13.

[52]Marc Leepson, "Advances in Agricultural Research," Editorial Research Reports 1 (1981), 373.

[53]Ibid., p. 373 from "The MacNeil/Lehrer Report," PBS-TV, March 19, 1981.

[54]Ibid., p. 373.

[55]Ibid., p. 374 from an Interview, April 28, 1981. The Cell Culture and Nitrogen Fixation Laboratory is located at USDA's Agricultural Research Center in Beltsville, MD.

[56]Ibid., p. 374.

[57]Ibid., p. 374.

[58]Ibid., p. 374.

[59]Ibid., p. 376.

[60]Ibid., p. 376.

[61]Ibid., p. 376.

[62]Ibid., p. 377.

[63]Ibid., p. 377.

[64]Ibid., p. 377.

[65]The Cornucopia Project is the first attempt by a non-governmental organization to make a systematic study of America's food system.

[66]The Cornucopia Project (Emmaus: Rodale Press, 1980), pamphlet 2, p. 2.

[67]Ibid., pamphlet 2, p. 2.

[68]Ibid., pamphlet 2, p. 2.

[69]Ibid., pamphlet 2, p. 12.

[70]Ibid., pamphlet 2, p. 12.

[71]Scott Johnson, "Starting Out: Young Farmers Face an Uphill Struggle," Colorado newspaper.

[72]Kurt Andersen, "A Bitter Harvest," <u>Nation</u> (October 1982), p. 31 quoted by Billy Fulton.

[73]Rick Potburg, "Starting Out: Young Farmers Face an Uphill Struggle," Colorado newspaper.

[74]Kurt Andersen, "A Bitter Harvest," p. 28 quoted by Herb Steffen.

[75]State of the Farm Economy. <u>Hearings before the House Committee in Agriculture</u>, U.S. Congress. House Committee on Agriculture (Washington, D.C.: GPO, 1982), p. 126.

[76]Ibid., p. 127.

BIBLIOGRAPHY

Andersen, Kurt. "A Bitter Harvest." Nation, October 1982, pp. 26-33.

Bell, Daniel. The Coming of Post-Industrial Society: A Venture in Social Forecasting. New York: Basic Books, Inc., Publishers, 1973.

Berardi, G.M. "Socio-Economic Consequences of Agricultural Mechanization in the United States: Needed Redirections for Mechanization." Rural Sociology 46 No. 3 (Fall 1981) : 483-504.

Daly, Patricia A. "Agriculture employment: has the decline ended?" Monthly Labor Review 104 (November 1981) : 11-17.

Dentzer, Susan, "Gloom in the Grain Belt." Newsweek, June 1980, pp. 66-67.

"Farm Economy, Depressed by Surpluses, Isn't Expected to Perk up Much in 1983." The Wall Street Journal, 9 December 1982.

"Farmers Start Reviving Penny Auctions, A 1930's Tactic for Preventing Foreclosure." The Wall Street Journal, 10 September 1982.

"Foreigners Buying Farms in Georgia." York Daily Record, 9 October 1979.

Fuel from Farms. Oak Ridge: Solar Energy Research Institute, 1980.

Higgs, Robert. The Transformation of the American Economy, 1865-1914: An Essay in Interpretation. New York: John Wiley & Sons, Inc., 1971.

Hoiberg, Eric O. and Gordon L. Bultena. "Farm Operator Attitudes Toward Governmental Involvement in Agriculture." Rural Sociology 46 No. 3 (Fall 1981): 381-390.

"Krueger Reports from Washington." Frederiscksburg

Standard, 26 April 1978.

Leepson, Marc. "Advances in Agricultural Research." Prepared by the Editorial Research Reports 1 No. 19 (1981): 371-388.

National Farmers Union; Washington Newsletter. 29 No. 25 (June 1982)

Nellis, Micki. Makin' It on the Farm: Alcohol Fuel is the Road to Independence. Iredell: American Agricultural News, 1979.

Poole, Dennis L. "Family Farms and the Effects of Farm Expansion on the Quality of Marital and Family Life." Human Organization 40 (Winter 1981): 344-349.

S. Edward Murphy; "American Agriculture Survival Kit." Compiled 1979.

"Starting Out: Young farmers face an uphill struggle." Clipping from a Colorado Newspaper.

Sweet, William. "The Plains States: World's Breadbasket." Prepared by the Editorial Research Reports 1 No. 19 (1980): 367-384.

The Cornucopia Project. Emmaus: Rodale Press, 1980.

Toffler, Alvin. The Third Wave. New York: William Morrow and Co., Inc., 1980.

U.S. Congress. Senate. Committee on Agriculture, Nutrition, and Forestry. Farm Structure: A Historical Perspective on Changes in Number and Size of Farms. 96th Cong., 2d sess., 1980.

U.S. Congress. House. Committee on Agriculture. State of the Farm Economy. Hearings before the House Committee on Agriculture. 97th Cong., 2d sess., 1982.

U.S. Congress. House. Committee on Banking, Finance, and Urban Affairs. Farm Income and Family Structure in the United States. 95th Cong., 1st sess., 1979.

"What's the Future for Maryland Farming?" Lancaster

Farming, 18 December 1982.

Williams, Anne S. "Industrialized Agriculture and the Small-Scale Farmer." _Human Organization_ 40 (Winter 1981): 306-312.

Witts, David A. _The Power of Parity_. Midland: Northwood Institute Press, 1979.

THE EFFECTS OF POST-INDUSTRIAL TECHNOLOGY UPON DEFENSE

Elaine Hesser

Introduction, Purpose, and Scope

War continues to be a reality of life. Armed conflict (here synonymous with "war") ranges in scale from the Beirut massacre to the battle of Gettysburg. It is as old as Israel, and as modern as satellites. It is as remote as Moscow, and as close as the draft registration card at the local post office.

If we accept war's reality and proximity as a "given", it is only logical to study the ways in which it is fought. Post-industrial technology affects warfare, just as it touches all aspects of life. It is my contention that the weaponry made possible by this technology will not, in the near future, drastically alter our capability to attain a decisive victory in battle, although its development will be necessary if we (the United States) are to have a credible defense. It is my further contention that the "human factor" is and will continue to be the most important determinant in any armed conflict.

In order to demonstrate these facts, this paper will first describe in detail three of the United States Department of Defense's latest weapons systems. The advantages and disadvantages of these systems will be discussed and some facts about other systems and equipment presented to reinforce the arguments both pro and con. Following this presentation of cases for and against development and employment of post-industrial weapons, an analysis and some conclusions will be offered.

This study will not provide a comprehensive picture of the United States' present or future military capabilities. It is intended only as an overview of the type and extent of post-industrial technology involved. Only after understanding the functions and capabilities of a technology can one discuss its effects and future prospects. Yet even the most responsible analysis may contain elements of subjectivity. Clearly the subject of this paper is open to debate.

Some brief definitions and explanations are helpful in understanding the terms and weaponry involved. "Post-industrial weaponry" will mean weaponry designed to kill or defend through its quality, not its quantity. In general, weapons of the past were designed to put out as much firepower as possible--large artillery pieces, automatic rifles, and so on. As we move out of the industrial era into post-industrial technology, greater accuracy becomes possible; therefore, less ammunition (and theoretically, less cost, in the long run) is necessary. Fewer men and weapons should be needed to counteract the same amount of opposing force.

Lasers are the first type of weapons to be discussed. Actually, lasers can be used as part of a guidance system, detection system, or as weapons in themselves.

Tied directly to lasers are satellites, which will be the second type of system discussed. "It has been estimated that of all man-made objects placed in space, 90 percent were for military purposes--put there by the two superpowers and NATO."[1] Under these circumstances, no study would be complete without a look at the effects of satellite technology.

A third kind of post-industrial technology being developed by our armed forces is remote control technology. In fact, a "drone" vehicle, called an RPV (Remotely Piloted Vehicle), has been used by the Israelis against the Arabs. This is one of the most exciting post-industrial advancements, because it is a step toward the "unmanned" battlefield. According to James W. Conan:

> The advent of these precisely controlled robots of the sky truly promises to eliminate humans from combat, to make of warfare a match-up of machines and electronic devices which the military and civilians alike will watch on the tube, complete with instant replays and halftime highlights. Should we live so long.[2]

This is perhaps an idealistic assumption, but let us leave the analyses for the end of the study. For now, it is time to examine some of the "space-age weapons" of the Pentagon, beginning with lasers.

Long Ago, In a Galaxy Far, Far Away: The World of Lasers

Lasers are beams of low-frequency light which can be concentrated in various intensities to do almost everything, from guiding a missile to removing a cataract and performing other microsurgery, to destroying helicopters. Laser weapons are not as new as one might think. The military use of lasers can be traced to the 1960's in the Vietnam conflict.[3]

As a guidance system--or rather, as part of one--a laser beam is emitted from a friendly soldier's position. The projectile being fired "locks into" the reflection of the beam from the target, and follows it to the target. This is an improvement over weapons such as the older, wire-guided anti-tank missiles. Since it is wire-guided, the operator must remain in position after it is fired, until it reaches the target. Since a good marksman takes approximately three seconds to aim at and shoot a target, remaining exposed for this length of time is regarded as undesirable. Also, the wire can be tangled or broken, leaving the missile flying an erratic path away from the target.

Laser-guided weapons can also be used against aircraft. One such weapon under consideration by the Army and Marine Corps is the FIM-92A "Stinger Alternate". This is an approximately 35-lb., one-operator weapon. Because of its use against aircraft, it is less likely to face the problems of thick underbrush or trees blocking the laser beam. This is a very expensive weapon, which cost approximately $45,705 in 1979.[4]

Of course, anything which can block a beam of light can also block a low-power laser beam. Also, a laser-guided weapon can be "confused" by playing another beam on a different target. One way to avoid this would be to "code" a laser's pulsations or vibration frequency. This would be a ". . . new dimension \in} the military art of codebreaking."[5]

Lasers can also be used as weapons in themselves at higher intensities. In fact, the Department of Defense requested $435 million for fiscal 1983 to fund ". . . the first U.S. long-range space \laser} system . . ."[6] The estimated cost over the next 20 years is up to $300 billion for the development and deployment of planned laser systems.[7] The United States is not alone in its development of this field, either: "There is no question that the Pentagon's growing enthusiasm for lasers is influenced by Soviet progress in the field."[8]

In the defensive mode, lasers can be used to "craze" enemy sensors or detection systems. As discussed above, enemy laser-guided missiles can be confused by more than one beam being reflected in the area of the target. In the "Buck Rogers" department, lasers which can melt metal are under development. These high-powered, high-energy weapons:

> . . . abruptly will upset the balance of today's offensive and defensive tactical and strategic weapons, superseding them as the penultimate defender and destroyer, capable of turning men into messes of mush, their machines into molten metal.[9]

These weapons can be mounted on aircraft (an Air Force KC-135, for example), and used to defend U.S. helicopters and airplanes. Another use of such lasers is in Anti-Ballistic Missile (ABM) systems.

These systems are designed to stop missiles by shooting them down in mid-air. So far, the systems which have been developed have not proven to be effective. Laser technology would change this, because ". . . a missile, ten miles away and moving at twice the speed of sound, could be identified, struck and destroyed before it could move one inch."[10] This has been tested on actual (non-nuclear) missiles, traveling between 450 and 780 miles per hour, with 100% accuracy.[11] The effect of such a weapon deployed in large enough numbers would be to make ballistic missiles obsolete as weapons of attack. If one superpower were to develop this capability before the other, it would seriously upset what some call the "balance of terror" by giving that nation an uncontestable first-strike capability.

Finally, lasers can be used to communicate. This would be particularly effective in communicating with nuclear submarines, a use being developed by the Department of Defense.[12]

83

Further limitations and advantages of lasers will be discussed in later sections; however, lasers and satellites are a combination which is quite effective. That will be the topic of the next section.

Lost in Space: Satellites

Since lasers can be fouled by haze, clouds, and terrain, an excellent place to mount them would be in space, where these problems are virtually non-existent. Also, outer-space mounting would increase the reaction time available in the event of a missile launch.[13] Satellite-based warfare capability is not a thing of the distant future. According to an article in Business Week, Richard D. DeLauer, Undersecretary of Defense, stated that ". . . by next year, the Soviets will be able to deploy a laser space station that could threaten U.S. satellites."[14] By 1990, he continued, U.S. bases, material, and weapons could also be targeted by Soviet satellites in space.[15]

The United States' work in space lasers does not fall under any single branch of the Armed Forces, but under the Defense Advanced Research Projects agency, or DARPA. DARPA is currently testing a space laser called "Talon Gold". Although it is subject to disruptions from the satellite's vibrations, it may be tested on a space shuttle by the late 1980's.[16]

In addition to their use as weapons bases, satellites also serve a vital communications function. They are used in our early warning systems, and to police the SALT agreements.[17] In addition to being extremely important to our defense, communications satellites are also terribly vulnerable. There are no armed guards floating around them with M16 rifles to ward off attackers.

Do the Soviets have the capability to disable our satellites? Consider the following:

Lasers of relatively low power would suffice to blind the infrared sensors of many early warning satellites. When one of our MIDAS satellites over the Indian Ocean was blinded in October of 1975, and two U.S. Air Force satellites over the Soviet Union were blinded the following month--one of them for more than four hours--<u>they</u> <u>may</u> <u>well</u> <u>have</u> <u>been</u> <u>the</u> <u>targets</u> <u>of</u> <u>illumination</u> <u>by</u> <u>Soviet</u> <u>lasers.</u> (Emphasis added)[18]

Also, satellite communications ". . . are subject to Soviet eavesdropping."[19]

Acknowledging all of these vulnerabilities of our own satellites in combination with the development of Soviet satellite capabilities (and "killer satellites"), the defense of space stations has become a concern.

As with the development of the task came the development of the anti-tank weapon, so with the development of the satellite comes the advent of the anti-sattelite weapon. In fact, over fifteen years ago, two bases were built in the Pacific to fight Soviet satellites. Their subsequent discovery and the furor raised over the bases-which were armed with nuclear missiles caused their closing in 1975.[20] Other possibilities for anti-satellite weaponry include "space mines" (just like the underwater mines used in the World Wars, these would explode on contact with an enemy satellite), or a "mini-bomb" that would zero in on a sattelite after being launched from an F-15 airplane ("mini" in this case refers to a weight of a mere four pounds).[21]

Other defenses of existing and future satellites will include placing spare satellites, operational on command, in orbit. Filters can be placed on satellites to guard laser blinding, and anti-jamming and encryption devices to aid in communications security can be added.[22]

All the possibilities of "space wars" have not yet been explored. Some ideas and questions that leap immediately to mind are that the more war that is fought in space, the less that is left to be fought on earth. On the other hand, space at the peak orbital band is limited; who will legislate which superpower will have access to this limited space? What effect would an electromagnetic pulse from a nuclear weapon have on satellites in space? What effect would this use of space have on our upper atmosphere? These are only some of the questions which must be answered before we see "space wars" or even space defense become a reality. The technology is nearing completion; is the knowledge for its employment ready?

Drone Bees: Remotely Piloted Vehicles

Remotely Piloted Vehicles, are in general, unmanned, "drone" airplanes. They can be equipped with television cameras, sensors, or radar, and were first used as reconnaissance vehicles. During the El-Al battle in the Arab-Israeli war, they were flown about by the Israelis to confuse Arab radar.[23]

The Israeli version of the RPV is called the "Scout", and has a rather remarkable viewing capability. At a height of 1000 meters (approximately 3/5 of a mile), it can view an area of 19.3 square miles.[24] This information is then transmitted by the RPV to a controller at the rear, who is far safer than any battlefield reconnaissance team that may have been sent out to accomplish the same mission. The information can also be recorded aboard the RPV for future use. In the most recent Arab-Israeli conflict, the Israeli attack on Syria in Lebanon, "The use of remotely polited vehicles enabled combat commanders at all levels to have a near real-time (in time to influence this battle) projection of the fighting in Lebanon as it was taking place."[25]

Another use the Israelis found for the RPV was as a "decoy". In order to use its radar-guided missiles, Israel first had to trick the Arabs into turning on their radar. Rather than risk an airplane and crew, or waste a missile, Israel sent unguided RPVs. Once the radar was turned on, the guided missiles could be released, acquire the target, and destroy it.[26]

Also, RPVs can be equipped with electronic jammers, or laser designators which take the place of the soldier standing exposed to fire while keeping his laser beam on the target.[27] The missile can then "home in" on the laser from the RPV without the risk to the individual soldier.

RPVs are rather small, from 8-12 feet in length, and can stay aloft for 3-5 hours. Their maximum altitude is approximately 10,000 feet. They cruise at the same speed as the average automobile.[28]

Another use for an RPV, suggested by Conan, would be a kamikaze type of mission.[29] One imagines the RPV carrying an explosive payload too small to warrant the use of a guided missile. An RPV could also fly low enough to escape detection by radar.

To summarize, the RPV contributes mainly to battlefield intelligence. It provides "real time" information, and can also record this information on videotape for future use. As a weapon, it can perform "kamikaze" missions, or replace the infantryman in laser-guided weapons systems. It can also serve as a decoy for enemy radar or "confuse" and jam enemy electronic equipment. An additional use could be to transport small replacement parts for battlefield computers, or even small arms. In brief, it replaces soldiers and larger, more expensive vehicles, in many hazardous missions.

It should be possible (at least one imagines it should) to develop RPVs with a capability to fight off other RPVs, which would take us one step further toward the unmanned battlefield described by Conan.

Procrastinate Now, Pay Later: Why Favor These Advances?

One reason offered for the outlays in military research and development is that the military gives rise to ". . . a technological advance that is unmatched in the civilian field."[30] Some of these advances may later have a positive effect on civilian technology and manufacturing. One very old example comes from the development of the rifle. Eli Whitney developed the concept of interchangeable parts in rifles. This made them easier to repair, because a handmade part was not necessary each time a part was broken or damaged. This idea of course spread to industry, and later led to the assembly line concept developed by Henry Ford. The airplane was also developed in large part by the military. Therefore, it can indeed be concluded that one reason for favoring technical and scientific advancements in the military is that these eventually will be beneficial for the civilian industries as well.

A second reason for favoring the type of advancements discussed is put forth by Daniel Bell: "The rated power of a country no longer rests on its steel capacity but on the quality of its science and its application, through research and development, to new technology."[31] Obviously, this applies to all areas of a country's interest, but particularly to the military. The late General Creighton

Abrams, in 1974, echoed this thought:

> When I entered the service \in 1936} it was
> largely a horsedrawn army. The battlefield has
> changed. We have had to assimilate radar, jet
> aircraft, nuclear weapons . . . to mention only a
> few. Every lesson we have been able to dig out of
> the Mideast war fully reinforces our faith that we
> are headed in the right direction . . . We must
> assure that American soldiers are not surprised on
> some future battlefield by a technologically
> superior enemy.[32]

Also, the geographical location which deTocqueville
once thought to be so advantageous has now become a
liability of sorts. The world has been shrunken by
interdependence and advanced communications, yet the United
States remains relatively isolated:

> \U.S. strategy} . . . must begin by confronting
> the harsh reality of geography: the United States
> is at a distinct disadvantage when it comes to
> leaning on other nations. Merely by massing its
> troops along . . . its borders, the Soviet Union
> can influence events in all three of the areas
> most vital to U.S. interests--Western Europe, the
> Middle East and the Pacific environs of Japan and
> the Koreas . . . In order to influence events in
> countries like West Germany or Saudi Arabia \the
> U.S.} must be able to project power over thousands
> of miles, with all the attendant logistical
> problems.[33]

So far, that projection of power has only been
available through nuclear weapons. Space and long-range
guidance systems for conventional weapons could change
this.

In more specific terms, the use of such items as
lasers, satellites, and RPVs has been proven effective in
both Lebanon and the South Atlantic. Clarence Robinson
called Israel's surveillance capabilities "pivotal" in the
Lebanese/Syrian conflict.[34] Since the Syrians used Soviet
equipment and tactics, this may be considered something of
an indicator of this technology's effectiveness against the
Soviet Union.

The USSR is perceived to have a superiority of quantity. Therefore, it is the role of the United States to match that superiority with greater quality. According to U.S. News and World Report,

> Even before the Mideast and South Atlantic conflicts, the so-called quality vs. quantity debate was tilting in favor of the high-tech 'quality' side . . . Now, the Pentagon intends to argue . . . that the U.S. should press ahead with a wide array of 'miracle' weapons. Defense experts claim that . . . equipped with such arms \the U.S.} can defeat a larger Soviet adversary.[35]

Another advantage touted by supporters of post-industrial weaponry is that it would save Western European property and cities from damage. "With an arsenal of precision-guided, electronic weapons, the North Atlantic Treaty Organization (NATO) could carry the battle deep into Soviet territory instead of being pinned down on Western soil."[36]

Obviously, the perception of the Soviet Union as a nation equipped with larger numbers of men and machinery is one major motivating force behind these developments. As Harold Brown, ex-Secretary of Defense, stated: "Given our disadvantage in numbers, our technology is what will save us."[37] This statement sums up rather neatly this particular set of arguments in favor of "high-tech" weaponry.

Other motivating forces deal with specific weapons. The laser, for example, is seen as an excellent weapon. It is not an indiscriminate killer, as a nuclear or even conventional warhead is. It will not kill large numbers of non-combatants. It can be directed at individuals, pieces of machinery, or whatever is necessary, ". . . only if they themselves are threatening."[38]

Additionally, laser guidance and optics can provide the infantryman with a "one-shot kill" capability. Ammunition can be heavy and cumbersome to carry. Since maneuver and mobility are two keys to success in conflict, increasing them by decreasing the weight a soldier must carry is certainly an advantage. Furthermore, the one-shot kill capability should save money in the event of a war, because less ammunition would be necessary. Other technological "wonders" (for example, a laser-guided missile), are far cheaper than, say, a tank. Therefore, we would not have to build more tanks to battle tanks, only more of the accurate anti-armor weapons which are now available. These would, of course, be less expensive.

Finally, weaponry such as an RPV would keep men and women off the battlefront. This would not only save such incidentals as salary, combat pay, disability pay, and veterans' benefits, but also the priceless commodity of human life. This life is precious both in itself and in its productive contribution to other sectors of the economy.

Satellites have the capability to remove the battle from our very planet. Imagine the savings in farmlands, factories, and irreplaceable structures and relics! There is no guarantee that satellites would not be targeted on ground sites, but if both superpowers deployed satellites at the same time, it is more likely that other satellites would become primary targets.

While computers can be used to integrate satellite systems and serve as part of advanced guidance systems, they are also useful by themselves. For example, the Pentagon is a great producer of "worst-case scenarios", in which everything that can go wrong, does. In these scenarios, there is a hypothetical situation--for example a Soviet attack on Western Europe--and then an attempt to generate responses to and probable outcomes of this attack. The computer brings these scenarios to life, making them more realistic and providing better training than merely moving plastic tanks across map sheets. The "Janus" program uses a $2.45 million computer to simulate in detail ". . . any 15-square-mile slice of the earth . . ."[39] It is an elaborate, computerized version of war games which date to the late 1700's, in which the "players" use war tanks, personnel, artillery, chemicals, and tactical nuclear weapons.[40] The computer can even calculate the effects of weather, and of fires caused by nuclear blasts. An M.I.T. professor who is ". . . an expert on the psychological impact of computer games . . .", stated that the computer ". . . could heighten the revulsion \to nuclear war}. The computer is confronting us with . . . the brute fact that we are playing with the survival of the planet."[41]

This is but one way in which post-industrial technology
can help prevent a nuclear war. It can also add to the
credibility of the U.S. deterrent. The laser-ABM systems
are a good example. Were the United States to develop this
defensive capability before the Soviet Union, our security
against a first strike would be assured. This would be a
great deterrent to a Soviet first strike, and thus a
nuclear war.

The Soviets' main conventional strength lies in their
tank force. An effective anti-tank weapon would deter the
Soviets from attempting to use those tanks. One such weapon
is the neutron warhead. This is a warhead which can be
fired from conventional artillery, and which limits the
blast area to a few hundred yards; however, opponents of
this weapon fear that it would lead to escalation to a
strategic nuclear war. If a viable, armor-piercing weapon
(the VIPER is the latest attempt) is developed, the state of
the art of guidance systems will be combined with it to be a
strong deterrent against an armor vehicle attack.

Why, with all these apparent advantages, are we not
wholeheartedly embracing technology as our defense? If, as
a 1974 Senate Armed Services Committee report states "'. . .
defensive electronic warfare equipment may well determine
the ability of our forces to survive an enemy attack.'"[42],
why do we only read and hear of B-1 bombers and MX missiles?
Obviously, there must be disadvantages to the use of
advanced electronics or other such equipment. Why would
some favor the old Sherman tank over M-1 Abrams or the B-52
over the new, advanced B-1? Obviously, this weaponry must
have some disadvantages, which will be discussed in the
following section.

Murphy's Law #3--The more complicated a machine is, the
more likely it is to break down: The Arguments Against
Post-Industrial Weaponry

Murphy was right about a lot of things--the toast falls
jelly-side down, an elephant was a horse created by
committee, and of course, Law #3, above.

Nowhere has this been more publicized than where it
concerns the M-1 Abrams tank. The tank has a new type of

stronger armor, can shoot while moving, travel at speeds in excess of fifty miles per hour, and has an on-board computer tracking and aiming system. It can also breach obstacles to a height of four feet and travel up steeper hills than other armor vehicles. Ostensibly, possession of such a vehicle is an unquestionable advantage. However, it has been described in somewhat less affectionate terms:

> The tank often proved 'allergic' to dust, a commodity often found on battlefields. Its transmission failed so often that one Senate tank expert observed, 'General Motors would have issued a recall notice for all of them'.[43]

The three space-age weapons described in this study also have their drawbacks. First of all, the laser can be stopped by haze or smoke, not to mention brush or buildings. Since smoke grenades, which are normally employed as signal devices, are in ready supply, this makes the laser quite vulnerable. Intentional smokescreens aside, there is also the smoke from fires, artillery, and explosions of all types. Furthermore, although substantial research has been completed, much extensive work remains to be done on laser weaponry. For example, in order to increase the power of a laser beam, it is "bounced" back and fourth between mirrors; when the power level becomes great enough (over 1 million watts), vibrations which upset the alignment of the mirrors begin to occur.[44] Once scientists learn how to stop these vibrations in aircraft, a way must also be found to prevent them in outer space. Further problems in outer space include "\designing} . . . a weapon that must be boosted into space and have enough fuel to produce pulse after pulse of power."[45] The expense of such weapons is also prohibitive. One estimate states that

> If the U.S. deploys all the laser weapons now under development, the cost could be gigantic--$300 billion over the next 20 years.[46]

This sort of feat is highly unlikely, but Senator Malcolm Wallop of Wyoming has proposed something almost as fantastic. His plan for the defense of the United States calls for 24 laser stations orbiting in space; the testing for this sort of project will not ". . . be possible until 1987 at the earliest."[47] As for the use of lasers as weapons in themselves (other than in outer space ABM systems), ". . . so far they have shown no superiority over conventional artillery."[48]

The satellites themselves pose even more problems. As mentioned earlier they can be "blinded" by enemy laser illumination. Even though ". . . satellites are relatively inexpensive in comparison to other long-distance communications", they are subject to Soviet "eavesdropping".[49] There is also the danger of Soviet "hunter-killer satellites". Furthermore, the satellites would have to be deployed in rather large numbers (See Sen. Wallop's plan, above), which raises the questions of cost and airspace. Even DARPA admits there are great difficulties involved.

> DARPA officials are optimistic that the larger battle station they are planning will be able to shoot down enemy satellites, submarine launched ballistic missiles, or flights of bombers. But such objects will be sitting ducks compared to countering land-based intercontinental missiles, which could be launched by the hundreds of thousands at the U.S. and must be destroyed in the early moments of flight.

> 'That would be a stressful mission.' admits DARPA's Tanimoto. 'We would need orbiting laser battle stations in multiples of 10 . . .'"[50]

Multiples of 10? The Wallop plan had only called for 24; current research and development funds are already set at $2.4 billion for high-energy lasers alone.[51] None of these figures take into account communications satellites and their maintenance. The cost for DARPA's "shield" concept was estimated to be "'. . . at least $200 billion, including use of a space shuttle to assemble stations in space and maintain them'", according to a "former high-ranking Defense Department official."[52]

So far, the RPV has exhibited none of these problems. However, to outfit it for employment on "kamikaze" missions--one use suggested by Conan in Superwarriors--carrying explosives would be to put them under the terms of the SALT II treaty. Specifically, in Article II, paragraph 8, which discusses cruise missiles:

. . . <u>unarmed</u>, pilotless, guided vehicles shall not be considered to be cruise missiles if such vehicles are distinguishable from cruise missiles on the basis of externally observable features (emphasis added).[53]

RPVs are "easily distinguishable" from cruise missiles; arming them would not help the cause of defense if they would then be counted as "cruise missiles".

In general, "high-tech" weaponry has received mixed reviews--from Conan's <u>Superwarriors</u>, which is irrepressibly optimistic, to James Fallows' article in <u>The Atlantic Monthly</u>. While Fallows admits that ". . . technical changes have repeatedly altered the nature of war," citing the machine gun (WW I) and the tank (WW II).[54] Fallows says that the theory of technology as our saviour in the realm of defense is ". . . wrong in its implication high technology always, or even usually, increases the military usefulness of a weapon."[55] By using simple mathematics, Fallows refutes the argument that technology is the key to our problem of numbers against the Soviet Union:

The Soviets now add about 500 tactical fighters to their force each year, compared with our average of 250. If a sensible plane could be built for $5 million instead of $25 million, then it would cost about $2.5 billion to match the Soviet output.[56]

Do such comparable pieces of equipment exist? One example might be the Fairchild Republic F-105 Thunderchief airplane, a piece of "industrial" equipment. The plane was used in Vietnam, where ". . . it was found to be one of the simplest and most reliable aircraft to fly . . . Severely battle-damaged F-105s have continued to fly for hours with all their oil lines shot away."[57] It had some problems--it did not turn quickly, and lost speed when weighed down. Yet when it returned from a bombing run, it was fast enough to outrun and outfight a Soviet MIG-21 or -23 plane.[58]

This plane can be compared to the General Dynamics F-16 Air Combat Fighter (see Table 1, below). This plane is highly maneuverable, and can carry the very accurate

Sidewinder Air-to-Air missiles. It has also been equipped with several early warning systems and a "remarkable guidance for its air-to-air missiles."[59]

Table 1: Comparison of F-105 Thunderchief
and F-16 Air Combat Fighter[60]

	F-16	F-105
Weight (loaded)	33,000 lbs.	54,000 lbs.
Maximum Speed	1,400 mph	1,485 mph
Operational Range	1,300 miles	1,840 miles
Ceiling	60,000 ft.	N/A
Armament	20mm cannon	22mm cannon
Ordnance	2 Sidewinder missiles, 10,200 lbs. conventional bombs (5 total)	13,000 lbs. conventional bombs (12 total)
Price	$12,322 million	$2.287 million

Source: Gervasi, Arsenal of Democracy II

Who is to make the decision as to whether the advanced guidance system and Sidewinder missiles are worth the additional $10 million per plane? The Congress and Department of Defense fight these battles regularly. Everywhere one turns, there is someone with an opinion on high-tech weaponry. Newsweek writers Melinda Beck and John Lindsday call attention to the appeal of high-tech weapons, as opposed to conventional forces:

> . . . the Pentagon must correct the glaring weaknesses in its 'general purpose' forces, including personnel, operations, maintenance, spare parts and training--the items that ensure 'readiness' for conventional war. These items are more expensive than nuclear missiles. They lack

the glamour of aircraft or new bombers.[61]

Indeed, one reason offered for the high-tech explosion in defense is the appeal these weapons have for Defense officials. A physicist who works at both IBM and Harvard, Richard Garwin, states that procuring mines for use in wartime to slow the Soviets in their passage to the sea would be a good investment. "'But no one can command a mine . . . You don't get promoted for procuring them. There's no glamour to them.'"[62] All this leads to the conclusion that another drawback of post-industrial weaponry is that it can draw money away from less "glamorous" but necessary or useful ideas.

If those are the major reasons for procuring high-tech weapons, then they are not good ones. This is another of the major arguments against post-industrial weapons. They are not the products of rational planning, some say, but of glamour-grabbing and gadget-hunting. What will be the outcome of all of this? Where do the arguments, both pro and con, leave the defense planner? This will be the topic of the final section.

Launching Pad for an Odyssey: Where do we go from here?

Some view the United States as a David, facing a Goliath-like Soviet Union. One cannot argue that the tendency of our Department of Defense is to sometimes present the Soviets as "ten feet tall". Our strategy, then, is not to get David more stones, but to computerize the guidance system on his sling.

Others, like Ruth Leger Sivard, argue that we have a deterrent, and that it is enough. She argues that "\Defense} can create more vicious weapons of revenge, but it is now powerless to protect populations against them," and ". . . neither distance nor oceans can any longer guarantee sanctuary for non-combatants."[63]

The weapons, when they work, provide a macabre promise of quick, "cleaner", kills. In the long run, they could remove soldiers from the battlefield and the battlefield from the earth. At worst, they are vulnerable to attack, over-sensitive for battlefield conditions, and sap money from more worthwhile projects in large sums. Still, in an era in which we are turning to the computer more and more

frequently for education and business, it seems only logical
to turn to it and its high-tech cousins for the defense of
the nation, as well.

The thought of an unmanned battlefield holds forth
great appeal, but there is a question as to the utility of a
war fought without hurting anyone, in space, without
destroying any property. This may appear to be a highly
paradoxical statement, but the purpose of fighting any war
is to eliminate the enemy's will to fight. The conqueror
then imposes his will on the conquered. The only way in
which the unmanned battlefield would result in that outcome
would be if it placed a drain on each adversary's economy.
Eventually, one nation would literally "go broke" and
surrender. Undoubtedly, this would be a serious drain on
the victor's economy also, and the "winners" would then have
to resurrect two failing economies at once. In a world
moving from bi-polar to multi-polar in terms of military
power, one envisions the uninvolved nations circling the two
dying combatants like buzzards, waiting to come in and take
over both. Even if an economic recovery were somehow
possible, it would have to wait until the defeated nation
were occupied and a proxy government installed, involving
still more expenditure on the part of the victor.

In light of these difficulties, it could be suggested
that computers be used to simulate space battles, rather
than actually to fight them. This would reduce war to
nothing more than a grandiose video game which could effect
an outcome only if it were understood that the simulated
capabilities actually existed and would be used as they were
in the simulation. It then could be argued that to save
both sides great economic hardship the outcome must be
accepted by both parties.

What does this concept of war do to the theory of
deterrence? After all, the primary objective of a
government should be to deter war. Obviously, if it becomes
no more hazardous or risky than a game in an arcade, there
is no real deterrent. War becomes a game of skill, in which
the stakes are the only frightening factors; the only
deterrent is the threat of loss. In the interim between the
present and the era of the unmanned battlefield, however,
the story is a little different. The main deterrent of
conventional war between nuclear states is the fearful
specter of escalation to a nuclear war, but the advent of

97

laser ABM systems changes this by theoretically rendering nuclear missiles useless. At the same time, the capability for a war fought over long distances with greater accuracy is increased by long-range computer guidance systems. If the nuclear deterrent were removed, would a nation with such long-range conventional power be tempted to use it? Nuclear holocast is a horrifying thought; however, conventional was is not a desirable alternative. I believe that without the nuclear deterrent nations would be more likely to engage in conventional warfare, but I would not go so far as to say that conventional war would become inevitable.

I also do not believe that post-industrial weaponry makes victory a "sure thing" for either side. Advantage in the military, as in all things, is relative. In the case of the two superpowers, developments have been moving at a more-or-less equal pace. Only when one side develops a capability matched neither in quality nor quantity by the other is there a distinct advantage. In order to illustrate this, and to move to my final point (which will become clear later) let me present the following scenario. While it will not be entirely accurate in terms of detail, an imaginary battle scene of, say 1995, can be created in one's mind. RPVs are patrolling the border between East and West Germany. They carry video cameras, and each side--Warsaw Pact and NATO--has its own surveillance team of RPVs tirelessly monitoring the scene. It is the Spring, and has been a very tense period of strained East-West relations. Andropov has recently died, and the younger Soviet party members are fighting among themselves for power. Fearing that perceptions of in-fighting and weakness will bring on an attack from the West, they decide to surprise NATO by invading West Germany.

Their tanks, with newly-improved armor, roll across the border. U.S. soldiers, at a base some distance from the border are alerted by the RPs, and immediately send out several more to maintain reconnaissance. Satellites are moved into position for a better overall view, but are crippled by Soviet satellites. The NATO forces move into position, and although they are outnumbered, hold their own through the use of the VIPER II, a powerful and accurate anti-armor weapon that is the son of the VIPER of the early 1980's.

98

The Soviets send in RPVs which randomly play laser beams on trees, rocks, etc., confusing the VIPER II. Fortunately for NATO, the fields are muddy and difficult to negotiate with the Soviet tanks. The Abrams tanks are speeding at 35-40 miles per hour from the U.S. base to support the infantry's delaying action.

When it becomes clear that the Soviets do not wish to negotiate, NATO wishes to end the conflict as quickly and with as little loss of life as possible. All the possible outcomes and current information from the RPVs are fed into a computer at NATO headquarters, and it decides that the best way to do this is to knock out the Soviet control bases, rendering their RPVs useless and crippling intelligence capacity and computers. Before sending out aircraft, they send out some RPVs to activate Soviet radar. Missiles then "ride" the radar emissions to their source, knocking out the radar towers. Unfortunately, they are located at some distance from the command posts, which are not damaged. A squadron of F-16 fighters, now protected from detection, is sent in to the headquarters, located by a combination of recorded satellite pictures and RPV information. The Warsaw Pact's command post is destroyed.

At this point, the battle's outcome will depend largely on the Soviet leaders. They can surrender, control the fighting from another location through use of satellites, or escalate and attempt to wipe out the NATO command post, and so on, ad infinitum.

This scenario, while it is lacking in detail, illustrates the improbability of a decisive victory through high technology. Its outcome brings me to my final point: although machines and computers can _fight_ the battle, _people_ still must control it. A machine can be programmed to choose an option or aim a missile. It cannot say "It's time to stop and negotiate," for what military leader would program a computer to recommend surrender? No matter what machines or weapons we create, none is unconquerable by itself. We hope to _deter_ conflict; in case of a conflict, we hope to win, of course, but victory is never guaranteed.

Turning to technology will not create an unqualified U.S. superiority. Post-industrial weaponry is not the final answer to deterrence, any more than the machine gun, airplane, U-boat, or tank was. People are the answer. All the satellites in space cannot help if the willingness to communicate is lacking.

At the same time, we cannot ignore the advances made in military technology. We cannot fight the U-boat with the Nina, the Pinta, and the Santa Maria. In all, it is a matter of degree. Mass--that is, sheer numbers, cannot be sacrificed for maneuverability--if we need a minimum of tanks to deter or destroy a Soviet onslaught, we cannot stop paying for them in order to build one supertank.

The principles of war--mass, unity of command, surprise, maneuver, objective, and so forth--are to be taken together. They create a delicate balance, which cannot be disturbed if one desires a victory.

The answer, then, lies in moderation. Procurement must be a careful process based on rational planning. The other necessities of government must be provided for in the budget. Priorities must be set and maintained. Humans must be rational, for in the end, it is the human, not the computer, who controls the outcome.

Endnotes

[1]Cees Hamelink, "Imperialism of Satellite Technology", cited by John Wicklein in Electronic Nightmare: The Home Communications Set and Your Freedom, (Boston: Beacon Press, 1981), p. 161.

[2]James W. Conan, The Superwarriors: The Fantastic World of Pentagon Superweapons, (New York: Waybright and Taller, 1975), p. 304.

[3]Ibid., p. 253.

[4]Tom Gervasi, Arsenal of Democracy II, (New York: Grove Press, Inc., 1981), p. 224.

[5]Conan, Superwarriors, p. 278.

[6]"Laser Weapons: From science fiction to fact", Business Week, July 26, 1982, p. 66.

[7]Ibid.

[8]Ibid.

[9]Conan, Superwarriors, p. 253.

[10]Gervasi, Arsenal, p. 30.

[11]Ibid.

[12]William J. Cook, "The Dazzle of Lasers", Newsweek, January 3, 1983, p. 39.

[13]Gervasi, Arsenal, p. 31.

[14]"Laser Weapons", Business Week, p. 66.

[15]Ibid.

[16]Ibid, p. 68.

[17]Gervasi, Arsenal, p. 31.

[18]Ibid.

[19]Wicklein, Electronic Nightmare, p. 161.

[20]Gervasi, Arsenal, p. 31.

[21]Ibid.

[22]Ibid.

[23]Conan, Superwarriors, pp. 316-317.

[24]Clarence A. Robinson, Jr., "Surveillance Integration Pivotal in Israeli Successes", Aviation Week and Space Technology, July 5, 1982, p. 16.

[25]Ibid.

[26]Ibid., pp. 16-17.

[27]Ibid., p. 16.

[28]Ibid.

[29]Conan, Superwarriors, pp. 325-326.

[30]Ruth Sivard, "World Military Expenditures 1980", World Politics 81/82, Chau T. Phan, ed., (Connecticut: The Dushkin Publishing Groups, Inc., 1981), p. 65.

[31]Daniel Bell, The Coming of Post-Industrial Society: A Venture in Social Forecasting, (New York: Basic Books, Inc., 1973), p. 389.

[32]General Creighton Abrams, quoted by Conan in Superwarriors, p. 287.

[33]Peter McGrath, "Where to Cut Defense", Newsweek, December 20, 1982, p. 26.

[34]Robinson, "Surveillance Integration", Aviation Week and Space Technology, p. 16.

[35]Robert S. Dudney, "Lebanon, Falklands: Tests in High-Tech War", U.S. News and World Report, August 18, 1982, p. 24.

[36]Ibid., p. 25.

[37]Harold Brown, quoted by James Fallows, "America's High-Tech Weaponry", The Atlantic Monthly, May 1981, p. 29.

[38]Conan, Superwarriors, p. 277.

[39]Philip Faflick, "Brutal Came of Survival", Time, August 16, 1982, p. 59.

[40]Ibid.

[41]Ibid., citing M.I.T. Sociology Professor Sherry Turkle.

[42]Senate Armed Services Committee (1974), cited by Conan, Superwarriors, p. 253.

[43]McGrath, "Cut Defense", Newsweek, p. 28.

[44]"Laser Weapons", Business Week, p. 66.

[45]Ibid.

[46]Ibid.

[47]Cook, "Dazzle", Newsweek, p. 39.

[48]Ibid.

[49]Wicklein, Electronic Nightmare, p. 161.

[50]"Laser Weapons", Business Week, p. 68.

[51]Cook, "Dazzle", Newsweek, p. 39.

[52]"Laser Weapons", Business Week, p. 68.

[53]"Text of SALT II Treaty" (1979), as reported by DuPre Jones, ed., in U.S. Defense Policy: Weapons, Strategy, and Commitments, 2nd ed., (Washington, D.C.: Congressional Quarterly, 1980), p. 39-A.

[54]Fallows, "High-Tech Weaponry", The Atlantic Monthly, p. 29.

[55]Ibid.

[56]Ibid., p. 30.

[57]Gervasi, _Arsenal_, p. 80.

[58]Ibid.

[59]Ibid., pp. 86-87.

[60]Ibid., pp. 81, 89.

[61]Melinda Beck and John J. Lindsday, "Defending the United States", _Newsweek_, December 20, 1982, p. 23.

[62]Richard Garwin, Cited by Fallows in "High-Tech Weaponry", _The Atlantic Monthly_, p. 30.

[63]Sivard, "World Military Expenditures", _World Politics_, pp. 65, 66.

Bibliography

Beck, Melinda, and Lindsday, John J. "Defending the United States". _Newsweek_. 20 December 1982, pp. 22-23.

Bell, Daniel. _The Coming of Post-Industrial Society: A Venture in Social Forecasting_. New York: Basic Books, Inc., 1973.

Conan, James W. _The Superwarriors: The Fantastic World of Pentagon Superweapons_. New York: Waybright and Taller, 1975.

Cook, William J. "The Dazzle of Lasers". _Newsweek_. 3 January 1983, pp. 36-40.

Dudney, Robert S. "Lebanon, Falklands: Tests in High-Tech War". _U.S. News and World Report_. 18 August 1982, pp. 24-25.

Faflick, Philip. "Brutal Game of Survival". _Time_. 16 August 1982, p. 59.

Fallows, James. "America's High-Tech Weaponry". _The Atlantic Monthly_. May 1981.

Gervasi, Tom. _Arsenal of Democracy II_. New York: Grove Press, Inc., 1981.

Lapp, Ralph E. "Hydrogen Bomb". _The World Book Encyclopedia_, Vol. VIII. Chicago: Field Enterprises, Inc., 1961.

McGrath, Peter. "Where to Cut Defense". _Newsweek_. 20 December 1982, pp. 24-31.

Robinson, Clarence A., Jr. "Surveillance Integration Pivotal in Israeli Successes". _Aviation Week and Space Technology_. 5 July 1982, pp. 16-17.

Sivard, Ruth. "World Military Expenditures 1980". _World Politics 81/82_. Chau T. Phan, ed. Connecticut: The Dushkin Publishing Groups, Inc., 1981.

"Text of SALT II Treaty", _U.S. Defense Policy: Weapons_,

Strategy, and Commitments, 2nd ed.   Washington, D.C.:
Congressional Quarterly, pp. 36A-49A.

Wicklein,   John.   _Electronic   Nightmare: The   Home
    Communications Set and Your Freedom_.   Boston: Beacon
    Press, 1981.

DEMOCRATIZING THE AMERICAN CORPORATION

Amy J. Rowe

We are witnessing the beginning of a potentially major transformation in the traditional American corporation. This transformation involves a complete redefinition of the nature and purposes of the corporation as it relates to social, political and economic issues. Such redefinition begins with the decision-making process or governance system of the corporation. In an attempt to explore the question of who should govern the corporation, I will examine the current trends of corporate reforms which move toward a more public and democratic form of corporate governance.

The large, private corporations, collectively called the Fortune 500, clearly represent the "predominant power in the society."[1] These economic institutions touch our lives daily in our various roles as consumers, employees, managers, stockholders or citizens. They began to flourish in the United States roughly one hundred years ago; throughout that time, they thrived, requiring only minor modifications. Until recently, there was little criticism of these private corporations and their decision-making executives, because the economic goal of increasing output of goods was congruent with the major consumer values of the society.[2]

Within the last decade, the public's confidence in the ability of large corporations to meet the expectations of society has diminished greatly. "Latent mistrust has grown to the point at which lack of confidence in business motives has become the overwhelming popular response to the role of the large corporation in the United States."[3] Sociologists have attributed the recent challenge to corporate legitimacy to a basic divergence between the traditional economic values held by the corporation and the new social values held by the general public. The central economic value of material accumulation has been challenged by a concern for the quality of life and the potential of human resources. As Daniel Bell explains in his book, The Coming of the Post-Industrial Society, these new social values reflect a transition to a more affluent and economically advanced society.

The desire and capability of groups to assert affluence have been improved by advances in educational opportunities and the opportunity of security about food and shelter, to spend life trying to satisfy higher level aspirations.[4]

The original autonomy of corporate decision makers is now being challenged by the public (consumers, employees, and citizens) who believe they have a claim on corporate behavior. The corporate decision-making system must be restructured in response to the public's desire for greater participation in decisions that will affect their lives.

The 19th Century Corporation

To understand more fully the recent debate over the governance of the corporation, it is necessary to examine briefly the original or traditional goals and responsibilities of this institution. The large business corporation emerged as "a new social invention" in response to a very strong societal goal--industrial growth and material accumulation.[5] A Polish sociologist, Henry Sienkiewicz, writing during the late 19th century commented:

It was a very one-sided civilization... the universally accepted opinion was that money alone constitutes the worth of a man and that material gains and enjoyment are the sole objectives worth striving for.[6]

The business corporation as we know it today began as a single-purpose institution whose success was measured solely by productivity and profitability. Alfred P. Sloan of General Motors, one of the originators of the present corporate form, was explicit in stating the corporation's philosophy:

We presumed that the first purpose in making a capital investment is the establishment of a business that will pay satisfactory dividends and preserve and increase its capital value. The primary object of the corporation, we declared, was to make money not just motor cars.[7]

All aspects of corporate policy were subordinate to the effect they would have on the rate of return on capital invested. There was little concern and interest in how management made decisions, unless the decisions were reflected in the price of the stock.[8] The corporation had no other role than that of economizing or securing the best allocation of scarce resources among competing ends.[9] However, as Daniel Bell explains in his book, The Cultural Contradictions of Capitalism, society is not a holistic system with one central unifying goal, for example, economic growth[10].

Bell describes modern society as consisting of three distinct realms--the techno-economic structure, the polity, and the culture. He explains that each part is ruled by contrary principles. The techno-economic structure deals with the production and allocation of goods and services. The polity is concerned with equality, social justice, legitimacy, democracy, and proper governance. Culture encompasses questions of one's human existence and self-realization[11]. Although this description of society is rather theoretical, it is useful in understanding what created the tension between corporate goals and the public's expectations.

At the height of the Industrial Revolution, these three realms of society were unified around the goal of economic growth. "The able business executive was a pillar of the community because business achievement was considered a high form of success, worthy in itself and crucial to social well-being."[12] Economic progress equaled social progress. But as we moved into the latter portion of the twentieth century, the economizing goal no longer was able to meet the other needs of society represented by the polity and culture. The public expectations of the corporate role began to change and questions of corporate governance, legitimacy and reform were raised.

Galbraith's Influence

One cannot discuss changes in the American corporation without introducing one of the major prophets and catalysts of corporate reform, Dr. John Kenneth Galbraith. As a leading social critic, economist, and reformer, he revealed his unconventional examination of the firm and the power of the corporation in his trilogy: <u>The Affluent Society</u> (1958), <u>The New Industrial State</u> (1967), and <u>Economics and the Public Purpose</u> (1973). Galbraith defined the industrial system today as a "few hundred technically dynamic, massively capitalized and highly organized corporations."[13] This definition differed greatly from the classical theory of the small, numerous firms which had no power over the market.

In the classical market, the consumer was sovereign; individuals expressed their desires to the producer by the way they distributed their income for goods and services. Galbraith asserted that because of the increased size, specialization, and capital investment of the modern corporation, it is forced to diminish the instability inherent in the market system. Corporations minimize the risks to their investment through formal planning and control. Galbraith describes this corporate strategy:

> Prices must be under control...effort must be made to ensure that the consumer responds favorably to the product...the firm is required with increasing capital...to control...the social environment in which it functions.[14]

Galbraith cites various examples of the tools used for market control. He describes the intensive advertising and carefully tested sales techniques as a method of creating stable demand for a good. Such demand management shifts "the locus of decision in the purchase of goods from the consumer where it is beyond control, to the firm where it is subject to control."[15] Social imbalance results from the corporation's power over the market. Corporate control fosters a bias in demand toward private goods at the expense of much-needed public goods and social investments.[16] This led Galbraith to advise that the corporation must be reformed to align it with the public purpose.

Galbraith's desire was to emancipate the public from the industrial system through an understanding of the true corporate forces and an awareness of what the corporation does not provide. It is interesting to note that Galbraith addressed his analysis to the general public, rather than his economic colleagues, in his attempt to educate. Many of his books were bestsellers; for this reason, many view Galbraith as an early catalyst to recent corporate reform movements.

Who Should Rule? -- Market, Management or Public

The realization that the economic goals of the private corporation are not congruent with the goals held by many people is the basis for the growing debate over the role of the corporation. In order to understand the directions current transformations in the corporation could take, it is necessary to explore the current attitudes of economists and business leaders regarding the question of corporate governance. These attitudes can be divided for the sake of clarification into three doctrines: classical, managerial and public consent.[17]

The classical doctrine is also known as the "invisible-hand" or "hand-of-the-market" doctrine. The contemporary spokesman for this group, Milton Friedman, states in his book Capitalism & Freedom:

There is one and only one social responsibility of business--to use its resources and engage in activities designed to increase its profits...if businessmen do have a social responsibility other than making maximum profit...how are they to know what it is? Can self-selected private individuals decide what the social interest is?[18]

Ideally, the free and competitive market is perceived as the sole and ultimate governor of corporate decision making. Morality, responsibility and conscience reside in the invisible-hand of the market, not in the hands of organizations nor their managers.[19]

The second doctrine is the managerial or "hand-of-management" and takes a position directly opposed to the classical doctrine. "The managerial [doctrine] expands the boundaries of executive discretion and hence managerial power and responsibility."[20] Unlike the other two doctrines, managerial seeks to align the corporate values with the public expectations from within, forgoing the need for heavy external controls. The executive's role includes a conscious effort to make decisions that balance the claims of shareholders, employees, customers and the general public. Managerial proponents see the corporation as having the potential to become an active, moral agent in society. Business leaders are increasingly adopting this position as a way to avoid more government regulation.[21]

However, critics of this attitude are "unable to see the engines of profit regulate themselves to the degree that would be implied by taking the principle of moral projection seriously."[22] Others add that corporate executives achieve their positions of power because they know how to sell or manage, but not solve social problems. Although the managerial doctrine is supported by well-meaning business leaders, many critics doubt that, unpressured, the corporation will reform itself to the extent its constituents (consumers, stockholders, employees and general public) may desire. After all, it was "the failure of most employers on a voluntary basis to provide safe working conditions that led to the stringent piece of federal legislation - the Occupational Safety and Health Act (OSHA) of 1970."[23]

A third doctrine has been emerging which recognizes the importance of the public voice in corporate decision making. It is called the "doctrine of public consent" or the "consent doctrine".

It represents an effort to base the legitimacy of the corporation on the principle of popular sovereignty; in effect it seeks to apply to the corporation the legitimating concept of John Locke that a government rules only with the consent of the governed.[24]

This doctrine has evolved as a result of powerful citizen pressures, which originated with civil rights movements of the sixties and subsequent activities in the seventies, particularly by the Project of Corporate Responsibility Group founded by Ralph Nader. The recent eruption of direct citizen protests resulted from the realization that modern corporations possess the power of governments and therefore should be treated as governments.[25] Proponents of this doctrine object to the fact that corporate decision makers impose their decisions on the public. The public consent doctrine seeks to remodel the governance system of the corporation so that it resembles more closely a political model in which the governed select the governors and the governors are accountable to their constituencies.[26]

This topic of the politicalization of the corporation is addressed in depth in a recent book Economic Democracy: The Challenge of the 1980's, by M. Carnoy and D. Shearer. The authors suggest a strategy for a new economy which would involve the transformation of the economic decision-making power from the few to the many.[27]

The Trend of the Future:
Corporate Governance by Public Consent

The classical, managerial, and public consent doctrines represent the three basic opinions regarding how economic goals should be aligned with societal goals. In essence, they reflect the three approaches to corporate governance; decision making governed by the market, by management, or by the public through the political body.

The question remains, however, which of these indicates the direction that corporate governance is taking? Daniel Bell deals explicitly with this question in his chapter entitled "The Subordination of the Corporation".

> What is evident everywhere is a society-side uprising against bureaucracy and a desire for participation, a theme summed up in the statement that 'people ought to be able to affect the decisions that control their lives' ... the politicalization of decision making ... in the decades to come will spread into the complex organizations.[28]

Similarly, Herman Kahn, author of The Future of the Corporation, states:

> Certain striking cultural changes will lead to greater dependence on consensual techniques and even participatory democracy rather than classical use of rules, orders, directives and top-down command and control.[29]

Today the governance of the corporation is a major issue of debate, since it is the governance system which provides the legitimacy to those who manage a corporation.

The statements by Bell and Kahn reveal that the future trend in corporate governance is toward public consent. Such a transformation involves a major institutional change, since this approach would ultimately shift the decision-making authority from the once autonomous institution, ruled by owner, manager, and stockholder, to its constituents. The rest of this paper will examine reforms in corporate structures and management styles that reflect this transformation.

The Emergence of the Public Voice

On April 17, 1980, Ralph Nader and a large coalition of labor and public interest groups launched what they called a decade-long attack on corporate America. Their objective was to "make corporations more accountable to their basic constituencies."[30] On that day, christened Big Business Day by Nader, an important bill was introduced to Congress by New York Democrat Benjamin Rosenthal, Chairman of the House Consumer Subcommittee. The bill, called the Corporate Democracy Act, would apply to companies with more than $250 million is assets or more than five thousand employees. These qualifications include the top eight hundred United States corporations. According to Rosenthal, the goal is to "democratize corporate decision making by creating tough new standards for corporate governance..."[31] The major provisions include:

1) a majority of outside (nonshareholder) directors

2) committees of outsiders to deal with auditing, compen-

    sation and nomination of directors

3) board members with special responsibility of investi-

    gating employee well-being, consumer protection,

    environmental problems, community relations, and

    shareholders rights

4) notification by stockholders of major corporate

   decisions

5) 24 month prenotification of plant closings

6) a list of the twenty largest shareholders with

   addresses.[32]

Supporters of this bill believe the existing system of corporate governance has failed in checking the activities of the corporation. Although this bill did not pass, many of its proposals have been adopted by corporations. The introduction of such a bill was very important because it was one of the first formal measures aimed at transforming the governance structure of the corporation.

What is Wrong With the Present System of Corporate Governance?

The present corporate governance system which is in question is based on a model designed to insure democratic governance. Abram Cheyes, author of "The Modern Corporation and the Role of Law", described the nineteenth century corporation as a republic in miniature, noting that an analogy can be made between the state and the corporation. "The shareholders were the electorate [and] the directors the legislature, enacting general policies and committing them to the officers for execution."[33] However, a brief examination of the weaknesses that have developed in this model will clarify why there is a need to set new, more democratic standards of corporate governance.

The electoral process involves the election of directors by the shareholders. Theoretically, the directors who then manage the corporation are accountable to the shareholders. The shareholder is assumed to be able to make an informed judgment about the ability of the director and vote accordingly. However, in reality, this electoral process is meaningless. In most corporations, "the chief executive officer dominates the process, selecting candidates that are assured election through their close ties with the corporation."[34] Since the cost of

115

challenging management's candidate for director is exorbitant, dissatisfied shareholders would rather exit by selling than voice their disapproval. "Management tends to promote exit in order to avoid development of democratic voice mechanisms that could threaten the managers' hold on their positions."[35]

Contrary to the model, the directors do not manage the corporation. In most cases, they make no major decisions; their decision-making power is transferred to the full-time managers. "Over the years the boards of directors ... as bodies of independent significance, have tended to wither away, so that the full-time managers of our largest corporations have been self-selecting, self-perpetuating and self-policing."[36]

Proposals to correct the flaws in the current corporate governance system range from small adjustments to comparatively radical reforms such as those included in the Corporate Democracy Act. The major focus of these reforms is to restructure the Board of Directors giving it more decision-making authority. The extreme proposals suggest that the boards should be composed of only one managerial director; the other directors should be representatives of the workers, the public, and other constituents. Traditionally, the Chief Executive Officer (CEO) is the Chairman of the Board, but it has been proposed that these two roles should be divided.[37] The CEO should be concerned with the traditional economic goals of profit and productivity, while the Board of Directors should function as an internal regulatory agency.

One innovative proposal called for the establishment of a quality-of-life advisory committee to the Board of Directors. The committee would be composed of representatives of employees, consumers, environmentalists, affected communities and other interested groups which would advise management on social concerns.[38] Such restructuring of the Board would provide for more public input in corporate decision making.

Democratizing Investment

One area which clearly reveals a trend toward increased participation in decision making is in pension fund investment. This issue is important since pension funds are now the largest single source of investment capital in this country.[39] Originally, pensions were considered a gift given to the employee by the employer. However, in the Inland Steel Case (1949), the Supreme Court held that pension contributions by employers are to be viewed as part of the employees wages; therefore, once a contribution is negotiated, that money no longer belongs to the employer.[40] However, this ownership of the pension fund means little, because the employee cannot use it. Until very recently, and in most cases even today, the worker has had no say in how the funds are invested.

Unions have recently begun to object to the ways these funds are being used. Their opinion is that "it is logical that these funds be invested in ways that benefit the union members."[41] For example, a union would not want its funds being invested in companies known for violation of the National Labor Relations Act or OSHA standards. Unions are now negotiating for two rights: 1) the right to channel a percentage of their pension fund into investments that serve social purposes; and 2) the right that no funds be invested in ways that subvert union principles.[42] One author pointed out that most Americans cannot afford direct stock ownership, and that by participating in the decisions of their pension funds, they have a greater stake in the economy.[43] This recent trend toward employee control of pension fund investments is one determinant in Peter Drucker's theory of the emergence of an "employee society".

Democratizing the Workplace

Peter Drucker, a leading management expert, recently published a book entitled Managing in Turbulent Times, in which he elaborates on two new realities with which management must contend. The first is the transnational integration of business; the second is the worker's search for "a place in the power structure."[44] This second issue clearly reflects the trend toward a more democratic, participatory form of corporate governance, and therefore warrants some attention.

Drucker states that society in the developed countries

has become an employee society, which he sees as a possible threat to the autonomy of the corporation.[45] "The status, function, power and responsibility of the educated, employed middle class are going to be central social issues during the next one hundred years..."[46] He bases his hypothesis on two facts. First, through their pension funds, the employees own almost one-third of the investment capital of all large American businesses. Second, today's employees tend to be "middle class in [their] education, [their] knowledge of the world and [their] expectations."(47)

Although employees have the property and the knowledge, they are not encouraged to take responsibility through the exercise of power. Drucker states: "Knowledge has to be endowed with responsibility or else it becomes irresponsible and arrogant."[48] Finally, Drucker adds that unless employees are integrated into the decision-making process, they will not support the enterprise or its management. Drucker suggests, for example, that employees should be given responsibility for the affairs of the plant, designing programs and setting goals for their own improvement.

This evolution toward a more qualified employee society has resulted in many alterations in the traditional management style. The traditional decision-making system was highly centralized with all decisions coming from the top-level managers. The management-employee relationship was predominantly adversarial. Management reforms have fostered increasing decentralization of the organizational structure, i.e. project teams, task forces and quality-control circles. The decision-making process, therefore, is also less centralized and is based on a more informal, consensus-seeking system.[49]

Many of these reforms toward a more participatory management style have come from the Japanese, whose traditional organizational structure was built on cooperation and consensus. Their motto is "Zen in keiei" which means "Make every man a manager".[50] The quality-control circle (QC circle) which originated in Japan to combine the concerns of workers and supervisors to improve job-related quality control is now being used throughout the United States.[51] Lockheed's space and missile unit was the first to adopt the QC circle in 1974. By 1981 over one hundred firms had adopted this idea. Although the name is often changed, as in the case of TOPS

(Turned Onto Productivity and Savings) or IMPS (Improved Methods and Products Seekers), the program is just as effective.[52]

The cited proposals and actual reforms in corporate governance structures, investment strategies and management styles clearly indicate the increasing involvement of employees, consumers, and shareholders in the decision-making process of the corporation. However, many radical political economists share the belief of Charles E. Lindblom, a Yale political scientist. He states that "the large private corporation fits oddly into the democratic theory and vision...indeed it does not fit".[53] Advocates of this belief feel corporate reforms thus far have fallen far short of creating a truly democratic institution in which the ultimate decision-making power is held by the people who will be affected by the decisions.

If one were to project this trend of increasing public participation into the future, it is questionable whether the private corporate structure would prevail. Two alternative corporate forms are suggested by the authors of Economic Democracy; although their suggestions are unconventional, as well as idealistic, they warrant examination. The two forms are 1) a competitive public enterprise, and 2) an industrial producer cooperative.

Competitive Public Enterprise

Public corporations do exist in the United States, but their aim has been to complement and aid private corporations, rather than to compete with and possibly reform them.[54] Under this proposal, competitive public enterprises would be created through a new government holding company which would buy the necessary number of shares to control at least one major firm in each major industry, i.e., auto, computer, chemical. The directors of the board would be non-management members representing consumers, labor and government. These companies would set standards in social responsibility, safety, product quality, and labor relations. The idea is that private corporations would emulate these standards in order to remain competitive.[55] Of course, this theory assumes that the public corporation would be meeting the true needs and expectations of the public and, therefore, they would be successful. A second major task of these public corporations would be to invest in new public enterprises.

This selective infiltration of an industry by a public enterprise should not be confused with nationalization of an entire industry in which the government would have total ownership and control. [56]

> What we are seeking over the long run is not greater government ownership but greater democratization of economic decision making. Public enterprises are only a means to that end, not an end in itself.[57]

Producer Cooperatives

The second alternative for a more democratic form is the producer cooperative (co-op) in which the employees both own and manage the organization. All the decisions are made by the members. "In capitalist societies, the producer cooperative represents the most complete form of structural reform of the traditional capitalist relationship within the enterprise."[58] The seventies have seen renewed interest in co-ops. This structure tends to be limited to small retail cooperatives, although a few medium-sized industrial producer cooperatives do exist.

One of the largest industrial cooperatives in the United States today is Hyatt-Clark Industries Inc. (HCI) which became a worker-owned company on October 30, 1981.[59] I had an opportunity to interview the Vice President, Mr. Peter Wallack, who explained how the transformation occurred and how such a system functions. Originally, HCI had been a division of General Motors called the New Departure-Hyatt Bearing Division, General Motors Corporation and manufactured automobile roller bearings. In August 1980, GM told Hyatt-Clark that it was going to close the plant because of its high labor cost and GM's decreasing demand for its particular type of bearing. (The bearing manufactured by Hyatt-Clark cannot be used in front wheel drive automobiles).[60] Although many prospective buyers studied the possibility of a purchase, no offers were made. Therefore, in December 1980, GM announced that the plant would be closed in July 1981. When the union heard this news it asked if an employee group could buy the plant, and shortly after plans for the purchase began.

The union's first attempt to raise the needed capital funds failed, so a combined management and labor committee called the Job Preservation Committee was formed. After much solicitation for support, the group was able to raise $125,000 for a feasibility study by convincing 850 employees to contribute.[61] Although the majority of employees supported the effort, there was a dissident group that felt that the threat of a shutdown was only a ploy by GM to get the workers to accept a wage cut. Regardless of this group, on October 30, 1981, the employees formally purchased the plant and it became an independent corporation. All those who had contributed were hired by the new HCI.

The major difference between the structure of the previous subsidiary and the employee-owned corporation is that HCI now has its own Board of Directors. Unlike the GM Board of Directors, there are three union representatives on the Board, along with three management and seven outside directors (bankers, lawyers).[62] Having union representatives on the Board is unique and significant since the worker now has a direct voice in the decision-making process. To those who favor a more democratic and participatory corporate form, this representation is a major strength. Mr. Wallack explained that a certain percentage of the profit goes into an equity fund, and upon retirement, regardless of position, all employees share equally in this equity fund depending on their length of employment.[63]

When asked about the workers' attitude now, Mr. Wallack stated that the change in morale and enthusiasm for work was remarkable.[64] There is none of the loitering which used to be a problem and absenteeism has dropped sharply. According to the new labor contract, salaried employees are now permitted and encouraged to assist labor in their jobs if the opportunity warrants it. Most union contracts prohibit this. According to Mr. Wallack, this change has improved the labor-management relationships greatly.[65] The traditional adversarial relationship between these two groups has eased. HCI, barely a year old and not yet making a profit, is an experiment which is receiving a great deal of public attention. If it is able to succeed, perhaps other corporations will see this as a viable alternative to divestiture. As the authors of Economic Democracy state:

> The key to positive results in worker owned ...
> firms is participation in decision making.
> Participation does not guarantee financial
> success, which often depends on market conditions

beyond the worker's control; but it does improve
productivity, help workers solve production and
financial problems, and increase their commitment
to each other[66] ... Worker-controlled firms act
as a model for democratic production.[67]

Worker Self-Management in Yugoslavia

The most highly developed example of such employee
ownership is the worker self-management socialism of
Yugoslavia. When this method was implemented in 1950, it
was perceived as an embodiment of Marxist socialism.
However, the Yugoslavs' interpretation went beyond
nationalization which would simply replace the capitalist
elite with a government elite as controller of the
decision-making power.[68] "Transcending both private
capitalism and state capitalism; self-management socialism
was to institute direct democracy in economic matters with
decision-making power as the exclusive prerogative of
individuals directly affected by decisions ..."[69]

In this Yugoslavian model, ownership and control of
production enterprises is in the hands of representives of
the community in which the company is located. A
capitalistic board of directors or communistic state
planning council is replaced by a social council composed of
representatives of local trade unions, employers,
professional association and residents.[70] This council
makes all the major policy decisions for the firm, including
hiring, investment and regular firm operations. Unlike most
American forms of worker participation in which the worker
is still accountable to management, in the Yugoslav model
management answers to the workers, consumers and residents
of the social council. Although there are problems with
this model, its structure suggests some interesting
possibilities for "democratizing" the American corporation.

Conclusion

During the Industrial Age corporations enjoyed an
autonomous, sovereign existence. Their decisions were made
relatively independently of the consumer, employee and
public citizen, although these groups were greatly affected
by corporate actions. However, as we have progressed into
the Post-Industrial society the "tolerant and benign
attitude toward the corporation has receded."[71] The many
who are affected by corporate decisions are demanding the

ability to participate in the decision-making process. Recent corporate reforms indicate a rapid movement toward governance by public consent. It is questionable as to what ultimate point this may be taken in the future; perhaps businessmen will eventually become civil servants chosen by the public techniques of election.

ENDNOTES

[1]Daniel Bell, The Coming of the Post-Industrial Society. (New York: Basic Books, 1973), p. 270.

[2]Ibid., p. 297.

[3]T. Bradshaw and D. Vogel, Corporations and Their Critics, (New York: McGraw Hill, 1981), p. xvi.

[4]W. Dill, Running the American Corporation, (New Jersey: Prentice Hall Inc., 1978), p. 17.

[5]Bell, The Coming of the Post-Industrial Society, p. 276.

[6]Henry Sienkiewicz, quoted in Neil Chamberlain, Remaking American Values - Challenge to a Business Society, (New York: Basic Books, Inc., 1977), p. 41.

[7]Bell, The Coming of the Post-Industrial Society, p. 278.

[8]Dill, Running the American Corporation, p. 39.

[9]Bell, The Coming of the Post-Industrial Society, p. 275.

[10]Daniel Bell, The Cultural Contradictions of Capitalism, (New York: Basic Books, Inc., 1976), p. xi.

[11]Ibid., p. 10-13.

[12]Bradshaw and Vogel, Corporations and Their Critics, p. xv.

[13]John K. Galbraith, The New Industrial State, (Boston: Houghton Mifflin, 1967), p. 9.

[14]John K. Galbraith, Economics and the Public Purpose, (New York: Houghton Mifflin, 1973), p. 39-40.

[15]Galbraith, The New Industrial State, p. 205.

[16]John K. Galbraith, The Affluent Society, (New York:

The New American Library, 1958), p. 201.

[17]L. Silk and D. Vogel, Ethics and Profits: The Crisis of Confidence in American Business, (New York: Simon & Schuster, 1976), p. 133.

[18]Milton Friedman, Capitalism & Freedom, (Chicago: The University of Chicago Press, 1962), p. 133.

[19]Kenneth Goodpastor and J.B. Matthews, "Can a Corporation Have a Conscience," Harvard Business Review, Jan.-Feb. 1982, p. 136-137.

[20]L. Silk and D. Vogel, Ethics and Profits: The Crisis of Confidence in American Business, p. 134.

[21]W. Greenbough, "Keeping Corporate Governance in the Private Sector," Business Horizons, Feb. 1980, p. 72.

[22]Goodpastor, "Can a Corporation Have a Conscience," p. 137.

[23]J. McKie, Social Responsibility & the Business Predicament, (Washington: Brookings Institution, 1974), p. 324.

[24]L. Silk, Ethics and Profits: The Crisis of Confidence in American Business, p. 136.

[25]D. Vogel, Lobbying the Corporation, (New York: Basic Books, Inc., 1978), p. 8.

[26]Bradshaw, Corporations and Their Critics, p. 223.

[27]M. Carnoy and D. Shearer, Economic Democracy: The Challenge of the 1980's, (New York: M.E. Sharpe Inc., 1980), p.

[28]Bell, The Coming of Post-Industrial Society, p. 368.

[29]H. Kahn, The Future of the Corporation, (New York: Mason & Lipscomb, 1974), p. 205.

[30]A. Reilly, "Assault on Corporate America," Dun's Review, April 1980, p. 104.

[31]Ibid., p. 106.

[32]Ibid., p. 112.

[33]Abram Cheyes, "The Modern Corporation and the Role of the Law," in The Corporation in Modern Society, ed. Edward Mason, (New York: Atheneum, 1966), p. 39.

[34]Dill, Running the American Corporation, p. 65.

[35]Ibid., p. 67.

[36]Bradshaw, Corporations and Their Critics, p. 225.

[37]G. Rosen, "Can the Corporation Survive," Dun's Review, August 1979, p. 41.

[38]Bradshaw, Corporations and Their Critics, p. 228.

[39]E. Glover, "Social Investment: Directing Union-Negotiated Pension Funds to Social Purposes," Employees Benefits Journal, Sept. 1982, p. 13.

[40]Carnoy, Economic Democracy: Challenge of the 1980's, p. 103.

[41]Glover, "Social Investment: Directing Union-Negotiated Pension Funds to Social Purposes," p. 14.

[42]Ibid.

[43]Ibid., p. 30.

[44]Peter Drucker, "Managing for Turbulent Times," Industry Week, April 14, 1980, p. 54.

[45]Peter Drucker, "Managing for Turbulent Times," Industry Week, May 12, 1980, p. 48.

[46]Ibid., p. 44.

[47]Ibid.

[48]Peter Drucker, Managing for Turbulent Times, (New York: Harper & Row Publishers, 1980), p. 190.

[49] Business Horizons, February 1980, p. 67.

[50]Ibid., p. 64.

[51]Ibid.

[52]Ibid., p. 68.

[53]Dill, Running the American Corporation, p. 18.

[54]Carnoy, Economic Democracy: Challenge of the 1980's, p. 62.

[55]Ibid.

[56]Ibid., p. 84.

[57]Ibid., p. 85.

[58]Ibid., p. 142.

[59]Peter Wallack, interview held in Westfield, New Jersey, January, 1983.

[60]Ibid.

[61]Donald Warshaw, "GM Turns over Clark Plant to Worker-Owned Company," The Star-Ledger, Oct. 31, 1981, p. 1,4.

[62]Wallack.

[63]Ibid.

[64]Ibid.

[65]Ibid.

[66]Carnoy, Economic Democrary: Challenge of the 1980's, p. 182.

[67]Ibid., p. 194.

[68]M. Schrenk, C. Ardalan and N. El Tatawy, Yugoslavia -- Self-Management Socialism--Challenges of Development,

(Baltimore: The Johns Hopkins University Press, 1979), p. 3.

[69]Ibid.

[70]Ibid.

[71]Bell, <u>The Coming of the Post-Industrial Society</u>, p. 272.

BIBLIOGRAPHY

Athos, A. "Is the Corporation Next to Fall?" Harvard Business Review, Feb. 1970.

Andrews, Kenn R. "Can the Best Corporation Be Made Moral?" Harvard Business Review, May-June 1973.

Bander, Edward J. The Corporation in a Democratic Society. New York: H.W. Wilson Co., 1975.

Bauer, Raymond and Dan H. Fenn, Jr. "What is a Corporate Social Audit?" Harvard Business Review, Jan.-Feb. 1973.

Bell, Daniel. The Coming of the Post-Industrial Society. New York: Basic Books, Inc., 1973.

Bell, Daniel. The Cultural Contradiction of Capitalism. New York: Basic Books, Inc., 1976.

Blake, David H. Social Auditing: Evaluating the Impact of Corporate Programs. New York: Praeger Publishers, 1976.

Bradshaw, Thorton and David Vogel. Corporations and Critics: Issues and Answers to the Problem of Corporate Social Responsibility. New York: McGraw Hill, 1981.

Burford, Victor and Alison Esposito. "Today's Executive: Private Steward and Public Servant" Hardvard Business Review, March 1978.

Carnoy, Marin and Derek Shearer. Economic Democracy: The Challenge of the 1980's. New York: M. E. Sharpe Inc., 1980.

Chamberlain, Neil. Remaking American Values - Challenge to a Business Society. New York: Basic Books, Inc., 1977.

Chamberlain, Neil. The Limits of Corporate Responsibility. New York: Basic Books, Inc., 1973.

Chamberlain, Neil. The Place of Business in America's Future. New York: Basic Books, Inc., 1973.

Dill, William R. Running the American Corporation. New Jersey: Prentice Hall Inc., 1978.

Drucker, Peter F. "Management Loses its Power Base" Industry Week, May 12, 1980.

Drucker, Peter F. "Managing for Turbulent Times" Industry Week, April 14, 1980.

Drucker, Peter F. Managing for Turbulent Times. New York: Harper & Row Publishers, 1980.

Drucker, Peter F. Toward the Next Economics. New York: Harper & Row Publishers, 1981.

Estes, Ralph. Corporate Social Accounting. New York: John Wiley & Sons, 1976.

Friedman, Milton. Capitalism and Freedom. Chicago: University of Chicago Press, 1962.

Galbraith, John Kenneth. The Affluent Society. New York: The New American Library, 1958.

Galbraith, John Kenneth. Economics and the Public Purpose. New York: Houghton Mifflin, 1973.

Galbraith, John Kenneth. The New Industrial State. Boston: Houghton Mifflin, 1967.

Glover, Eugene. "Social Investment: Directing Union-Negotiated Pension Funds to Social Purposes" Employees Benefits Journal. Sept. 1982.

Goodpastor, Kenneth D. and John B. Matthews, Jr. "Can a Corporation Have a Conscience?" Harvard Business Review. Jan.-Feb. 1982.

Greenbough, William. "Keeping Corporate Governance in the Private Sector". Business Horizons. Feb. 1980.

Hession, Charles H. John Kenneth Galbraith and His Critics. New York: New American Library, 1972.

Inglehart, Ronald. The Silent Revolution: Changing Values &

Political Styles Among Western Publics. New Jersey: Princeton University Press, 1977.

Jacoby, Neil. Corporate Power & Social Responsibility. New York: MacMillan Publishers, 1973.

Kahn, Herman. The Future of the Corporation. New York: Mason & Lipscomb, 1974.

McKie, J. Social Responsibility & the Business Predicament. Washington: Brookings Institution, 1974.

Nader, R. and M. Green. Corporate Power in America. New York: Grossman, 1973.

Novak, M. and J. Cooper. The Corporation: A Theological Inquiry. Washington, D.C.: American Enterprise Institute for Public Policy Research, 1981.

O'Keim, Gerald and Barry D. Baysinge. "The Corporate Democracy Act" Business Horizons. March-April 1981.

Reilly, Ann. "Assault on Corporate America" Dun's Review. April 1980.

Rosen, Gerald. "Can the Corporation Survive?" Dun's Review August 1979.

Schiff, Frank. Looking Ahead: Identifying Key Economic Issues for Business & Society in the 1980's. Washington, D.C.: Committee for Economic Development, 1980.

Schrenk, M., C. Ardalan and N. El Tatawy. Yugoslavia -- Self-Management Socialism - Challenges of Development. Baltimore: The Johns Hopkins University Press, 1979.

Silk, Leonard and David Vogel. Ethics and Profits: The Crisis of Confidence in American Business. New York: Simon & Schuster, 1976.

Toffler, Alvin. The Third Wave. New York: Bantam Books, 1980.

Vogel, David. Lobbying the Corporation: Citizen Challenges to Business Authority. New York: Basic Books, 1978.

INDUSTRIAL POLICY FOR A POST-INDUSTRIAL SOCIETY:

AN ALTERNATIVE FOR THE UNITED STATES

Mary Louise Biunno

Since the 1930's, and especially since 1945, there has been a growing awareness among economists, sociologists, and others, that the United States is evolving into an essentially different society under a "mature" industrial economy. The market is no longer "free," nor is there perfect competition. There has also been a boom in technology and information in the Post-War era that was unthinkable in the thirties. Now, in a post-industrial society, we are discovering that even Keynes's economics of interventionism is faltering. We need something more, to take account of recent technological developments, and to guide our economy for the greater social good. In our American laissez-faire psyches, at least one remedy, industrial policy, may have a chance of acceptance although the governmental coordination of policies affecting America's industries causes shudders among conservatives.

This paper will suggest specific elements of an industrial policy and will consider various questions, such as: Who will do the necessary long-range coordination? What bodies will determine priorities? and What will be the possible goals or directions? Naturally, criticisms arise in this controversial area and a sampling of the most pressing will be explored, along with a brief analysis of these arguments. First, we must return to the beginning in order to understand the situation in which the United States finds herself in this last quarter of the 20th century.

Section I
Understanding our Dilemma

The United States, along with other advanced industrial societies in Western Europe, finds that--economically, socially, politically--things are changing. Students of society, such as Daniel Bell, Alvin Toffler and Herbert J. Muller, attribute these changes to revolutions in knowledge and technology and their implications, such as a service economy. However, each, while realizing that there is no single answer, places emphasis on either knowledge or technology.

Daniel Bell believes that the post-industrial society is based on a preponderance of information or knowledge, and is characterized by a service economy. The service economy involves a shift from goods to services as its predominant output. The large amount of information is due to computers having provided the link between formal economic theory and a large data base. Out of this has come econometrics and the policy orientation of economics. "The concept 'post-industrial society' emphasizes the centrality of theoretical knowledge as the axis around which new technology, economic growth, and the stratification of society will be organized."[1] This intellectual technology, growing up from, and concerned with, information, is Bell's axial principle. The universities and research centers will share common economic ground with science-based industries. There is so much information about economic, political and social aspects of society that increasing amounts of coordination are required in order to make sound decisions. With respect to economics, specifically, Bell suggests that there will be increased governmental intervention in the economy. He observes that Keynes provided a theoretical justification for intervention while Kuznets and Hicks gave government policy a strong framework through the creation of a system of national accounts. Providing a meaningful classification for the great collection of statistics about the economy, the system of national accounts is useful in explaining important economic fluctuations and interrelationships.

Inevitably, complex societies become planning societies even though each individual may act rationally. Without coordination, individual choices result in irrational collective decisions. Technology, created by knowledge, enables us to gather and analyze this information for the greater social good. Further, the ethos of post-industrial society is communal, concerned with social goals which should be defined by national policy. Cybernetics, the complex of sciences dealing with electrical communication and control systems, also makes possible more output with less labor, so that post-industrial society provides more leisure time. Hence, rooting social character in work is no longer present to as great an extent.[2]

133

So, we can see that the changes in man's values and the way he spends his time--structural changes in society--are based on the change in the character of knowledge in respect to exponential growth in science and the rise of "intellectual technology."

Roger M. Troub and G. J. Wijers, both economics professors, share with Bell his emphasis on information. Troub refers to the U. S. as a service economy and also as an information society, the service activity involving decision making, or transfers and generation of information. This portion of the service economy is the information economy--Information for the purposes of design, planning, and decision making.[3] As knowledge accumulates, potentials for greater control over interacting systems grow. Transforming these potentials requires more complex planning, but does not necessitate centrally designed and executed packages. As the network of technology and social organization become more complex, the society's structure must change and planning processes are necessary to facilitate this change.[4]

Wijers emphasizes the importance of the information structure in the economy. This structure comprises mechanisms and channels for the collection, transmission, processing, and analysis of economic data, and largely determines the effectiveness of the economic system, which in turn affects all of life. The various actors within the economic system are influenced by the degree of uncertainty they face in making decisions. Naturally, reducing the uncertainty increases the chances of decisions being in line with the objectives. A good information system should also coordinate the many interdependent decisions in an economic system.[5]

The alternative emphasis in the development of post-industrial society and its implications is on technology. Herbert J. Muller, Alvin Toffler and others believe that new technology is the hallmark of the new age. Since 1945, new industries such as space sciences, computers, and semiconductors, have sprung up, and many radical changes and technical marvels have occurred. Events like the Cold War, the Atomic Age, and the rise of the non-Western world, occurred because technology was present. "All these changes were due to technology. In alliance with science, it is now unmistakably the basic determinant of our history."[6] Technology fostered the development of a new type of economics after World War II. With the advent of the computer, it became possible for the first time to process large amounts of data and for industrial executives to uncover significant interactions among the many economic variables of an expanding area of business activity.

Muller and Toffler, like Bell, see cybernation giving us more leisure time to enjoy. The technological revolution has created spectacular economic growth, resulting in what Muller (following John Kenneth Galbraith) calls the "affluent" society. It is this society created by technology that is now demanding some kind of social and economic planning. Now that we have a higher standard of living and our concern with work as a means of human expression is decreasing, we can worry about such matters as the externalities, such as pollution, created by technology. The service economy, too, is expanded by computerization because it creates the need for more brain workers. With an increasingly complex society and ever more sophisticated technology, more professionals will be needed. Where these professionals and technocrats have gone is where affluence has followed, while places involved in passé, industrial era industries have languished--Sun Belt versus Detroit. Electronics will bring major shifts in economic power and social political alignments, and these leaping advances make necessary some form of guidance.[7] "As technological progress brings individuals and nations into closer contact, there is a tendency toward planning and a different economic system."[8]

Both of these emphases on the origin of the changes occurring in society result in the one conclusion: we are in an era characterized by increased leisure, increased concern for fellow man and the environment, a rise in white-collar service jobs, growing amounts of information and communication with the rest of the world affecting all structures of society, social, political and especially economic. Whether knowledge led to computer development or computers led to a preponderance of information is difficult to conclude; they seem caught in a circle and dependent upon each other. We must be satisfied with the recognition of a profound transformation in industrial society and keep an open mind to the importance of both emphases for an understanding of modern society.

The trend toward a post-industrial society is accompanied by an ailing economy. "All the high-technology nations are reeling from the collision between the third Wave (post-industrial society) and the obsolete, encrusted economies and institutions of the Second (Industrial Era)."[9] In the United States we have begun to realize that Keynesian economics, and certainly classical economics, no longer function well for our society. Indeed, during the most recent period, beginning in the early 1970's, recession and inflation have gone hand in hand, giving rise to the trauma of "stagflation."

Keynes formulated a theory appropriate to the time in which he lived. The years between 1919 and 1939 primarily witnessed the problems of a depressed post-war period. He wrote about unemployment, low production and depressed demand while having little to say about the problems of a full employment, affluent society, the roles of large corporations and economic interest groups, the individual consumer, the social costs of continuous ʼgrowth, or the impact of technology. For him, the future was more of the same goods and services, with no qualitative difference. Keynes did provide the impetus for government intervention into the economy, but the scope of his interests was too narrow to explain the new problems that arose after 1945. Simply indicating fiscal and monetary policies as methods by which to maintain an economy at full employment, left his theory vulnerable to attack. This fine-tuning is still in use today, but it does not appear to be working effectively. Keynes's position was that full employment was possible without inflation. At that time, 1936, inflation was not a danger, but today, wages, which affect prices, do not fall in the slump. The labor unions maintain and increase these wages in industries whose productivity is not rising. At the same time, these firms yield to the union demands for higher wages, because of the workers' cry that they are being treated unfairly if they are forced to take a wage-cut.[10] "While there are many different varieties of advocates of post-Keynesian economics, their one common trait is that they assert that 'Keynesian' economics is not prepared to cope successfully with the current era of stagflation and chronic unemployment in the advanced industrial economies."[11] In Western Europe, Scandinavia and France, Keynesian interventionism has been replaced by indicative planning.

Although Keynes made great contributions to pure theory, modern day institutionalists, called neo-institutionalists, consider the broader area of the evolving industrial system. The market is no longer consumer want-serving, but is consumer want-creating, the government is a factor mediating among economic powers, and the free market system is declining. Neo-institutionalists do not wish to abandon markets (because they do many things well), but to design the markets as tools for particular purposes in particular settings. This overall design and location in the set of planning processes will result in increased utility of the market, rather than just leaving it up to the "invisible hand."[12] The Post-Keynesians simply see the growing disparity between theory and reality as reflecting the growing impotence of abstract market forces on modern economic conditions, thereby attacking not only Keynes but the classical economists.[13] Veblen, the institutionalist, would do away with all classical economic principles, wherease today's neo-institutionalists accept established theory for what it is worth and then go beyond to include not only decisions in the market, but the guidance of the larger evolving economic system. Ayres, a neo-insitutionalist, would like to see the market system placed within an institutional framework. This would better reflect the maturing economy of the post-industrial society which is no longer a competitive or theoretical system. People choose among alternative uses of scarce resources within the framework of a real world of historical time and geographical place.[14] There is capital immobility, unpredictable consumer behavior, technological revolution, and international instability. A highly uncertain economy such as this requires a drastic change in theory, although neo-institutionalists do use Keynesian-Marshallian concepts for short-run economic problems where the equilibrium concept has some relevance.[15]

The continuance of Keynesian interventionism as our official national economic policy adds to the stagflation problem by resulting in a hodgepodge of public policies and loss of competitveness in the international market. We are afraid to come right out and accept an industrial policy for fear of being called socialists, but are instead, content to muddle through. According to Robert Reich, the United States has responded inadequately to its emerging and declining industries, because the programs have not been

137

based on agreement between government and business to achieve positive economic adjustment. Programs have not been related to the commercial strategies of competitive firms within the international economy.[16] The United States is also sorely lacking in acquiring strategic business information, negotiating adjustment agreements, and coordinating policies. There is not even an agency responsible for gathering information about world market trends, the strategies of our trading partners, and the long-term outlook for particular U.S. industries. On the other side of the coin, American business managers have been reluctant to give out information, and have relied on the element of surprise. Nevertheless, the United States could establish a panel of business analysts to monitor the public consequences of declining and emerging industries, in view of the fact that the bargaining arena in Washington is filled with independent lobbyists, lawyers and related professionals pursuing competing goals.[17] The relief measures are ad hoc and produced by the political pressures of the moment without much thought about long-term effects. The only time policies are coordinated in when they are in opposition to each other and even then, action may not be taken.

This piecemeal effort is also affecting our position in the international market. Lester Thurow points out that for the first time in many decades America is in competition with countries which are her technological and financial equals. IBM is not at all assured of dominating the computer market in the future. New technologies and foreign compeitors are about to take over the industry, and the United States must be made more efficient.[18] Japan subsidizes her microcomputer industries and they are becoming quite strong. "Our companies are losing their share of the world market. Twenty years ago ... American industry accounted for 22 percent of the world's manufacturing exports. Today, that number is around 17 percent, with the U.S. having lost market share in 78 out of 92 separate manufacturing categories."[19] Gunnar Myrdal, too, criticizes and expresses concern for the U.S. economy. In the recession of 1969 and 1970, the low growth rate of the U.S. was due to the failure of the American government to adopt activist programs in industry similar to Scandinavia, Western Europe and France. This trend has only deepened today.[20]

As a nation, we must make a concerted effort to do something about international competitiveness. However, we have not yet awakened to the fact that we have become a dependent economic nation. Isolationism is wrong and will, in turn, erode the foundation of our military and diplomatic strength. All nations must trade, and the degree to which each nation succeeds or fails on this trade front greatly affects its economic health, its political structure and its standard of living. Calling for government to "get out of our hair" may be counterproductive to rebuilding international competitiveness. "It is clear to me that our present reactive, ad hoc system isn't working. If our traditionalists rule out even a debate on an industrial policy, then, by God, we'll all deserve what we get."[21]

Our present situation in the United States today paradoxically consists of a restructuring of society for the "post-industrial" era of high technology, information, service sector growth, and greater concern for the environs, and a rickety economy plagued by inflation, unemployment, loss of competitiveness and overall turmoil.

How can such a situation be remedied? One can guess that the solution lies somewhere in the realm of industrial policy. However, the various elements, ends, and implementation of such a policy need further consideration.

Section II

Looking for a Way Out: Institutionalization of Industrial Policy as a Future Alternative

Simply defined, industrial policy is concerned with the relationship between state and business and with the indistinct line between public and private enterprise. It is a complex of government policies, beyond fiscal and monetary attempts to steer the economy, that directly affect different sectors of business. Industrial policy is selective in that it bears directly on one or more sectors of business. The problem is one of effective resource use. Growth areas must be identified, which means picking out activities in industry and trade--based on population changes, trends in world demand and trade, etc.--that hold out the greatest prospect of rapid expansion.[22] It is an institutional capacity to view industries as a whole and to fashion policies that complement and support one another; it is a systematic and complete process for asking and answering complex social, economic and political questions which affect industry.

In its functioning in the economy, industrial policy would not replace the market system, but would include it in its framework. As Ayres explained, the market would be set in such a framework to reflect the non-competitive, but real industrial system in the Western world, and would be a subsidiary part of a developing, evolving economic system inside a larger social system. Direction provided by the industrial policy in terms of the acceleration of the overall pace of adjustment within the national economy, and a coordination of the paces at which firms, workers and communities adjust, would enhance the adaptation of the greater social system to post-industrial society.

This type of policy regarding industry has already been used in the United States, and hence, the groundwork is laid. All that needs to be done is to muster up a well-publicized call for industrial policy. "It is pointless to debate whether we should have an industrial policy. We have one. The issue is whether it will be open, rational, and useful in solving the problems of the new era."[23] It is almost inherent in the modern world. It is simply that the orthodox conservatives have not wanted to believe it, and have focused on the wrong issues. There is no denying the large role the government plays in the economy. The existence of the Council of Economic Advisers, the Federal Reserve System, and Congressional proposals in 1968 and 1969 for a Council of Social Advisers, a Commission on National Goals, and an Office of Program Analysis and Evaluation, all show that the trend is toward social and economic guidance in the United States. In July 1982, Governor Kean of New Jersey selected a twenty-one member commission on science and technology to explore ways to attract high technology business to New Jersey. Kean said that as we enter the final decades of this century, business will become based on advanced technology and, hence, it would be a strong force in our economy. The plan to attract such industries was formulated to enable New Jersey to remain viable. [24]

Another even more recent development in the trend toward industrial policy came in October 1982 when President Reagan signed the Export Trading Company Act. This bill puts into words the far-reaching views of the senators, both democratic and republican, who designed it. These senators see that the industrial course of the United States is changing, and through this bill they hope to promote the export of goods and services. Alleviating the depressing effect of American antitrust laws and permitting commercial banks to own equity in the trading companies should produce this desired effect of revitalizing exports. An idea such as this one is modeled after the successful Japanese general trading firms such as Mitsui and Mitsubishi which allow for joint exportation efforts with otherwise competing firms and a much closer link between banking services and trading policies.[25] This Act is at least part of what could eventually evolve into a more comprehensive industrial policy.

An obvious example of government's existing ability to successfully intervene is in the aerospace industry. It is the industry's major buyer; there is not a demand for such products in the private sector. However, our government's research and development (R & D) support is declining, and aerospace industry spokesmen are concerned they will lose their foreign markets to other countries whose governments are increasing R & D support. In West Germany there is emphasis on private industry but the Ministry for Education and Science makes grants for up to one half R & D cost while the industry has the rights to research results. "The Aerospace industry believes that the government should adopt an overall policy that promotes ... R & D and the export of high-technology products."[26] Viewed in this light, there is a striking similarity between America's defense-related and aerospace programs, and other nations' industrial policies designed to accelerate the development of emerging industries.

A final example demonstrating how important a move into the emerging area can be for the social well-being of society is the example of Lowell, Massachusetts. This small town, which had been a booming center for textiles in the Industrial Era, has fallen into decline and its inhabitants were unemployed and hopeless until Wang Laboratories entered the scene with a high technology factory and changed the face of the community. The people are now employed, and Lowell is a veritable fairy tale example of how emerging industries are so vital to economic health.[27]

Indeed, it seems as if we are headed in the direction of an industrial policy. The interventionist trend that started in the New Deal era may come to fruition in the post-industrial society if our leaders take cognizance of individual indicators and are not afraid to face a reality already created. Businesses themselves plan, and use long-term GNP projections as reference points for individual plan projects. Looking ahead, "... it may not be too hazardous a prediction to say that a future development in the U.S. is likely to be in the direction of more comprehensive...'indicative' planning."[28]

In order for industrial policy to be effective and accepted by government and business, there are some necessary elements which future policies should contain. Of primary importance is that policies have purposes, goals, or objectives and that an effort is made to achieve these ends. This does not mean that a central planning authority must determine the purposes. Rather, individuals and groups in different settings can set goals for the policy. Gerhard Colm's objective, as it appears in Allan Gruchy's book, is the successful operation of a free market enterprise economy in which public policies supplement market forces. Private enterprises remain independent but the coordination between government and business is within the common frame of reference--the growth of the economy as a whole.[29] Although this sounds similar to Western European countries such as France, or even East Asia's Japan, it is necessary to keep in mind that neither of these models can be grafted onto our society due to our laws, size, pluralism, and love of an adversarial system. Business and government are not natural partners here, and our political system is a highly contentious one in which disorder and opportunism have taken firm root. West Germany and Japan, on the other hand, have needed government to get started, and so have developed close ties between government and business. A United States industrial policy will have to be uniquely American.

The most controversial of the elements industrial policy should contain is this second one under consideration--the avoidance of protectionism. Liberals such as Senator Edward Kennedy (D. Mass.) believe an industrial policy should prop up failing smokestack industries, such as steel and autos, and it is precisely this thrust which laissez-faire proponents oppose. However, industrial policy can be different. We should be

142

forward-looking--looking into those industries with the greatest potential for growth. Protecting steel or autos only postpones the kind of revamping needed to adjust to economic realities.[30] The one problem democracy poses for industrial planning is that it is difficult to get political support not to subsidize a failing industry. Lobbyists from these ailing industries always garner support, because they are so much bigger and more powerful than the emerging industries. If bailing out a Chrysler or Penn Central prevents the loss of thousands of jobs and hundreds of small businesses, there will be political pressure to save the white elephants. A more sensible industrial policy, rather than bail-out, would be to encourage capital to shift from making passenger automobiles the country does not need, to making other things the country does need, in Detroit.[31]

Robert Reich sees these protections as perpetuating the downward cycle by retarding economic growth through the delay in the movement of resources toward more productive uses. Government must enter into a type of contract where assistance is exchanged for agreed shifts in private sector resources. Hence, the assistance acts like a bargaining chip to prod industries into adjusting. Subsidies to steel makers could be made contingent on steps by the industry to reduce excess capacity and relocate and retrain their workers. Industries should also be required to specify the investment strategy they will pursue, and they need not reinvest in their original product. Diversification into a more competitive industry may be a far better adjustment strategy.[32] It seems only fair that if government gives something it should receive something in return, help for the nation's overall economic health.

There are many policy instruments, such as indirect aid, which includes general subsidies and educational measures; direct aid, encompassing directly-financed R & D and social experiments, (for example, in the development of solar energy); and indirect specific aid in the form of subsidies, tax relief, and technology transfer programs in particular areas, sectors of industry or geographical regions.[33] Much of this is already in use, the aerospace program being one example and the use of the Domestic International Sales Corporation (DISC) tax incentive program in the computer industry another example. This DISC program gives tax relief on up to fifty percent of a company's income. The purpose is to expand exports and improve the balance of payments so that U.S. manufacturers can enter overseas markets in the face of competitors who are subsidized by their governments.[34] While this tax incentive program is a commendable effort, it is not enough. It should include every industry facing competition overseas, and be openly acknowledged as industrial policy. Otherwise, the program will probably never expand into something more, and may be allowed to slip away altogether.

Along with this important goal of protectionism avoidance, industrial policy should be holistic but should also consider local and regional levels. This third goal is important if the policy is to gain political support and the cooperation of the areas affected. Encouraging capital mobility between industries is sound economics, but it is much less wise to encourage mobility between regions and communities, because this necessitates worker relocation. The result of this relocation will be the abandoning of the schools, hospitals, government buildings and public utility installations -- public works -- similar to those in Detroit and Youngstown which only had to be reproduced in San Diego and Houston. Some industries require certain locations, but most do not. This separation of industrial from geographical mobility is the only way in a democracy to avoid poor socialism.[35] Individuals should be able to exert their effect on goal selection and policy performance in order for the final draft to gain their assent. This holistic-local approach may also be able to breach the gap between the sometimes inadequately understood government declarations, because individuals will have had a say in the policy's formulation from the beginning. However, these local policies must be integrated with the development of the entire country or else the hodgepodge of policies that exists today will only become exacerbated.

Naturally, this holistic approach will require the combination of information, and feedback, from the private sector industries. This is the fourth element of industrial policy. As was said earlier, it is difficult to procure information from industry, but while an industrial policy should not manage our economy, it can help generate the information that is needed for a better "trialogue" among business, government and labor. Through this action, the policy will provide better guidance to a major sector of our society that has shunned any such policy but which needs help in keeping in balance with economic reality.

In the United States, industry duplicates R & D needlessly. For every corporation to rediscover what others have already learned represents a waste to them and to society. Instead, the same base technology can be used to promote competition by many different applications. High-technology companies have cooperated and exchanged information over the years, but none of the methods sufficiently eradicates unwarranted duplication. In addition, small and large companies can cooperate. Since small companies produce 24 times more innovations per dollar than large ones, a large company can realize additional income from past investment by making available to the smaller, its underused technology and personnel.[36]

Effective solutions will require government and industry working cooperatively and supportively together. We must switch from an adversarial relationship to one of mutual support, but, of course, this may be difficult as it requires a change in our basic ideology. Feedback is important as it opens the door to public criticism, evaluation, and the participation of all interested parties. The means-end problem, too, must be considered when using fiscal, monetary, subsidies, and tax incentives policies. Their effectiveness in getting economic behavior to move in the desired direction must be monitored and reviewed in order to avoid misdirection and the resultant waste of time and energy. This is not to suggest non-Western socialist planning, because in the Eastern communist nations, one of the major limitations of planning is that there is no real feedback. In democratic industrial policy there is nothing more important than providing for feedback so that the general economic health , and the interest of the general public are served. With feedback, the system as a whole will be better understood.

Finally, the fifth element of an industrial policy is the inclusion of provisions for the workers, in order to reduce structural unemployment and to gain support from people in the lower three-quarters of the income distribution in America. A strong national capacity for innovation depends on deliberate government intervention in industry to support education and training, as well as R & D. In this way, the transition out of declining industries into emerging ones can be effected. Most workers need assistance in retraining and relocation, because they do not know where jobs are located and for what jobs retraining should be sought. This assistance must be designed to function not as a prolonged type of unemployment compensation, like protectionism, but in conjunction with industrial adjustment. By tying these programs to industry adjustment, the social costs of economic change are internalized.[37]

If such elements as the five mentioned above are taken into consideration, the ends of this type of system--achievement of coordination, restoration of competition, control of inflation and unemployment, and the social good--will be served. The achievement of coordination will enable us to see more clearly the correct industries in which to become involved. This would, in turn, create more job opportunities, and the retraining programs would further reduce unemployment. By having jobs in an area in which the productivity is high, workers would receive better wages. As long as productivity keeps pace with the increased wages, it will not create inflationary pressure. At the same time, workers receiving higher wages would likely cease demanding increased pay, another contributing factor to inflation.

Coordination and the use of incentive-creating tools, such as tax breaks, discussed above, would encourage entrance into emerging industries. This would be good for the country as a whole, because we would not only compete more effectively in the world market, but would regain the edge we have lost.

Finally, the concern with the social good, a characteristic of post-industrial society, would be served through the resulting expanded economic growth and the maintenance of the "affluent society's" standard of living while making society more affluent for more people by reducing inflation and unemployment. People, on the whole, would feel healthier psychologically in a revitalized economy than in the highly unpredctable one of today with its "fun-house-like" future. There would be more willingness to accept the other phenomenal changes accompanying post-industrial society, instead of retreating into a reactionary position out of fear. Post-industrial society's advent would be facilitated, and the anticipated increase in leisure time accompanying it would be better enjoyed if there were improved economic health.

146

"Industrial policy is an effort to get out ahead
of our coming problems. By thinking about
standards and objectives of industrial health,
assessing the foreign strategy, and contemplating
the allocation of federal resources affecting
industry, industrial policy make(s) it possible to
lock the barn door before the horse is out. In
fact, if the economy's bottom line is not healthy
enough there will not be adequate productivity
growth, adequate innovation or, adequate capital
formation. Our industry will continue to
atrophy."[38]

So, we are in the process of change. Let us accept it
gracefully as part of the new society. It is evident that
Keynesian economics is not enough to cope with present day
economic problems, and as such, there should be no fight to
give up the past.

"One thing seems clear. Continued human progress
requires (1) more accurate comprehension of what
is happening to us, (2) of what can happen to us,
(3) of what should happen to us, and (4) the
creation and use of the required conceptual and
organizational technology--the primary tools of
envisioning, formulating, evaluating, learning
and fashioning the human role structures needed
for pursuit of the better."[39]

Organizational technology should be an important
concern, for industrial policy is just a dream if there are
not concrete methods of implementing it. As will be seen,
this can be done by existing structures or by the creation
of independent agencies. While there is a need for an
industrial policy, there is no process or entity responsible
for devising one. The Council of Economic Advisers has been
mentioned by many as one body capable of doing the job. It
employs the techniques of input-output analysis and linear
programming so that we have an economic table to plot
various combinations of resources and requirements in
accordance with different value assumptions. However, still
lacking is an adequate forecasting model of the
economy.[40]

There are even private plans for the nation's largest

corporations and an effort has been made to coordinate them through forecasting and other services provided by groups such as the Conference Board, Data Resources, Inc., Chase Econometrics Associates, and the Wharton Economic Forecasting Unit. What is missing, according to Alfred Eichner, is some organ of government that is capable of fusing the results of these private efforts into a comprehensive policy or plan. Neither the Council of Economic Advisers, the Office of Management and Budget, nor the Congressional Budget Office can develop such a plan because they operate within a narrow political constituency with a meager staff and resources. Eichner argues that what is required, is a new body--a planning secretariat which would function as a technical arm of a social and economic council on which would be represented, along with key public officials, all the private interest groups whose support is essential. The secretariat would permit government to pursue maximum growth without fear of inflation.[41] This, of course, is more comprehensive and akin to a plan than to an industrial policy, but it is a possibility for the future.

Other alternatives for the more immediate future, and ones that might be more readily accepted given the present structure and ideology of our nation, involve organized labor, America's high-technology industries and individual state action. At least, labor can provide political support for such a policy if nothing else. It has an interest in industrial policy because it has been battered by unemployment, plant closings, and foreign competition. In fact, some elements within the labor movement have begun to develop their own proposals for industrial policy. The Machinists and the Industrial Union Department of the AFL-CIO have issued comprehensive approaches to the achievement of full employment, worker relocation, and creation of long-term high paying jobs. However, there is little evidence of a larger effort to push industrial policy as of now.[42]

The new high-technology companies will be the first to propose reforms since they appear to have the most to gain from adjustment and the most to lose from protection. They are aware that they compete directly with firms overseas which are beneficiaries of public policies.

While Governor Kean is interested in attracting

148

emerging industries to New Jersey, California is offering a
$22-million package of subsidies aimed at invigorating the
state's high-technology firms. Minnesota and North
Carolina, too, are financing micro-electronic research
centers, and Ohio is establishing a $5-million fund to aid
high-technology companies.[43]

More speculative alternatives, aside from the planning
secretariat mentioned, involve the creation of independent
agencies. These alternatives include a System of Social
Accounts, a high-technology working group, a Community of
Neo-institutional economists, a department for International
Trade and Industry, and a group of independent experts, to
mention a few.

Daniel Bell's system of Social Accounts would put
economic accounting into a broader framework that would be
useful in clarifying policy choices. The accounts would
measure the utilization of human resources in our society in
four areas: (1) the measurement of social costs and net
returns of innovations in order to see if they are worth it;
(2) the measurement of social ills; (3) the creation of
working budgets in the areas of defined need, such as
housing and education; and (4) indicators of economic
opportunity and social mobility.[44] This, again, is more
of a plan for society, but it has its merits in considering
economics as part of the real world rather than isolating
it.

An even more speculative approach than Bell's, is Allan
Gruchy's proposal that the neo-institutionalists not only
educate the public concerning effectiveness of an economic
policy, but actually develop a following with a sense of
community that would recognize the advantages of the
neo-institutionalists' proposals, and that would provide
extensive support for these proposals.[45]

G. J. Wijers proposes a combination of industrial
policy institutions staffed by independent experts able to
operate outside the normal bureaucratic government
framework. They would combine the available technical
knowledge and experience into a strategic industrial policy
to be submitted to the political system for approval and to
be returned to the experts for execution.[46] This approach
might gain wider acceptance than the previous two, because
it could be carried out within our present framework of

government, Congress possibly being the body to which the experts would submit their industrial policy.

Robert Reich's idea for the creation of a single bargaining arena for allocating costs and benefits of adjustment would, at the least, provide government with broad-based consensus about adjustment policies. This single arena, as contrasted to the multiplicity of those existing today, would provide a focus for an ongoing national debate about economic change. This was the idea behind the establishment of the Special Trade Representative in the White House; however, its powers are too advisory, its jurisdiction limited, and the bargaining over which it presides too covert for the Representative to become a base for building consensus about industrial policy.[47]

Finally, the two most feasible alternatives for the present are Frank Weil's suggestion for the creation of a Cabinet department for International Trade and Industry and Norris' high technology group. Responsibility for gathering and developing a reliable and specific information base about all industrial sectors would rest in this Cabinet department. It would also shape trade policy, direct import and export controls, and report to the President and Congress on the state of health of U.S. industry.[48] Clearly, this alternative would fall within our existing structure, but would simultaneously provide a real change that is not so drastic as to scare away moderate conservatives.

Norris' high technology working group would operate in the absence of an industrial policy, and thus may be a necessary first step along the road to acknowledged industrial policy. This group would work under the Department of Commerce to prepare a list of existing and potential high technologies that are critical to the American competitive position. It would also provide recommendations to accelerate progress or fill any gaps. This type of work has been done by industry in the area of military technology, so there is a precedent.[49]

This brings our discussion of the alternative for the future-- industrial policy--to a close, but naturally such a controversial subject involving and affecting so many does not end here. There are criticisms and analysis to be considered, and this is the topic of the next section.

Section III
Analysis and Criticisms

Critics' main concern is that industrial policy will lead to imperative planning of the Communist sort, where bureaucrats substitute their judgment for the market. People lack any clear idea about the nature of such a proposal, thus leaving it prone to caricature and retarding political discussion of the subject. Although government would be promoting market forces under an industrial policy, people do not differentiate between this reality of adjustment and the horrible vision of communism in which our freedom would be reduced. Furthermore, since these undemocratic forms of policy receive worldwide attention, no one takes into account the feedback and participation by the public, in a visible arena, advocated by Western Industrial policy. Scandinavia, West Germany, and Japan are not communist, and yet they have industrial policy. We certainly have learned not to follow the Communist model because its gross inefficiency is evident. By the same token, we cannot graft the Western European systems onto our own unique culture, but they can at least hold out the hope of industrial policy without the specter of communism.

Another criticism is that the issue of industrial policy may wane in the face of economic recovery, and thus, it cannot be terribly important. However, everytime the economy goes sour, the issue resurfaces. This alone should hint at its potential worthiness. Additionally, we must consider whether we want to maintain this economic instability. Not only is it a sign that the present economic policy does not work well, but it is inconsistent with our concern for the social good. One day there may be no economic recovery, and then the United States will face a crisis once more as it did in the 1930's. It is worthy to note that an economics of interventionism--Keynesian--and Roosevelt's program of extensive government involvement helped to bring the country back to its feet at that time. Now, we are ready to go beyond the degree and kind of interventionism of Keynes, and this should be done to avert, rather than to remedy a crisis.

Finally, the most serious criticism, in that it may have some substance, is that we may have difficulty obtaining and coordinating the plethora of information

151

generated. The question is whether the information
structure of the economic system can be perfected to provide
these industrial policy boards, agencies, and the like,
with a reliable picture of technological and market
developments. Of course, one answer to this is that the
other Western countries seem to be operating fairly well and
perhaps we could use their system or technique of gathering
information. Also, since there will be much information,
the national policy can be general while leaving the details
up to the governance of the state. However, smothering
amounts of details will not be needed either, since the
market mechanism with its allocative role will not be
dispensed with. This aspect eliminates the need for the
fixed prices and wages found in communist countries because
price theory as developed by standard economists will
remain.

We do need this policy, and ways can be devised to
remedy a possible information overload if we truly wish to
do something about our economy as it stands. Times are
changing and "the world is being shaken up like a
kaleidoscope. Nothing remains the same but for a few
seconds. Tomorrow, looking at the tube, reading the paper,
we will be living in a different world, and after that, in
yet another world, and after that..."[50] Who knows? Yet,
in order to make reality resemble something more than a
hectic, jumbled string of isolated events, an economic
policy that organizes the situation and takes stock of what
is happening is necessary not only to provide stability for
the society, but to administer effective remedies, such as
avoiding protectionism, providing for the workers and
relying on feedback, so that the changing times can keep on
changing, but for the better. Putting this policy into
operation will be tricky and will require some study of the
various structures that will be most helpful, a sampling of
which was discussed. Taken as a whole, however, this is the
most sensible way to go, because many of the requisite
elements are already in place. All that needs to be done is
the prompt activation of the policy that may open up a
radically different future, domestically and
internationally, for the United States,--that is, industrial
policy.

Conclusion

The road to post-industrial society will undoubtedly be

152

fast-paced and filled with many new developments and much strife, tension, and debate, but we must accept this, and accept the fact that we are headed toward a new frontier. We can no longer afford, in the face of challenge, to revert back to outdated, "Industrial" methods of existing. Acknowledgment of the panorama unfolding before us is crucial to our continued growth and survival as a nation and if this includes the alteration of our economic system, then so be it, because economics is inextricably intertwined with our individual lives, as well as with the larger structure of the society. Our older leaders may perhaps see the new light, but certainly, the younger leaders and leaders-to-be will realize that this post-industrial society needs an industrial policy.

Endnotes

[1]Daniel Bell, <u>The Coming of Post-Industrial Society:</u> <u>A Venture in Social Forecasting</u>, (New York: Basic Books, Inc., Publishers, 1973), p. 112.

[2]Ibid, p. 462.

[3]Roger M. Troub, "A General Theory of Planning: The Evolution of Planning and the Planning of Evolution," <u>Journal of Economic Issues</u> 16 (June 1982), p. 381.

[4]Ibid, pp. 386-7.

[5]G.J. Wijers, "Institutional Aspects of Industrial Policy," <u>Journal of Economic Issues</u> 16 (June 1982), p. 589.

[6]Herbert J. Muller, <u>The Children of Frankenstein: A Primer on Modern Technology and Human Values</u>, (Bloomington & London: Indiana University Press, 1970), p. 87.

[7]Alvin Toffler, <u>The Third Wave</u>, (New York: Bantam Books, 1980), pp. 138-141.

[8]Allan G. Gruchy, <u>Contemporary Economic Thought: The Contribution of Neo-Insitutional Economics</u>, (Clifton, New Jersey: Augustus M. Kelley, 1972), p. 298.

[9]Toffler, <u>The Third Wave</u>, p. 14.

[10]Sir John Hicks, <u>The Crisis in Keynesian Economics</u>, (New York: Basic Books, Inc., 1974), pp. 70-2.

[11]Allan G. Gruchy, "Planning Contemporary Insitutional Thought," <u>Journal of Economic Issues</u> 16 (June 1982), p. 373.

[12]Troub, "A General Theory of Planning," p. 388.

[13]Alfred S. Eicher, ed., <u>A Guide to Post-Keynesian Economics</u>, (White Plains, New York: M.E. Sharpe, Inc., 1979), p. 11.

[14]Gruchy, <u>Contemporary Economic Thought</u>, p. 301.

[15]Ibid, pp. 302-3.

[16]Robert B. Reich, "Making Industrial Policy," Foreign Affairs, 60 (Spring 1982), p. 870.

[17]Ibid, pp. 870-4.

[18]Lester C. Thurow, "A New Era of Competition," Newsweek, 18 January 1982, p. 63.

[19]John M. Henske, "The Changing World Environment for International Trade," Vital Speeches of The Day, 15 October 1982, p. 28.

[20]Gruchy, Contemporary Economic Thought, pp. 193-4.

[21]"The U.S. Needs an Industrial Policy," Fortune, 24 March 1980, p. 152.

[22]Eric Roll, The World After Keynes, (New York: Frederick A. Praeger, 1968), pp. 74-6.

[23]Jeff Faux, "Who Plans?," Working Papers, November/December 1982, pp. 12-13.

[24]Tom Johnson, "Kean selects a blue-ribbon panel to lure high-tech firms for Jersey," The Star Ledger, 30 July 1982, sec. 1, p. 3.

[25]Yoshi Tsurumi, "Export Trading Company Act of the U.S.: The Beginning of a New Industrial Policy," Pacific Basin Quarterly 8 (Fall 1982): p.p. 3-4.

[26]Kenneth E. Knight, George Kozmetsky, and Helen R. Baca, Industry Views of the Role of the Federal Government in Industrial Innovation (Austin, Texas: The Graduate School of Business, The Univ. of Texas at Austin, 1976), pp. 53-6.

[27]Fox Butterfield, "In Technology, Lowell, Mass., Finds New Life," New York Times, 10 August 1982, sec. 1, p. 1.

[28]Roll, The World After Keynes, p. 62.

[29]Gruchy, Contemporary Economic Thought, p. 270.

[30]Art Pine, "Debate Heats Up Again On Industrial Policy," Wall Street Journal, 11 October 1982, sec. 1, p. 1.

[31]Faux, "Who Plans?," pp. 14-15.

[32]Reich, "Making Industrial Policy," pp. 852-59.

[33]OECD, Technical Change and Economic Policy (Paris: OECD, 1980), p. 99.

[34]Knight, Kozmetsky, and Baca, Industry Views, pp. 81-3.

[35]Faux, "Who Plans?," p. 15.

[36]William Norris, "Keeping America First," Datamation, September 1982, p. 283. The only drawback to this plan is that the large companies could buy out the smaller ones and stifle innovation.

[37]Reich, "Making Industrial Policy," pp. 855-863.

[38]"The U.S. Needs an Industrial Policy," pp. 149-52.

[39]Troub, "A General Theory of Planning," p. 389.

[40]Bell, The Coming of Post-Industrial Society, p. 336.

[41]Eichner, ed., A Guide to Post-Keynesian Economics, pp. 180-81.

[42]Faux, "Who Plans?," p. 16.

[43]Reich, "Making Industrial Policy," p. 881.

[44]Bell, The Coming of Post-Industrial Society, p. 326.

[45]Gruchy, "Planning in Contemporary Institutional Thought," pp. 378-79.

[46]Wijers, "Institutional Aspects of Industrial Policy," p. 593.

[47]Reich, "Making Industrial Policy," p. 876.

[48]"The U.S. Needs an Industrial Policy," p. 150.

[49]Norris, "Keeping America First," p. 283.

[50]John Friedmann, Retracking America, (Garden City, New York: Anchor Press/Doubleday, 1973), p. 108.

Bibliography

Bell, Daniel. <u>The Coming of Post-Industrial Society: A Venture in Social Forecasting</u>. New York: Basic Books, Inc., Publishers, 1973.

Butterfield, Fox. "In Technology, Lowell, Mass., Finds New Life." <u>New York Times</u>, 10 August 1982, sec. 1, p. 1.

Eichner, Alfred S., ed. <u>A Guide to Post-Keynesian Economics</u>. White Plains, New York: M.E. Sharpe, Inc., 1979.

Faux, Jeff. "Who Plans?." <u>Working Papers</u>, November/December 1982, pp. 12-16.

Friedmann, John. <u>Retracking America</u>. Garden City, New York: Anchor Press/Doubleday, 1973.

Gruchy, Allan G. <u>Contemporary Economic Thought: The Contribution of Neo-Institutional Economics</u>. Clifton, New Jersey: Augustus M. Kelley, 1972.

Gruchy, Allan G. "Planning in Contemporary Institutional Thought." <u>Journal of Economic Issues</u> 16 (June 1982): 371-79.

Henske, John M. "The Changing World Environment for International Trade." <u>Vital Speeches of The Day</u>, 15 October 1982, pp. 28-31.

Hicks, Sir John. <u>The Crisis in Keynesian Economics</u>. New York: Basic Books, Inc., 1974.

Johnson, Tom. "Kean Selects a Blue-ribbon Panel to Lure High-tech Firms for Jersey." <u>The Star Ledger</u>, 30 July 1982, sec. 1, p. 3-8.

Knight, Kenneth E.; Kozmetsky, George; and Baca, Helen R. <u>Industry Views of the Role of the Federal Government in Industrial Innovation</u>. Austin, Texas: The Graduate School of Business, The Univ. of Texas at Austin, 1976.

Muller, Herbert J. <u>The Children of Frankenstein: A Primer on Modern Technology and Human Values</u>. Bloomington &

London: Indiana Univ. Press, 1970.

Norris, William. "Keeping America First." Datamation, September 1982, pp. 280-87.

OECD. Technical Change and Economic Policy. Paris: OECD, 1980.

Pine, Art. "Debate Heats Up Again on Industrial Policy." Wall Street Journal, 11 October 1982, sec. 1, p. 1.

Reich, Robert B. "Making Industrial Policy." Foreign Affairs, 60 (Spring 1982): 852-81.

Roll, Eric. The World After Keynes. New York: Frederick A. Praeger, 1968.

"The U.S. Needs an Industrial Policy." Fortune, 24 March 1980, pp. 149-152.

Thurow, Lester C. "A New Era of Competition." Newsweek, 18 January 1982, p. 63.

Toffler, Alvin. The Third Wave. New York: Bantam Books, 1980.

Troub, Roger M. "A General Theory of Planning: The Evolution of Planning and the Planning of Evolution." Journal of Economic Issues 16 (June 1982): 381-90.

Tsurumi, Yoshi. "Export Trading Company Act of the U.S.: The Beginning of a New Industrial Policy." Pacific Basin Quarterly 8 (Fall 1982): 3-5.

Wijers, G.J. "Institutional Aspects of Industrial Policy." Journal of Economic Issues 16 (June 1982): 587-96.

POST-INDUSTRIALISM, INTERDEPENDENCE, AND THE THIRD WORLD

John N. Geracimos

Daniel Bell, in his book The Coming of Post-Industrial
Society, defines and describes the phenomenon of
"post-industrial society." He basically limits his
discussion to the advanced industrial countries, since those
areas are where the characteristics defined as "post
industrial" either presently exist or are showing signs of
emerging.[1] His study does not examine the effects of the
new "post-industrialism" on those areas of the world which
have not been able to achieve this level of social
development (or, vice versa, the effects of these non-post
industrial areas on the more advanced sectors). At the end
of his book, Bell briefly describes these interrelationships
as the "outer limit of our trajectory -- a problem for the
twenty-first century."[2]

The Western societies which are candidates for
membership in the post-industrial club do not, however,
exist in isolation; post-industrialism does affect the rest
of the world, whose responses echo back to change the
post-industrial world even further. It is not necessary to
wait until the twenty-first century to see that these
interactions are having serious consequences for
non-industrial, industrial, and post-industrial worlds
today.

In this paper, I plan to examine some of the results of
the interactions between the post-industrial and non
post-industrial worlds. My first section will be a brief
look at the effects of the developing world on the
post-industrial world. A short introduction is necessary to
"bring home" the conflicts discussed in detail in the rest
of the paper, to show that the effects of post-industrialism
on the rest of the world rebound to affect those of us
living in the post-industrial areas. The remaining sections
of this work will deal with the effects of
post-industrialism on the rest of the world, which is
predominantly the "developing" or "Third World."[3]

Most of the discussion will surround the changes in
the "international division of labor." This new division

160

gives the post-industrial areas primacy in science and information industries, while assigning a manufacturing role to the Third World. I will examine a number of the major political-socioeconomic effects of this new division on the Third World. Throughout the discussion, I will also try to identify post-industrialism within the Third World itself, as the change in the international division of labor causes a number of unexpected post-industrial characteristics within (and among) Third World societies. While these effects tend to be distorted versions of those occurring tn the already industrialized world, they do shed an interesting light on trends in the post-industrial world itself.

I

By the end of the century, the United States, Japan, Western Europe and the Soviet Union will take on aspects of the post-industrial society.
-Daniel Bell, The Coming of Post-Industrial Society[4]

The [technical] revolutions have given the Western countries an extraordinary advantage in high technology, and paved the way . . . for the transfer of a large part of the routinized manufacturing activities of the world to the less-developed countries.
-Daniel Bell, "The Future World Disorder: The Structural Context of Crises"[5]

The change in the international division of labor is having a large and often disruptive effect on the post-industrial societies. The traditional division of labor assured the industrialized societies the dominant role in the world economy. They imported raw materials from the developing world, and used mechanical means to change these material materials into manufactured goods. Primary goods tend to be low value-added products such a cotton or tin. Therfore, the periphery was not able to accumulate much capital from their sale. The manufacturing process was the step where the largest percentage of value was added, and this step was monopolized by the old colonial powers. The industrial countries were then able to sell these high-value manufactured goods back to the developing countries. Because they were exporting low-value primary goods and had to import all of their high-value manufactured goods, the Third World countries experienced a net drain of their resources, which was manifested in severe trade deficits and mounting debts. Technical innovation in the advanced capitalist countries (such as the synthetic substitution of many primary products) drove commodity prices down even further, exacerbating the unequal exchange. The dominance of the industrialized countries over international trade regulations such as the General Agreement on Tariffs and Trade (GATT) also helped perpetuate the inequality.

Starting in the 1960s, the industrialized countries began to feel the effects of a change in this arrangement. Manufactured goods bearing such labels as "Made in Japan" and "Made in Taiwan" began to permeate the Western markets. The first of these products were labor intensive goods such a textiles and shoes. However, the quality and quantities of the goods increased. By the early 1970s, the Japanese-manufactured goods being imported were equal in quality (if not better than) the goods produced in the West. The West grudgingly accepted Japan into the circle of advanced industrial nations. Other countries began exporting large amounts of advanced products to Western markets as well. These imports were often priced lower than similar goods produced in the core countries. Workers in the so-called "traditional" industries (steel, auto, etc.) in the West were being laid-off by the thousands, as Americans bought the cheaper Korean steel and Taiwanese televisions. The Third World production of steel rose from 2% of global output in 1955 to 11% in 1977.[6] By 1982, the United States showed a 1.9% negative GNP growth rate, while the new industrializing countries, despite the global recession, showed gains on the order of 5%.[7]

This shift of heavy manufacturing capacity to the Third World areas is a symptom of the changing international division of labor. This new division is theoretically to be composed of a post-industrial region which will specialize in high technonlgy products and services, and a traditional manufacturing sector in the Third World selling its products to the post-industrial countries. The change is proving to be very disruptive to the First World, as this area experiences growing unemployment resulting from the move of high employment industries, such as auto and steel

162

manufacture, to areas in Latin America and the Far East.

There are a number of reasons commonly given for the West's loss of comparative advantage in the traditional manufacturing sectors. Probably the major reason cited is the relatively higher wages and benefits of the Western (and now Japanese) workers. This "revolution of rising entitlements" effectively raised the cost associated with the post-industrial worker to the point where the businesses can no longer conpete with the lower cost labor in the Third World.[8] This welfare state mentality also significantly (along with other things as well) boosts government spending and increases inflation. Popular pressure to reduce inflation makes it difficult for the government to deal effectively with the simultaneous problem of unemployment.

A cursory glance at other domestic reasons for the loss in comparative advantage reveals such varied factors as:

-"Loss of dynamism of the investment process", resulting in less investment in the traditional industries.
-The "relative saturation of the markets for durable goods."
-Less technological innovation in the traditional industries.
-The vulnerability of the traditional sectors to primary good supply instability, especially petroleum.
-Too many government regulations.
-The rising costs of environmental and consumer concerns.
-Labor organization pressure.
-The "oligopolilization of societies;" the increasing number and size of large conglomerates raises costs and contributes to inflation.
-Post industrial value changes, such as the decreasing emphasis on the value of economic growth.[9]

Ironically, some of the very advances in communications, transportation, and information control with which the West hopes to maintain its global economic lead are being used to further the move of the sites of advanced manufacturing out of the developed world. As Bell observes, the third technological revolution is "tying the world together in almost real time," enabling Western multinational corporations to locate manufacturing operations in areas (tending to be in the developing world) where they can significantly lower costs and raise profits. This trend will be examined in greater detail in a later section.

A radical would see the loss of comparative advantage in a different light, as a symptom of the ongoing crisis of capitalism. Since about 1966, the global capitalist system has been in a crisis of accumulation, a downword swing in the fifty-year Kondratieff cycle.[10] As the profit rates of business fall, companies tend to seek opportunities to recoup their losses. The multinational corporations in this period increasingly begin investing in plants abroad. They find tht the can make more profits by investing in the low-cost Third World than in the traditionally industrialized countries.[11] Corporate operations in a capitalist mode of production have one goal, the increase of profits. Interest in employees extends only as far as it is necessary to extract profit from their labor. If a multinational can acquire the same labor outside the home country at a lower cost, there is nothing binding it to support its obsolete domestic workers.

There are probably as many different approaches to the policies the West should pursue in the face of this changing division of labor as there are reasons for the change. The typical reaction is to use the Western technological lead to offset the comparative advantage that the Third World possesses in labor-intensive industries. This soultion emphasizes the use of technology to increase the productivity of Western (and Japanese) industry.[12] However, technical solutions tend toward an increasing capital intensity of industry, which preserve the Western share in these industries, but do not necessarily preserve the position of the Western worker. It makes little difference to an unemployed worker whether his job is filled by a $50,000 IBM robot or an Indonesian worker making $1.00 per day. Also, once the markets eventually become saturated with a product, increasing productivity merely increases excess capacity and waste.

The Americans, and to some extent, the Europeans have another approach to the problem of the emerging international division of labor, protectionism. The

Americans, through multilateral agreements such as tariffs (GATT) and through bi-lateral agreements on quantitative restrictions (orderly market agreements) try to limit the import of products in which they lack a comparative advantage. This policy tends to preserve inefficient American industries, removing the incentive for them to modernize. It also has considerable effects on the developing nations, which will be dealt with later. Protectionism shows that "American industries [are] not responding to the challenge from the developing countries."[13] Tariff barriers also raise the specter of trade wars, in which a number of countries retaliate against each other's restrictions with other restrictions. In an increasingly interdependent world, such a policy could become economic suicide for many countries, depriving them of vital markets. Also, although American labor unions might favor such a policy, it is not likely that the American multinationals will appreciate it, as they increasingly depend on the international free flow of goods and capital.[14]

The Japanese seem to have found the most realistic of the solutions to the shift in the international division of labor -- accept it. Through the 1960s and early 1970s, the Japanese had been taking the productivity route. In the earlier boom days of the late 1960s, Japanese industry had been spending an incredible 25% of GNP on plant construction and improvement. Because of the saturation of its domestic and export markets, this percentage fell to 13-14% in the late 1970s.[15] They shut down many of their comparatively unproductive and superfluous textile mills and moved them to South Korea and Taiwan, where they could better take advantage of the lower costs. Japan also cut its steel output and shipbuilding; naval construction was reduced to 33% of the peak of the late 1960s.[16] Japanese businessmen realized the inevitability of the loss of comparative advantage in the traditional sectors, and strove to reduce its dragging presence in the Japanese economy. In the place of these older industries, Japanese business invested heavily in the new high technology industries. Today, the Japanese reorientation is visible in their overwhelming dominance of the American stereo and 35 mm. camera markets.

The Japanese also made moves toward the "post-manufacturing age." Because the slowdown in the rate of growth in the Japanese economy and export markets prevented them from investing in more productivity, Japanese business started increasing quality of life investments.

> We will invest to make the paint shop environmentally more pleasant. . . We will invest to meet auto-emission standards. . .We will invest to make our company hospital, which is already one of the most advanced, even better.[17]

This investment pattern is a fundamental reorientation away from growth measured in quantitative terms toward growth measured in qualitative terms.

The new international division of labor is forcing the West into a reappraisal of its concept of growth in a post-industrial society. The role of growth in a post-industrial society is a very touchy subject.[18] It has been decsribed as a "political solvent" which serves to satisfy the growing expectations of a growing population without causing the upheavals associated with redistribution. Others see the failure of growth as a major crisis which will increase the distance between the "haves" and the "have nots."[19] The opponents of quantitative growth say that it only serves to widen the gap between the rich and the poor in a society.[20]

Some writers seek to preserve the idea of growth by redefining it. One reinterpretation of growth revolves around the post-industrial idea of the primacy of information. Growth in this interpretation means increases in "the quantity of information exchanged by men and the quality of their communication."[21] This type of growth is not based on material property, and can exist without the limitations of goods growth.

Finally, Bell offers an interesting scenario of a "Headquarters Economy." Although the United States might have an unfavorable balance of trade in goods, it would be able to make up the lost revenue through "invisible" trade, such as that accruing from "interest, dividends, re-patriated profits, [and] royalties."[22] Even though the means of production would be located abroad, they still could be owned by businesses in the United States. He points out that the United States could exist as a "rentier" economy, living off the profits of its subsidiaries abroad.

Bell notes, however, that such an economy would have to increase at the expense of American labor, as many workers would be unemployed and have little access to the incoming profits.

Although the change in the internaional division of labor has by no means yet approached the extreme of the headquarters economy, it it is moving in that direction. It is not "an outer limit to our trajectory -- a problem for the twenty-first century," but a structural problem today (as the 219,000 workers who lost their jobs because their companies relocated in the Far East can attest.)[23] The problem of the international division of labor shift is relevant for study today. The rest of this paper will examine the problem from the other perspective, that of the developing countries.

II

> The developing countries behave as though they were facinated by the apparent success of Western civilization, and attempt to reproduce at home all the means of achieving it, especially wide-scale industrialization which they consider to be a token of accession to power.
> -Andre Danzin[24]

> The aura of prosperity was not limited to the major industrial countries. Large MNCs [Multinational corporations] brought prosperity to other countries such as South Korea, Taiwan, Brazil, Mexico. These developing countries which played by the rules -- exporting cheap raw materials and labor-intensive light manufactured goods, importing capital and consumer goods and limiting trade and investment restrictions - and which had certain natural advantages at home in terms of raw materials or a disciplined productive labor force, experienced rapid increases in per-capita income.
> -Gregory Schmid[25]

> Manufacturing will go east.
> -Norman Macrae[26]

A number of countries in the Third World today seem to be successfully following the Rowtowian model for development. Some of the societies are modern industrial societies and appear to be moving from the Rostowian "Take off" stage to the "Drive to maturity."[27] These countries are now competing with the traditional center in manufactured industrial goods.

Industrialization is not a new phenomenon in the Third World. In the first half of the twentieth century, a number of Third World countries (mostly in Latin America) implemented a policy of "import substitution." They sought to change the traditional relationship between the center and the periphery by producing domestically many of the goods that had been imported from the industrialized countries. This policy, however, offered a limited opportunity for growth, because it depended on an internal market created by steadily rising wages. Expectations rose faster than the ability of producers to raise wages; it became difficult for companies to make a profit and still maintain the high wages. The demand in these industries was limited to the home markets. Tariff barriers, designed to protect the growing infant industries, enabled relatively (by world standards) inefficient industries to exist, effectively preventing their access to the larger Western markets.

In the late 1960s, the internal pressures and limited nautre of the import substitution system caused a reorientation of economic strategy. The total imports of these countries had not decreased at all. The previously imported consumer goods were replaced by the imports of producer goods. These countries realized that their businesses could make better profits (thus their economies could grow faster) by aiming for the foreign rather than the domestic markets. By changing to an export orientation, Third World industrializers no longer found it necessary to maintain the higher wages of the import substitiution period, as the workers were producing for a foreign, not a domsetic market. The lower wages also enabled the Third World producers to compete more effectively on the global market, since their costs decreased.

The change to an export orientation in many countries engendered a dramatic rise in their growth rates. From 1953 to 1962, South Korea had been following an import substitution model for its economy. Its GNP growth during this period, despite much American aid, was only 4% per year. After 1962, however, when the country steered its orientation toward foreign markets, its GNP growth rate increased to levels greater than 8% per year.[28] By 1978, its growth was 12.5% per year.[29] Korea was not the only example of this spectacular growth. In the late 1960s and early 1970s, Hong Kong and Singapore also had growth rates of around 10%.[30] During this same period, Brazil had GNP growth rates averaging 12% per year, and by the late 1970s, was the eighth largest manufacturing country in the world.[31]

Table 1
GROWTH OF MANUFACTURED EXPORTS -- 1965-1974[32]

North America	9% per year
Western Europe	10% per year
South Korea	37% per year
Japan	16% per year
Taiwan	29% per year
Brazil	25% per year
Mexico	21% per year

Source: LaPalombara and Blank

All of the "newly industrialized countries," such as Brazil, Hong Kong, Mexico, Singapore, South Korea, and Taiwan, experienced growing per capita incomes, industrial employment, and shares of the global trade.[33] Their growth rates radiply outpaced the slowing rates of the traditional industrialized countries. It was estimated in 1980 that there were fifty-three Third World countries that were either producing or were planning to produce manufactured goods for export markets[34] In a 1975 meeting at Lima, Peru, the United Nations Industrial Development Organization (UNIDO) set a goal for the developing countries' production of over one quarter of the world's industrial output by the year 2000.[35]

There are a number of theories put forward as to why
these countries could grow as spectacularly as they did in
so short a time. One theory looks at common domestic trends
that occurred both in the industrialized societies of the
West and in the high growth societies of the East. However,
the most commonly held ideas ironically look to the West
itself as the source of the development.

Norman Macrae, and editor for The Economist, sees four
steps that are necessary for industrial development. He
argues that these steps have been followed (not as much
through planning as through circumstance) by Western Europe,
the United States, Japan, and the rapidly growing societies
in the Far East.[36] The first step is the elimination of
unemployment in the rural areas. This phase is usually done
while a society is in a feudal stage, before the capitalist
profit motive, with its idea of decreasing the number of
workers to decrease costs, takes hold. When "everybody [is]
somehow kept busy in the rural areas," there is little
incentive for them to flock to the cities as they do in many
developing societies today. In Taiwan, many of Chiang
Kai-shek's loyal followers were rewarded with land in the
country and established a type of paternal feudalism.[37]
In Korea, Park Chung Hee, of peasant background himself, had
his "political constituency in the countryside" and gave
large support to rural areas.[38]

The next step, according to Macrae, is that the
government encourages the success of the rural "squires," so
they can raise capital for later investment. In Korea, the
government encouraged this success with price supports for
high yield crops and tried to deep the terms of trade in
favor of the countryside.[39] As late as 1974, the annual
household income of rural Koreans was higher than that of
the of the urban citizens. This policy is visible in the
American experience of large government support of the
westward movements in the ninteenth century. In Malaysia,
this step was accomplished by other means. The government
had a bias against the Chinese who tended to live in the
cities and toward the Malays who lived in the countryside.
Consequently, Malaysian government policy tends to favor the
rural areas. Interestingly, Singapore and Hong Kong are
exempt from these first two stages because they do not
possess any significant rural area.

170

The next of Macrae's stages is one on which there seems to be most agreement among writers on export growth. He maintains that it is necessary to exploit the urban workers, to pay them less than their labor is worth. This stage also occurred in the currently affluent countries. In Great Britain, the living standards of the workers fell consistently from about 1815 to 1849, while those of the upper and middle classes rose. In Japan, the wage increased at a rate below that of the productivity increase until the late 1960s.[40] This step is hailed as necessary because it increases investment by making business more profitable. Macrae acknowledges that this step is potentially very disruptive, as it usually becomes necessary to decrease the rate of wage increase at a time of growing worker expectations. The consequences of this step will be examined in greater depth in a later section.

The next step is the overthrow of the rural "squires" after they have invested enough in industry for the economy to achieve a "take off." Macrae then sees much of the land being turned over to "entrepreneural small farmers."[41] This stage is also difficult to achieve as it muct occur at a time when the gentry class has a considerable degree of political power. Usually, this transition has occurred during a crisis period. In Japan, it was a result of the second World War; in the United Kingdom, it occurred because of the potato famine of the 1840s.

Macrae sees two other steps in the development process: the introduction of a "cost-conscious technology" and the non-interference of government in business. In Korea, the government watches business, but there is little interference. The government does seek, however, to prevent the businesses from becoming too politically powerful and to keep them "competitively at each others' throats."[42] Of big business in Korea, Macrae says:

> [It is composed of]. . . about a dozen cost conscious entrepreneurs, all competing with each other like crazy, all self-made industrial toughies in this generation, all now heading large and expanding conglomerates.[43]

These large conglomerates are also starting to spread abroad. Over 40,000 Koreans are working on projects for

Korean companies in the Middle East.[44] The extreme cost consciousness is evident in this description of a large Korean company's modernization technique:

> If the new technology puts Hyundai's [a large Korean conglomerate] existing technology out of date, as had happened in the "old" foundry shop built in 1969, then a new foundry shop is built next door, and on May 1 all workers will move into it.[45]

Macrae's theory, however, seems to leave out one of the most important factors in Third World industrial development: the rest of the world. These export-oriented economies are very fragile. The countries often possess few natural resources (epecially city states such as Singapore and Hong Kong) and are very susceptible to political and economic decisions made outside of their borders. Apprehension about the expiration of the lease on much of Hong Kong's territory in 1992 has caused its growth rate to fall from 11% in 1981 to 4% in 1982.[46] Taiwan also suffered a similar drop in growth following the American normalization of relations with China.

It can be argued that the prosperity of the four major growth areas in the Far East depends on politicial and geographical conditions over which they have no control. South Korea and Taiwan (and before its "liberation" - South Vietnam) are areas considered important to American security. It is America's interest to maintain wealthy, stable regimes in these countries to avoid "Communist infiltration." Economically, Singapore is stragically located at the end of the strait of Malacca, the major channel connecting the North Indian Ocean with the Southeast Asian area of the Pacific. It is the fourth largest port in the world.[47] Singapore's position as a major port and trading center makes it an ideal site for manufacturing.

Singapore's value as a trading center is probably surpassed only by Hong Kong. Hong Hong has traditionally (especially since the 1949 Chinese Revolution) been a major port of trade with China. Its central location in Southeast Asia and proximity to China make it a very prosperous port and an ideal site for manufacturing. Its political position as a colony of Great Britain effectively gives it a government with "no politics."[48] Its value as a port and manufacturing city is increased by its support of free trade:

> Hong Kong is one of the few territories, if not the only one, which has remained completely faithful to liberal economic policies of free enterprise and free trade. The government's role has always been and still is to provide a stable and secure framework within which industry can flourish with maximum efficiency and minimum interference. The government intervenes in economic processes only in response to the pressure of overriding social or world events.[49]

Consequently, Hong Kong has low taxes, few regulations, and no stipulations about the local reinvestment of foreign capital and the use of foreign labor.[50] The decline in its growth rate in the wake of the uncertainty about its post-1997 future points to the major role of foreign capital in Hong Kong's economic growth.

Most of the theories about the changing international division of labor concentrate on the role of foreign capital in the newly industrializing societies. Although industry generated by import substitution was able to create a substantial infrastructure and fulfill domestic demand, it was not able to comepte well in foreign markets. With the turn to export markets, Third World governments courted foreign capital to upgrade their industry. Many countries offered "Free Production Zones" which offered foreign businesses an area to establish operations with a minimum of taxation or regulation.[51] Multinational corporations tended to invest in areas which had already had undergone import substitution, as they were able to take advantage of the already well-developed infrastructure. Traditionally, Latin America has been the largest recipient of such investment, followed by Asian countries such as Taiwan, South Korea, Hong Kong, and Singapore.[52] However, the new technology resulting from the so-called "third technological revolution" has enabled companies to invest outside of the traditional areas.

The old production technologies were limited to those areas where a sufficient infrastructure existed to support them. The llimited existence of highway networks, port, power, and communications facilities limited industrial investment in the Third World to a few places. The new technologies, however, eliminate these bottlenecks. The distances from the sites to the headquarters no longer are significant. The biggest irony is that these technologies are the ones which the post-industrial societies are hoping to use to further their own development, but instead are being used to advance the industrialization of the Third World. The most important advances are in four major areas.[53]

One of these revolutions is in transportation. Containerization of air freight has enabled the materials necessary for an overseas plant to be shipped virtually anywhere in the world cheaply and quickly[54] This ability relieves the multinationals of much of their dependence on local infrasuructure, as they can easily fly anything that is needed into the country. Entire factories can be shipped into areas, such as Indonesia, that previously were unable to cater to multinationals because of their relative isolation.

Another of the third industrial revolution's technological innovations that facilitate the expansion of overseas industrial expansion os the revolution in communications. Satellite link-ups make possible instantaneous two-way communication between any two points in the world. Like the transportation revolution, this communications revolution removes the investment restraints due to the lack of infrastructural development. It is no longer necessary for a company to wait for (or to persuade) a country to establish its own communications system, as a dish on the factory roof would enable the plant management to have daily two-way real-time communications by voice, teletype, facsimile, or even video, with the home office in New York, Bonn, or elsewhere.

The new communications technologies, along with the new information processing technologies, enable multinational companies to maintain a much stonger control over their overseas operations. Computers allow huge amounts of information to be analyzed and processed, resulting in much more efficient decision making. Computers are now small and cheap enough to be installed in overseas branches. Information can be exchanged between computers in the overseas plant and the home offices via a satellite data-link.

One of the farthest-reaching of the new technologies is an extension of an old concept. Standardization of parts and assembly line techniques enabled the industrialized countries to make great production increases at minimal costs. By having a worker perform only one task in the production of a product, it was not necessary to hire only very skilled workers who were able to build an entire product. However, the increasing complexity of many products, especially in the electronics industry, made highly-trained technicians indispensible to the production process. The new production revolution is the extension of the techniques of assembly line production to processes that were the realm of the trained technician. This neo-Taylorization enables companies to use unskilled workers to produce highly complex products. The need for only unskilled labor in the production process encourages the companies to move their facilities to areas where such labor is in abundance.

Another production technique stemming from neo-Taylorization which encourages the move to Third World locations is component assembly. Each factory often assembles only a part of some product. These factories can be anywhere in the world, therby preventing any one country from gaining a monopoly on the technology and the production of the product. There would be little incentive for a country to nationalize a factory that only manufactures one specific component. The components are eventually shipped to an assembly plant to be put together into the final product. Assembly is a labor intensive process, giving the Third World the comparative advantage in such areas.

The combination of all of these technologies gives the multinational corporations unprecedented power. It allows capital to take the active role; it can now move to labor instead of waiting for labor to move to it. If labor starts demanding higher wages in a certain plant, the company can close it and move to an area where the labor will be cheaper. With its information processing and communications capabilities, the multinational can rapidly evaluate local conditions and use them to its best advantage. Its mobility enables it to relocate if necessary. There have been a number of moves from the traditionally-industrializing Third World areas, such as Mexico, South Korea, Taiwan, Singapore, and Hong Kong, to such "out of the way" locations as the Philippines, Malaysia, Indonesia, Thailand, and Sri Lanka.

These "out of the way" areas court multinational investment from other countries. They establish Free Production Zones (see above) to enable the companies to work with a minimum of restrictions. Developing countries see multinational companies as a way for them to acquire modern technology and compete on world markets. They look forward to the creation of new jobs that comes with the establishment of manufacturing plants. Third World countries look to the multinationals as a means of reducing their debt to Western banks. The companies resident in a country must pay taxes to the host country, providing revenue which is sorely needed to pay back large loans and their interest. Ironically, the multinationals are often perceived as a way for a country to increase its self-reliance.[55] Any revenue generated inside the country's borders reduces its dependence on outside, unstable sources of income. Many Third World countries (and especially the poorer Fourth World ones) ard dependent on the export of one or two commodities for their income, the prices of which fluctuate widely on world markets. The presence of an industry in their country is seen as giving it a relatively stable source of income.

It is difficult to determine which model for the change in the international division of labor more closely approximates reality. Third World industrialization is both a domestic and a foreign phenomenon. The degree to which a country is penetrated by foreign capital seems to be related to its degree of indigenous economic development. There appear to be a number of patterns for multinational investment in a country.[56]

One of these patterns reflects the scenario of the multinational's having the power to do as they wish in a host country, the Free Trade Zone mentality. This pattern is characterized by few restrictions on the multinational. Also, the companies are usually given incentives by the host government, such as low land prices and electric rates and a tax holiday for a certain period at the start of the venture.[57] The countries that follow this pattern of relations tend to be the poorer ones, since they often need

176

to accept any revenue they can acquire and have little power
to bargain for better terms. The developing countries which
fall into this pattern are located mainly in Asia (with the
notable exception of India), Africa, and Central America.
The number of countries in this category seems to be
expanding, as multinationals use the new technology at their
disposal to establish operations in previously
unindustrialized areas. Operations in these areas are
usually wholly owned by the multinational, seriously
limiting the abilities of the host countries to control and
profit from them.

Another pattern of multinational investment is similar
in many respects to the previous pattern. However, this
model is characterized by a higher degree of joint ventures
and reinvestment stipulations. The multinationals still are
able to operate with relatively few restrictions, but their
actions are better mediated by the government through joint
ownership. Also, there are some "local participation
quotas" which require the companies to use local labor and
reinvest some profits locally. This form of investment is
most visible in areas that have some degree of leverage,
often in the form of natural resources for which the
multinationals are willing to make some concessions.
Consequently, this model is most commonly found in the oil
rich areas of North Africa and the Middle East.

The third model of multinational investment takes place
in the most developed of the Third World countries, usually
those which have developed significant domestic industry
through import substitution. This relationship is most
commonly found in South America and India. In these areas,
joint interest investments predominate. In India, the
multinationals are effectively nationalized, as the
government in 1974 passed the Foreign Exchange Regulation
Act which required Indian control of 51% of the stock of all
foreign industries in the country. Those states which
follow this mode of interaction with the multinational
companies are able to dictate terms to the companies in many
areas such as labor use and local profit reinvestment. They
can forbid foreign involvement in strategic sectors of their
economies such as in defense.[58] The larger the native
industrial resources of a country, the more power its
government may have over foreign operations in the country.
In Brazil, where there are large numbers of joint
enterprises, aligining foreign capical with both the

government and Brazilian private industry is hailed as an ideal of this model of behaving more as a partner and less as a puppet of international capital. Brazil is said to have a "mature and sophisticated approach to the foreign investor."[59]

Although the more developed Third World areas wield more power over the foreign companies operating within their borders than countries with smaller economies, they also have the largest amounts of foreign capital invested in their economies. In Brazil, about half of large industry is owned by foreign capital; the other half is divided between the government and a few large domestic companies.[60] Because of the complexities of ownership, it is difficult to tell exactly which owners control which industries.[61] Generally, multinationals' participation tends to be the greatest in component assembly industries (especially in Mexico and the Far East), heavy industry and in processing of local commodities for export (especially in Latin America). Local ownership is concentrated in the traditional labor intensive industries.[62] Aparently, it is difficult to determine who is in control of the new labor intensive industries.(See table 2).

Table 2
GROUPS OF EXPORT INDUSTRIES IN THIRD WORLD COUNTRIES[63]
----

1. - Traditional labor intensive
       Labor intensive articles such as shoes (6% of total
         1972) and textiles (35% of total 1972)

2. - New labor intensive
       Consumer goods - sporting goods, toys, furniture,
         wigs, plastic goods

3. - Component manufacture and assembly
       Televisions, computers, stereos, test equipment,
         and other electronic products

4. Processing of domestic commodities (15% of 1972
       total) Meat packing, food canning, processing of
       minerals

5. - Heavy industrial and engineering (25% of total
       1972) Steel, automobile, chemicals

Source: Frank, _Third World_.

Regardless of the actual ownership of the industries,
the fact remains that a number of the Third World societies
are becoming industrialized, their economies growing at
often incredible rates. Their industries are increasingly
competitive and are having major effects on the
post-industrial societies. However, the displacements often
are more severe for the emerging industrial societies. The
next section will look at some to the effects of the
changing international division of labor on these new
industrial societies.

                            III

    The chip produced in the pleasant environs of
"Silicon Valley" in California has its circuitry
assembles in the toxic factories of Asia.
          -A. Silvanandan[64]

                           179

The firms have let us know that, in case of labor trouble or wage demands, they can stop production within a month and transfer to another neighboring country with a cheaper labor force in the Asian area.
   -A Malaysian labor leader[65]

For a vivid illustration of these facets of exploitation, picture a barefoot, ill-clothed man - or indeed a woman or child - carrying heavy loads on an unsafe scaffolding on a noisy, dusty, hot construction site for ten or twelve hours (plus two to four hours back and forth from home to work by bus or foot) for the pay of $1 per day.
   -Andre Gunder Frank[66]

No matter how impressive the government statistics seem, they don't always translate into a better life.
   -Shim Jae Hoon, New York Times[67]

As the American economist Milton Freedman once noted, "There is no free lunch." The rapid industrialization of the Third World discussed in the previous section could not have been a "free lunch;" it had to have had a price tag attatached. The price tag attached to the Third World industrial development can be measured both in monetary and in human costs.

The countries that are being hailed for their "miracle growth" are the ones that owe the most money to the Western banks. According to an Organization of Economic Cooperation and Development (OECD) report, the total Third and Fourth World debt in 1979 (excluding International Monetary Fund loans) was $397,617.6 million; the debt service alone was close to $74 billion.[68] By January 1983, this debt has risen to about $700 billion. The industrializing Third World countries owe a disproportionate share of this debt. Brazil owes the most, alone owing $87 billion to the West. (See Table 4).

Table 3
ESTIMATED DEBT OF SELECTED THIRD WORLD COUNTRIES[69]

Country	Total at end of 1982	Debt service payment due in 1983	Payment as a % of exports
Brazil	$87.0* billion	$30.8 billion	117%
Mexico	80.1	43.1	126
Argentina	43.0*	18.4	153
S. Korea	36.0	15.7	49
Venezuela	28.0	19.9	101
Isreal	26.7	15.2	126
Poland	26.0	7.8	94

* _Time_ estimates

Source: Morgan Guaranty Trust Company, cited in _Time_

This debt problem is being exacerbated by rising prices for oil. However, growing protectionism in Western markets is posing an even greater threat. The industrializing countries tried to increase their exports after 1973 to pay for their increasing oil import bill, but found many of their most lucrative markets blocked. One observer noted that "no one is screwing the poor of the world like the British worker," as the British government moved to protect noncompetitive English textile mills.[70] By 1977, the Western countries were actually importing less textiles and clothing.[71]

In order to pay for these debts, many countries are required to take out more loans. With these loans, notoriously from the International Monetary Fund, come stipulations about how the country should operate the economy. Usually, the stipulations involve austerity measures such as wage freezes and social services cuts. Reductions in these areas lead directly to a changing of the dollar cost costs of industrialization to the human costs; it is inevitably the people who must, one way or another, feel the effects of these reduction reductions. The bulk of this discussion will concern the human costs in the Third World, because all of the other costs can be reduced to

181

human terms.

A major human factor resulting from rapid industrialization is the disorientation of the labor force. The change to an industrial society involves a significant degree of de-skilling the native worker. The workers forget the stills used to survive in the pre-industrial environment. Over 90% of the labor used in the export industries are unskilled workers, who because of neo-Taylorization, learn nothing of value on the job.[72] Therefore, they are prisoners of the industrial age, unable to return and find employment in the rural areas. In the mid-1970s, Argentina, traditionally an exporter of foodstuffs, had to import wheat from Spain and cattle from neighboring Paraguay and Uruguay to feed its own population.[73] This effect is especially visible in the OPEC countries. In Kuwait, before the oil boom, the natives engaged in "fishing, pearling, pasturing, and a little trade."[74] After the oil boom, all of these activities were abandoned. Oil provided 99% of the government revenue, but only employed a few thousand people. Over 70% of the total working population were immigrants. Of the 30% of the workers who were natives, 75% of them were engaged in services, and actually had little to do. A UN agency estimated that the average Kuwaiti civil servant works about 17 minutes a day.[75] Also, many citizens of developing countries who do acquire marketable skills emigrate to the industrial world to take higher paying jobs. This "brain drain" represents a substantial loss of human capital from these societies.

Those people who are unable to emigrate from the developing countries are forced to endure high levels of social inequality. One of the causes of this inequality is associated with the increase of foreign investment and the resulting rise of the service sector. Multinational investment in a country tends to augment the percentage of people who are employed in the tertiary sector (characteristically, a post-industrial effect) and increase inequality.[76]

Although foreign companies often establish operations in Third World countries to trim labor costs, their factories tend to be capital intensive relative to the already existing economy. However, people in the countryside move to the urban areas in search of employment,

182

the "city lights effect."[77] Inevitably, more people migrate to the city than the industrial sector will be able to employ, so these people search for any work they can get, which, because of the limited nature of manufacturing in early industrializing societies, will tend to be in the service sector. In 1976 South Korea, over 29% of the population was engaged in service occupations, compared to only 21% in the manufacturing sector. Singapore's tertiary is proportionally much larger. Over 64% of the population is engaged in services; manufacturing only accounts for 27% of the total.[78]

The service sector is characterized by the widest polarization of income in society. At the top of the scale are the professionals: doctors, laywers, civil servants, etc. However, the great majority of service workers will end up as servants, janitors, and elevator operators. Service sector expansion can be interpreted as a drag on productivity.[79]

Interestingly, the growth of employment in the manufacturing sector had been shown to have a minor effect on inequality compared to that of the tertiary.[80] A large number of workers in the service sector, however, can have an effect on the wages of those in the manufacturing sector. A large pool of very poorly paid workers already present in the urban areas tends to have a depressing effect on the wages of all of the unskilled workers there.

The employment offered by the industrialization is usually for only a small portion of a country's population. Brazil, a country that is often extolled as an example of an export-oriented industrial country that also has a large domestic market, shows this low participation. The total labor force in Brazil is just above one third of the total population.[81] Industry, however, only occupies one eighth of the labor force (or one twenty-fourth of the total population). The largest industrial employer, the textile industry, only employs 15% of the industrial labor force. That means that Brazil's largest employing industry, the textile industry, only affects one one-hundred and sixtieth (or 0.625%) of the total population. As late as 1976 in Korea, 45% of the population still worked in agriculture, forestry and fishing, compared to only 21% working in the manufacturing sector.[82]

The industrial employment in many areas is often only temporary. The multinational companies which assemble high-technology microprocessor based products move their operations from country to country. Because of the neo-Taylorized technology, the assembly of electronic equipment is comparatively labor intensive, encouraging the companies to seek cheaper and cheaper labor sources in order to reduce costs. The assembly plants were first constructed in areas such as Singapore, Hong Kong, Taiwan, South Korea, Puerto Rico, and Mexico. However, rising wages in these areas prompted the companies to move to even cheaper areas. In 1972, companies moved to Malaysia, the following year, to Thailand, and in 1974 to the Philippines, Indonesia and Sri Lanka.[83] To prevent further corporate moves, the older Third World industrial areas have been encouraging the immigration of cheap labor from abroad to help keep their wage costs down. Hong Kong had been experiencing labor shortages in a number of industries, so it started negotiating with the Philippines for labor imports.[84] Businesses in the colony encourage Chinese to emigrate from the People's Republic. Because many of these refugees are considered to be in Hong Kong illegally, they are in no position to complain to the government about oppressive working conditions.

The most notorious employer of immigrant labor is Singapore. Because of its small size, it does not have a large native pool ot unemployed workers, a situation which would tend to drive up the wage rate. However, workers are imported from neighboring countries to keep the wages down. About 40% of the manual laborers in Singapore are from Malaysia.[85] There are also significant numbers of workers brought in from Indonesia, the Philippines, and Thailand. The guest workers are housed in the worst of accommodations and are not allowed to settle in Singapore; it is made clear that they are only in the city state to work, as they can be deported if they are unemployed. Their employment is subject to government control; they cannot change jobs without its permission. Strict regulations even govern their love lives. Foreign workers are not allowed to marry until they have worked for five years and show a "clean record." After that period, a couple may marry if it receives government approval. However, it must agree to be sterilized after the birth of a second child. This worker repression is accompanied by low wages and rather high prices. It costs about a day's wage to buy coffee and a sandwich in Singapore.[86]

Worker exploitation seems to be a major side effect of the changing international division of labor. The native populations of the industrializing states suffer many of the same types of hardships as the guest workers. When a country's industry is producing for the export market, there is little incentive to raise wages. Under import substitution, wages were increased to enable the workers to purchase the products that they produced. Under an export orientation, however, there is no need to create a domestic market.

An illustration of this different orientation can be seen by comparing the development of South Korea and Japan. Korea's economy at at the end of the 1970s was thought to be on the same order of the Japan Japanese economy of the early 1960s. Korea's 1977 percapita national income of $(1972 constant) 630 is very close to Japan's 1960 level of $(1972 constant) 611.[87] In 1961, Japan was exporting about 11,000 automobiles per year, about the same (if not less) than Korea's 1977 automobile export. In 1961-1963, Japan's home market for autos was about 300,000 cars per year. However, the Korean government limits the domestic sale of automobiles to about 10% of the Japanese figure, showing the strong commitment to overseas, rather than domestic markets.[88]

Production for export creates a type of cultural shock for the Third World worker. He becomes disoriented, as his production seems to have no purpose as seen through both "traditional" and "modern" perspectives of value.[89] In a traditional theory of value value, something has worth if it is useful. By this definition, his production is meaningless, as the worker produces goods which are of no conceivable use to him. Often, he is manufacturing products such as television tubes or speaker woofers, only components of a larger product, having no relevance whatsoever to his life.

The worker is also disoriented when he views the products of his labor from the "modern" exchange value perspective. In the West and Japan, a worker knows that if he has the desire, he can save his wages and eventually purchase the product of his labor. However, the products made by the average Third World worker are useless to him from this perspective. Because he cannot buy what he makes, his output is rendered vauleless from this perspective as well.

It is rather damaging foe a worker to toil to make some product which is worthless to him whether he holds either a modern or a traditional value definition. This dysfunction is exacerbated by the knowledge that somebody else must be gaining from his labor. Whether is is the domestic industrialists, the foreign multinationals, or the consumers in the post-industrial societies across the ocean, he knows he is not gaining very much from his production.

The working conditions in the export societies are intolerable by the standards of the Western industrial societies, and unimaginable by those of the post-industrial societies. Because the surplus of low-cost labor is what gives these areas their comparative edge in manufacturing, it is to the advantage of government and business to keep the costs associated with labor as low as possible. There is little incentive to raise wages, as that change would renter the goods uncompetitive. Like good capitalists, the Third World industries seek to maximize their profits by maintaining high productivity and low costs.

There are a number of methods employed by Third World industries to cut costs, all of which bear most heavily on the worker.[90] One of these methods is to increase the intensity of work. The laborers are encouraged to work at faster rates. The electronics industries of the Far East employ 70% to 90% women from 14 to 24 years old, since they will work for about 20% to 50% lower wages than men.[91] Women are also used because their smaller hands are rather adept at performing delicate operations with small components. A Malaysian brochure advertises:

> The manual dexterity of the oriental female is famous the world over. Her hands are fast and she works fast with extreme care. Who, therefore, could be better qualified by nature and inheritance to contribute to the efficiency of a bench assembly production line than the oriental girl.[92]

This employment in the electronics industry (one of the more "modern" manufacturing ventures) is considered by the manufacturers to be temporary work. The high pase of work wears out the workers rather quickly, resulting in worker turnover rates of 5% to 7% a month and 50% to 100% per year.[93] The very nature of the work causes short working lives. Much of the electronics assembly is done through microscopes. After a few years, the vision deteriorates to the point where a person can no longer continue working. Also, the chemicals which are used to bond the components to the circuit boards are rather dangerous. Workers often suffer "burns, dizziness, nausea, and sometimes even lose fingers."[94] The chemicals are suspected of causing cancer.

The high pace of work also causes many work related accidents. A Free Market Zone in South Korea records about 4500 accidents per year out of a work force of 24,000 (1970).[95] Long hours and a high intensity work environment are blamed for most of the accidents. The Brazilian Ministry of Labor estimates that gloves could reduce accidents by about 22%. However, factories tend to set quotas according to the abilities of the best workers in the plant. In order to meet this quota (and keep a job), the ordinary workers must work without gloves.[96] This high pace of work results in many injuries. When a worker is injured, he is fired.

The fate of the workers laid off is of little interest to the companies. The women who work in the factories often provide the only income for large extended families. When they are fired, the women are forced to seek new employment to support the family. Because they have been declared unfit for work in the manufacturing sector, the women tend to seek jobs in the service sector, often in such areas as prostitution and drug sales. Prostitution is encouraged by the large tourist trade in many of these countries. "Packaged sex tours" for foreign businessmen are offered to take advantage of the surplus women.[97] The businesses can continue the the high turnover rate as long as there are surplus workers in their countries (which is why Singapore and Hong Kong are so eager to import them). As long as there is a surplus of unemployed labor, it will be more cost effective for the industries to maintain the highest worker productivity and merely to replace the worker when he/she wears out.

Another means used by industry in the Third World to

increase the amount of labor extracted from each worker is to increase the amount of time each worker toils per day.(See Table 4). The figures in the table were officially reported to the International Labor Office (ILO) by the respective governments, the actual hours worked in many countries may vary considerably.

Table 4
REPORTED AVERAGE WORKING HOURS
PER WEEK IN MANUFACTURING[98]
(All statistics are for 1976
unless otherwise noted.)

Federal Republic of Germany	39.0-44.6
Singapore	45.7-56.2
United States	35.6-42.4
Mexico (1973)	39.1-48.2
Japan	37.5-41.2
Republic of Korea	48.3-58.6
Ecuador	37 - 58
U.S.S.R.	39.9-41.0
Guatemala (1975)	43.2-50.4
South Africa (1975) White manual workers	42.1-52.4
Egypt (1973)	47 - 62

Source: Yearbook of Labour Statistics 1977

In Hong Kong, there are no regulations dealing with working hours for males over eighteen years old.[99] In South Korea, a seven day, 84-hour work week is considered "not uncommon;" a sixty hour week is called "normal." In Singapore, some electronics and textile industries only have two shifts per day, consequently the workers (usually women) work twelve hours per day. In Brazil, the 1970 census stated that 24% to 29% of the workers in the "food and beverage, construction, and mechanical industries" toiled longer than fifty hours per week. An eleven to twelve hour day is considered normal in some industries. Sixty-six hour weeks were reported as "normal" for workers in the metallurgical industry of Sao Paulo. A labor union president overlooked sixteen hour days in the sugar mills in Pernamubuco. Because of the very low wages paid, he realized that the workers needed the long hours to earn enough wages to survive.

The most obvious way for an industry to cut labor costs is to pay the worker less wages. As mentioned above, high wages are not necessary to the companies to create demand (as in import substitution) since the demand is outside the borders of the country. National minimum wage laws are often unenforced, or are applicable only in the unionized sectors. Wages in Brazil are described as "steadily decreasing.[100] The industrial wages in Brazil are higher than the agricultural wages. This difference, however, merely serves to accelerate the movement of rural residents to the already overpopulated towns, further depressing wages. The degree of worker exploitation is illustrated by the comments of a Malaysian plant mamager:

One worker working one hour produces enough to pay
the wages of ten workers working one shift plus
all of the costs of materials and transport.[101]

The low wages of the Third World worker (see Table 5) have severe effects on the health and living conditions of the laborers' communities.

189

Table 5
AVERAGE HOURLY WAGE RATES IN SELECTED COUNTRIES[102]
(In 1976 U.S. cents.)

Republic of Korea	46
Singapore	62
Brazil	136
Textile industry	79
Hong Kong	69
Maritius	20
Mexico	144*
El Salvador	
Males	57
Females	50
Spain	203
Columbia	47
South Africa	
White manual workers	322
National average	134
Yugoslava	101
Japan	358
United States	560
Kenya	55*
Philippines (1975)	29

*An estimate based on a 45 hour work week.

Source: Yearbook of Labour Statistics 1977

The falling wages and worsening labor conditions following
the change in the international division of labor have

190

engendered a parallel drop in the quality of life in the societies. Infant mortality is considered a major component when determining the "physical quality of life" of a society.[103] In Sao Paulo Brazil, a major industrial area, infant mortality showed a significant rise during Brazil's 1968 to 1973 "economic miracle" period (see Table 6).

Table 6
INFANT MORTALITY IN GREATER SAO PAULO BRAZIL[104]
(Excluding the city center area.)

Year	Rate per 1000 per year
1950	148
1960	73
1961	65
1962-1965	67-69
1966	74
1967	77
1968	74
1969	88
1970	99

Source: Frank, Third World

The reduction of the physical quality of life resulting from the industrialization of the Third World is surely accompanied by a decline in the mental quality of life of the workers. The lack of purpose that was discussed above must be accompanied by a strong sense of frustration. The average worker sees his society grow, yet gains little himself; he works in the "modern sector," yet personally lives in the colonial past. The laborer is told to be patient, that the wealth accruing from his country's spectacular growth will "trickle down" to him, yet he sees his own condition worsening. This frustration is heightened by the realization that a few in the country are getting rich, but there is little he can do to join their ranks.

Those workers who do aspire to a better position for themselves in society are often suppressed by the government, arbitratily arrested, imprisoned, and subjected to indignities such as torture and long isolation. The governments of the export-oriented societies tend to be rather heavy-handed, authoritarian ones, often headed by the military. These tough governments are necessary to preserve the orderly business climate (the low cost labor environment) of the country and maintain high rates of national growth. The next section will examine these authoritarian governments in more detail, governments that ironically seem to possess some post-industrial characteristics themselves.

IV

Investors can expect that the lid will be kept on wages for some time. The Singapore Government and the organized labor movement under its firm control are making a special effort to restore the island's attractiveness as a low-wage center instead of allowing wages to rise above those paid in competitive manufacturing countries such as Taiwan, Hong Kong, and South Korea.
  -Far Eastern Economic Review[105]

I feel it [industrialization] is painful but necessary-- someone must be hurt; it is unavoidable.
  -Thai technocrat[106]

The dominant role of the military in politics is no longer "pathological" but normal. . .
  -Roger Benjamin[107]

In Japan, the decision process is talk, talk, talk, until you reach a consinsus. In Korea and in China, it is talk, talk, but then sombody on top makes a decision.
  -South Korean technocrat[108]

The changing international division of labor has caused major alterations in the nature of governments in the

Third World. Pressures of rising wage demands and threats of internal revolt and external encroachment into markets have often resulted in a swing toward authoritarian governments. These regimes, however, are not the typical "tin-pot banana republic" personal dictatorships which one expects to be so prevalent in Third World states. The new political arrangements can best be described as military-run corporate technocratic states. The basis for power is the military control of the internal political scene. However, many of the important decisions are mady by technocrats, highly trained experts who are predominantly committed to modernization through industrialization. The alliance between the military and the technocrats is the keystone of the arrangement, and seems similar to the relationship between a monarch and a prime minister. Like a king, the military provedes the source of power, but it is the technocrat, like the prime minister, who actually wields that power. The state is the leading device for the implementation of modernization, and tries either to eliminate or co-opt opponents to its programs.

The wage increases that resulted from the import substitution policies created a number of internal problems for the governments of those societies. They were faced with rapidly rising expectations among their people, yet could only afford a limited program to fulfill them. Through the change to an export orientation the government sought to increase the resources of society, but in order to do so had to curtail the rising worker demands. The democratic or populist regimes were unable to cope with the contradictions inherent in the change to a new economic orientation.[109] The military, never very great admirers of politicians, assumed the reins of government in these societies, seeing themselves in the role of the saviors of their respective nations.

The military governments moved against the perceived enemies of development, becoming preoccupied with internal rather than external security.[110] In the best corporate-fascist style, these regimes imprisoned opponents to their rule. There were "more political prisoners in Asia than in the rest of the world."[111] Indonesia is the worst offender, with over 55,000 to 100,000 political prisoners in at least their eleventh year of confinement. In June 1975, Indira Gandhi declared Emergency Rule, "detaining" over 40,000 people without trial.[112] Korea and the

Philippines reduced their constitutional safeguards and made extensive use of rule by presidential decree. India, Pakistan, Malaysia, and Singapore passed emergency laws providing for preventive detention. Korea had been accused of using torture in its treatment of political prisoners.[113]

The military governments were engaged in repression in Latin America as well. The four countries accused of the most human rights violations were Chile, Argentina, Brazil, and Uruguay, countries which had had the "highest socio-economic standard[s] in Latin America."[114] These countries were also among the most advanced with their import substitution policies before they changed their economic orientation toward exports.

The military also consolidated its control over organizations, especially political parties and unions. The control over the various institutions varied from country to country. In South Korea, the unions were subject to strict regulations, leaving them relatively impotent. Strikes were declared illegal; the peanalty for participating in a strike was a maximum of seven years in prison.[115] The government was to mediate all labor disputes, and acted very swiftly in cases involving a foreign-owned company. The government reserved the power to dissolve any union it saw as becoming a "threat to the safeguarding of public order," and had the important ones infiltrated by agents of the Korean CIA.[116] Similar restrictions governed the unions in Marcos' Philippines.

Suharto's Indonesia came even closer to a corporate organization of society. The press and labor unions were put under strict control. All of the traditional and left-wing political parties were eliminated, and in their place was created a single state party.[117] The army was unified and brought under tighter government control. Using his national party and unified army, Suharto was able to use "carrot and stick" policies to co-opt or eliminate political enemies. The large number of political prisoners in Indonesia suggests that the stick was wielded rather often. Business, however, was as unregulated as before, since foreign investment was seen as desirable for development.

The main element which differentiates these export regimes from previous corporate dictatioships is the overwhelming presence of the technocrats. Technocrats are non-partisan university-trained experts who "have been given extraordinary influence over the formulation and execution of policy becuse of their professionsl expertise."[118] They are "concerned with the application of modern transnational knowedge to developing societies."[119] Adjectives commonly used to describe technocrats include: rational, systematic, professional, impersonal, and modern.

The concept of technocratic rule seems to conjure up some of the images associated with rule in a post-industrial, more than a developing society. Bell's emphasis on theoretical knowlege and the importance (and power) of science in society could lead one easily to conclude a technocratic fate for the post-industrial governments. Most of the Third World technocrats have PhDs which they acquired at Western univeristies such as MIT, Columbia, and the University of Chicago.[120] This academic background, in the light of Bell's emphasis on higher education and theoritical knowledge, lead us to question whether the Third World technocrats are a post-industrial phenomenon in their own right.

Bell looked at the relationship between science and government in the United States and determined that the scientists take a backseat role to the politicians and the military in the political decision making process. Science as a political force reached its peak around the time of the development of the atomic bomb. He deccribes a struggle between civilian scientists and the military for decision making powers over the new technology from 1945 to about 1955, in which the military emerged victorious.[121]

He listed three reasons for the break-up of the scientists as a political force:

1. -The break-up of the closely knit elite groups of scientists that were formed during the second World War.

2. -The military was able to acquire its own laboratories and became more independent of the political stipulations of civilian science.

3. -Scientific research was able to acquire much

195

funding from private as well as public
sources, enablilng scientists to diversify
their connections.[122]

However, the conditions in many Third World countries were
considerably different from those in the United States.
Instead of breaking up, science (personified by the
technocrats), became aligned with the military governments.

The expmple of Indonesia's technocratic evolution can
serve to illustrate this point.[123] Before 1966, a group
of university professors taught economics at the Staff of
the Army Command and General Staff school. They instructed
the military leadership of Indonesia on theories on economic
development, mainly the idea that "economic "progress" was a
major ingredient necessary for national security. In 1966,
General Suharto, who had received economic training at the
army school, took control of the government from the
civilians as a result of crisis stemming from the changing
international division of labor. Because much of his
criticism of the civilian government stemmed from economic
grounds, he turned to his old mentors for economic advice,
naming some of them to posts in the government. They were
in a position to try their theories in a real country, and
quickly convinced Suharto of their views. "His dedication
to economic development became an obsession," and he
followed their advice for economic growth.[124]

This chain of events highlights a number of important
differences between the Third World and the American
situation that allowed technocratic rule to prevail in the
former and not the latter.

1. -There was no break-up of the university
    economists in Indonesia. The smaller numbers
    of professionally trained personnel in a
    traditional society often hostile to
    modernity helps to maintain the cohesiveness
    of the Third World scientific elite.

2. -The military was not able to maintain its
    independence from the technocrats to-be.
    Actually, they turned to them for advice,
    becoming dependent on their expertise.

3. -Although it is not readily apparent from the

Indonesian example, technocrats and the military share a number of common characteristics, the most important of which is the distrust of politicians and democracy.[125] Both groups greatly feel that politicians are corrupt and that democracy breeds anarchy, making effective rule impossible. Autocratic rule through the military gives the academics a chance to test economic theories without the meddling effects associated with politics.

This combination of military and technocratic rule is considered to be a form of new authoritarianism, more precisely defined as "bureaucratic authoritarianism."[126] This rule is impersonal and institutional. In time, the military rule might "wither away," after all of its changes have taken hold on the society (assuring the goal of economic growth). Less force will be needed as the people become acclimated to the new economic orientation.

The philosophy of the technocrats is generally in favor of the industrialization of the Third World. They are strongly in favor of modernization and see the state as the principal actor in society able to accomplish it (see Table 7). This statism fits in well with the corporate philosophy which tends to be the result of the suppression of dissidents. The technocrats admire labor unions in the context of "economic unionism." Under this concept, the purpose of the union is to rally workers to the banner of modernization.[127]

Table 7

TECHNOCRATS' PERCEIVED MODERNISING GROUPS IN SOCIETY[128]

Groups	% of those polled who named it
Government	55%
Private business	26%
Modernising entrepreneurs	21%
Modernising intellectuals	19%
External forces (foreign aid, business-	10%
Agriculturalists (large landowners/farmers)	10%

Source: Ilchman, et al.

However, technocrats do not like to become associated with the excessess often demonstrated by military rule. They are often criticized for their alliances with the autocrats and try to distinguish themselves from the dictators as much as possible, especially by maintaining "high levels of personal and professional integrity."[129] Technocrats tend to shy away from political office and personal gain. Because they attended many of the same schools, they try to maintain contact among each other and with colleagues at their home university, and are very sensitive to criticism about their relationship with the military. Their transnational experience also makes them more receptive and knowledgeable about foreign investors, dealing with them with "self-confidence," as "demands and concessions, are, more often than not, likely to be based on direct experience."[130]

198

Table 8
TRADITIONAL VALUES PERCEIVED BY TECHNOCRATS AS CONTRIBUTING TO "BACKWORDNESS"[131]

With Traditional Values	Named By	Compatible Modernization

Ascriptive values:
   Importance of family
   Respect for elders
   Caste and class
	66%	9%

Religious values	4%	6%
Conservation	26%	0%
Respect for law and authority	21%	71%
Regional and local identity	10%	0%

Source: Ilchman, et al.

This internatonal attitude tends to make technocrats less tolerant of traditional idiosyncrasies within the country. They see most of the obstacles to development as internal rather than external.[132] They typically scorn traditional values and tend to ignore local sentiment, their support of such values extending only as far. as they are seen as aiding progress (see Table 8). In an international survey of technocratic planners, 57% said that they felt that the population of the country was "unsympathetic" to modernization.[133] The technocrats are also rather unsympathetic toward nationalism. They are interested less in simple self-reliance and more in growth through international trade. They openly seek foreign aid and capital in the effort to modernize. Also, they have been active in denationalizing some enterprises that were nationalized by the civilian government.[134]

The overall success of the military technocratic alliance is mixed. It has been very successful in the growth area; most of the "miracle growth" economies of the late 1960s and early 1970s were run through this type of a system. However, the alliance reflects the major problem the the changing international division of labor, the neglect of the worker. The record of technocratic rule in social issues had been rather dismal, as inequality has generally increased under its tenure. "There is mounting evidence that from the standpoint of social goals, the technocrats have made no more impressive records thain their colonial predecessors."[135]

The technocrats also have not been very successful in curbing the excesses of the autocrats they serve. In the Philippines, the President's Executive Secretary was fired for trying to eliminate greedy men in the army.[136] Technocrats are limited by the nature of their knowledge. Their position is based on an institutionalized body of knowledge, not necessarily on any personal qualities. This type of knowledge makes them personally expendable, as it would be possible to hire someone else with a similar background to to the same job.

However, there are signs that the technocrats are acquiring power bases outside of the military. As more people in the society become educated, the more potential support the technocrats will have. Various technocrats have been associated with revolts at a number of universities, indicating both a greater political role and a more independent base of power.[137] This possession of an independent power base could enable the technocrats to survive as a politically relevant entity if the military turns its power over to the civilians. Because the military is not likely to return to the barracks until its growth orientation has been ingrained in the society (and in their political successors), the technocrats might find themselves welcome participants in the new civilian governments. Their momopoly on the modern, theoretically-based knowledgd in the society (that is, their post-industrial approach to industrial problems) could earn them significant positions in future non-military governments.

V

Several middle powers have emerged from the ranks of the Third World, distinguishing themselves from the majority of states, the very poor Fourth World.

-Timothy M. Shaw[138]

Independent development by way of a new economic
order it seems, will befefit those foreign
interests that are in a position to take
advantage of the new opportunities.
-Raul Fernandez[139]

The same period will be one of acute suffering for
truly peripheral areas, whose non-essential
exports will find a very weak world market and
whose internal food production may collapse
further.
-Timothy M. Shaw[140]

You start on drugs because the job's so boring,
hour after hour, and you don't even know what the
board is for. You take a "crank" (a
methemphetamine) and you feel a flash of
energy--zzt, zzt, zzt--- and do you work! Then
the technician stands behind you and says, "Hurry
up, you did 100 boards last night!"
-Marijane Esparza
ex-assembly line worker on a Silicon
Valley circut board assembly line.[141]

When discussing the new international division of
labor, it must be realized that the new industrialization
and economic miracles accrue only to a small minority of
Third World countries. About 73% of the growth of value
added of Third World manufacturing occurs in only ten
countries.[142] This concentration of wealth in the Third
World mirrors, on an international level, the inequalities
that exist within the industrializing societies. The
differences place the industrial Third World in a position
similar to that which was traditionally reserved for the
original industrialist powers.

The new world powers are able to engage in
sub-imperialism. While they still have some degree of
dependence on the West, they are able to exert their
infulence over the poorer Fourth World areas. This new
domination is even encouraged by such Western policies as
the Nixon Doctrine, which acknowledges the roles of U.S.
client states as the international policemen of their
regions. The Nixon Doctrine is the most visible in the

military realm, when it becomes U.S. foreign policy to help maintain large, modern military establishments in Third World countries such as Saudi Arabia, Iran (until the Revolution), Taiwan, and Isreal. However, the penetration is much more subtle in other areas.

Third World industrial countries are starting to make investments in the Fourth World. They are able to gain from the same unequal terms of trade that were used to exploit them. They have the additional advantage of first hand experience of foreign penetration in their economies, so they should know the most effective means of making a profit from investment in a less developed country. Singapore is the principal investor in Malaysia.[143] Third World foreign investment is not limited to the country's sphere of infulence. Brazilian companies have substantial investments in Nigeria.[144] The newly industrialized countries also use the underdeveloped countries as markets for their exports. Many of the import substitution products that were too inferior to compete in Western markets are dumped on Fourth World countries.[145]

Another way the "advanced" Third World abuses is poorer brethren is through migration of skilled labor to areas where they can make a better living. This phenomenon is analogous to the post-World War II "brain drain" to the Western countries. The OPEC countries are a notorious recipient of the new wave of emmigration. Much of the skilled labor in Pakistan has been siphoned to the wealthy, yet labor-poor regions in the Middle East.[146] Asian industrializers such as Singapore, Hong Kong, and South Korea also absorb a considerable amount of skilled labor from their neighbors.

The new international division of labor is splitting the Third World solidarity that existed in the wake of the 1973 OPEC crisis. The new industrial states realize that their interests are different from those of the Fourth World countries. The much-touted New International Economic Order has been criticized as an effort of the richer Third World countries to enlarge their share of world markets.[147] The poorer countries are increasingly being forced to look to themselves for development, as the Western powers are concentrating most of their efforts in the more strategic Third World industrial powers. The benefits of the post-industrial changes do not seem to be for them.

The changing international division of labor seems to be a very compartmentalizing phenomenon. Throughout the world, there are islands of varying degrees of development. In the countries called "post-industrial," there are still large groups of people who are unemployed or suffer from the same purposeless production complex as the Third World worker. In the newly industrial Third World, there are post-industrial areas as well. India and Brazil are developing post-industrial industries (aside from mere assembly) such as the computer industry.[148] These same few countries are also expanding their international service sectors. Disagreements are brewing in the international trade organizatons about the regulation of the international goods trade. "Industrial" Third World countries are approaching post-industrial size service sectors.[149]

Clearly, the post-industrial society is not a phenomenon which is restricted to the United States, Japan, and Western Europe, just as these countries are not devoid of their underdeveloped areas. The new division of labor, which in theory relegates the information based economies to the post-industrial world the the manufacturing industries to the Third World, has serious consequences for both worlds. Also, it must be noted that such a division can never be absolute. In almost any country, there will be enclaves of underdeveloped, industrial, and post-industrial societies. The changing international division of labor is merely the redistribution and changing concentration of these areas between one country and another.

ENDNOTES

[1]According to Bell, a post-industrial society manifests a number of observable characteristics. One of the easiest to verify through statistics is the shift from a manufacturing dominated to a service dominated economy. In the United States today, more of the labor force is engaged in service occupations than in manufacturing ones. This shift can be compared to the similar shift from agricultural employment to manufacturing employment during the Industrial Revolution. The industrial society still produces food, just as the post-industrial society will continue to produce goods, but with fewer workers involved in the process than before. The service orientation points toward the increasing importance of knowledge in such a society, the transition from a "goods" society to an "information" society. Bell emphasizes the importance of science and technology in a post-industrial society, writing of the "centrality of and codification of theoretical knowledge." Scientists and trained professionals would have the central roles in a post-industrial employment setting. See, Daniel Bell, The Coming of Post-Industrial Society A Venture in Social Forcasting (New York: Basic Books, Inc., Publishers, 1973), p. 117.

[2]Bell, p. 486.

[3]Throughout this essay, a number of terms will be used to describe the various areas of the world. The "First World" will mean the advanced capitalist industrial regions: Western Europe, Canada, the United States, and Japan. These areas are the ones that are considered moving toward a post-industrial society. The "Third World" will be those regions that are considered "developing countries." The term "Fourth World" will refer to the poorest areas of the world, to those countries which are unable to undergo development in the Western sense. Another terminology which parallels these definitions is the "center/periphery" description. Essentially, the "center" (or the "core") is the First World. The "semi-periphery" is the most developed of the Third World (for example: Taiwan, Korea, Brazil, etc.). The rest of the Third World and all of the Fourth World is the "periphery."

[4]Bell, p. 483.

[5]Daniel Bell, "The Future World Disorder: The Structural Context of Crises," Foreign Policy 27 (Summer 1977), p. 112.

[6]Raul Fernandez, "Third World Industrializaion: A New Panecea?" Monthly Review 32 (May 1980), p. 13.

[7]"Hooked on Growth," Time November 29, 1982, p. 59.

[8]Bell, "Global Disorder," p. 110.

[9]This list is not meant to be an exhaustive list of the reasons for the Western loss of comparative advantage in manufacturing, as that problem is not the focus of this paper. It is only presented to give an indication of the types of ideas prevalent in Western society regarding its position. The list was compiled from, Luciano Tomassini, "Industrialization, Trade, Trade, and the International Division of Labor," Journal of International Affairs 34 (Spring 1980), p. 149.

[10]According to neo-Marxists, Kondratieff long waves, with periods of fifty years, have generally characterized the global economic situation from the onset of the Industrial Revolution. The rising sections correspond to rising prices and profits (boom) and the falling sections to declining wages and profits and increasing unemployment (recession). It has been argued that political and economic repression increases during the downswings as businesses increase their efforts to compensate for their declining profit rates. Greater concentration of the means of production is also expected to occur during this period. As competition for the shrinking markets increases, the less efficient businesses are forced to sell out to the (usually larger) more efficient ones.

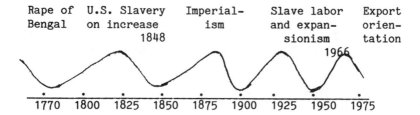

| Rape of Bengal | U.S. Slavery on increase 1848 | Imperial-ism | Slave labor and expan-sionism 1966 | Export orien-tation |

1770 1800 1825 1850 1875 1900 1925 1950 1975

See, Andre Gunder Frank, <u>Crisis: In the World Economy</u>, (New York: Holmes and Meier Publishers, 1980), p. 22-24. And by the same author, <u>Crisis: In the Third World</u>, (New York: Holmes and Meier Publishers Publishers, 1981), p. 160.

[11]Folker Frobel, Heinrichs Jurgen, and Otto Kreye, "Export Oriented Industrialization of Underdeveloped Countries," <u>Monthly Review</u> 30 (November 1978), p. 23.

[12]Andre Danzin, "Are Science and Technology Leading to a New Pattern of Development," <u>Impact of Science on Society</u> 29 (1979), p. 207. Also, Norman Macrae, "Two Billion People," <u>The Economist</u> 263 (May 7, 1973), p. 28.

[13]Henry Scott-Stokes, "Third World Industrializatizes, Challanging the West. . . " <u>New York Times</u> International Economic Survey, February 4, 1979, p. 13-14.

[14]Multinational corporations are "increasingly dependent on earnings from abroad." Bell lists a number of American corporations and the percentage of their profits earned abroad: IBM--50%; Goodyear--33%; Firestone--39%; Uniroyal--75%; H.J. Heinz--44%; Colgate-Palmolive--55%; the chemical companies -- Dow, Pfizer, and Union Carbide--30% to 55%; General Electric--20%; Ford (excluding Canada)--24%; and General Motors (excluding Canada)--19%. See, Bell, <u>Post-Industrial Society</u>, p. 484-485.

[15]Macrae, p. 32.

[16]Scott-Stokes, p. 13-14.

[17]Macrae, p. 32.

[18]This paper will not discuss the debate raging about

the "limits to growth" issue, as it is a large subject with limited bearing on the main topic of this paper. However, for a lively exchange on the issue, see, Jeremy Bugler, "Towards the No Growth Society," New Statesman, March 11, 1977, p. 312-316. Wilfred Beckerman, "What No-Growth Society?" New Statesman, March 18, 1977, p. 339-342. And, Jeremy Bugler, "The Left and Growth," New Statesman, March 25, 1977, p. 398.

[19]Bell, "World Disorder," p. 114.

[20]Bugler, March 11, 1977, p. 313.

[21]Bell would probably agree with this definition. Danzin, p. 268.

[22]Bell, Post-Industrial Society, p. 485.

[23]The figure of 219,000 is only for the period from about 1968 to 1973. Bell, Post-Industrial Society, p. 484. Thousands of others have been laid off since that time.

[24]Danzin, p. 209.

[25]Gregory Schmid, "Interdependence Has Its Limits," Foreign Policy 21 (Winter 1975-1976), p. 192.

[26]Macrae, p. 42.

[27]W.W. Rostow, in looking at the development of the advanced capitalist countries, formulated a stage theory which he expected Third World Countries to follow in their attempts to develop. He presents his theory as a capitalist alternative to Marxist stage theory, assuming that the developing countries would be able to repeat the steps taken by the West. Briefly, his stages of economic growth are: the traditional society, the precondition for take-off, the take-off, the drive to maturity, and the age of mass consumption. Rostow also examines the stage which would follow the age of mass consumption. Interestingly, he tends to take a rather optimistic view of this post-industrial phase. See, W.W. Rowtow, The Stages of Economic Growth A Non-Communist Manifesto, (London: Cambridge University Press, 1960), p. 6-10.

[28]Roger Benjamin, "The Political Economy of Korea," Paper--Department of Political Science, University of Minnseota, n.d.

[29]Shim Jae Hoon, "South Korea Pays a Price For Success," New York Times International Economic Survey, February 4, 1979, p. 59.

[30]Macrae, p. 52. Also, "Hooked on Growth," p. 60.

[31]David Vidal, ". . .But Brazil Finds It Costly," New York Times International Economic Survey, p. 13.

[32]Joseph LaPalombra and Stephen Blank, "Multinational Corporations and Developing Countries," Journal of International Affairs 34 (Spring 1980), p. 124.

[33]LaPalombra, p. 125.

[34]Frank, Third World, p. 101.

[35]This target has been regarded as being unrealisticly high. A United Nations Conference of Trade and Development (UNCTAD) meeting in Nairobi in 1976 better defined the target, determining that an overall Third World growth rate of 11% would be necessary (compared to the 1972 rate of about 6.6%). An International Labor Office (ILO) meeting in June 1976 decided that the target could be met if substantial redistribution would accompany high investment. Interestingly, Bell seems bothered by this combination of redistribution and growth: ". . . the key terms 'substantial income redistribution' and 'high levels of investment' have a menacing ambiguity." Bell, "World Disorder," p. 126-128.

[36]Macrae, p. 17-22.

[37]Ibid, p. 17.

[38]Ibid, p. 41.

[39]Ibid, p. 17.

[40]Ibid, p. 21.

[41]Ibid, p. 21-22.

[42]Ibid, p. 42.

[43]Ibid, p. 41-42.

[44]Frank, Third World, p. 115.

[45]Macrae, p.42.

[46]"Hooked on Growth," p. 59.

[47]A. Sivanandan, "Imperialism in the Silicon Age," Monthly Review (July-August 1980), p. 30.

[48]Macrae, p. 55.

[49]Industrial Developments in Asia and the Far East Vol. II. Selected documents presented to the Asian Conference on Industrializatio Industrialization in Manila from December 6 to 20, 1965, (New York: United Nations, 1966), p. 221.

[50]Macrae, p. 55.

[51]Frank lists a large number of countries that are establishing Free Market Zones or World Market Factories. They include: Bahrain, Hong Kong, India, Singapore, Syria, Taiwan, Indonesia, the Democratic Popular Republic of Yemen, Thailand, Western Samoa, Ivory Coast, Morocco, Swaziland, Egypt, Maritus, Senegal, Togo, Tunisia, Botswana, Ghana, Lesotho, South Africa, and Liberia. See, Frank, Third World, p. 100-101.

[52]LaPalombera, p. 121.

[53]Frobel, et al, p. 23.

[54]For a discussion of containerization and other third generation technical innovations in trade, see, Martin L. Earnst, "The Mechanization of Commerce," Scientific American 247 No. 3 (September 1982), p. 132-145.

[55]The four reasons for wanting MNC investment are from, Frank, Third World, p. 102.

[56]The investment patterns are from, LaPalombera, p. 130-131.

[57]Frank, World Economy, p. 326.

[58]LaPalombra, p. 130.

[59]Ibid, p. 135.

[60]Vidal, p. 13.

[61]Frank, Third World, p. 99.

[62]Ibid., p. 100.

[63]Ibid., p. 99.

[64]Silvanandan, p. 32.

[65]Folker Frobel, Jurgen Heinrichs, and Otto Kreye, The New International Division of Labor, cited in Frank, Third World, p. 145.

[66]Frank, Third World, p. 173.

[67]Hoon, p. 59.

[68]Geographical Distribution of Financial Flows to Developing Countries, (Paris: OECD, 1981), p. 231.

[69]The debt service payment includes the interest and the amortization that are due in 1983. See, "The Debt Bomb Threat," Time (January 10, 1983), p. 42-43.

[70]Bugler, March 11, 1977, p. 312.

[71]"Hurt the Poor and You Will Hurt Yourselves," The Economist 268 (September 9, 1978, p. 87-88.

[72]Frank, Third World, p. 104.

[73]James H. Street and Dilmus D. Street, "Closing the Technological Gap in Latin America," Journal of Economic Issues 12 No. 2 (June 1978), p. 482.

[74]Silvanandan, p. 29.

[75]Ibid.

[76]This theory is the subject of a facinating study: Peter B. Evans and Michael Timberlake, "Dependence, Inequality, and the Growth of the Tertiary: A Comparative Analysis of Less Developed Countries," American Sociological Review 45 (August 1980), p. 531-552. Some of their findings seem as if they could be relevant in post-industrial societies as well.

[77]Evans, p. 533.

[78]Yearbook of Labour Statistics 1977, (Geneva: International Labor Office, 1977), p. 324-355.

[79]Bell, Post-Industrial Society, p. 463.

[80]Evans, p. 546.

[81]The Brazilian employment statistics are from, Vidal, p. 13.

[82]Yearbook of Labour Statistics, p. 324-355.

[83]Silvanandan, p. 37-38.

[84]Ibid, p. 30.

[85]The guest worker restrictions were from, Silvanandan, p. 31.

[86]Ibid.

[87]The original figures for the per capita national incomes were: Japan 1960 -- $421, Korea 1977 -- $887. However, I changed these values into constant 1972 dollars. By comparing values of constant and non-constant dollar tables in the Statistical Abstract of the United States, I was able to determine a formula to convert a given year's dollars to constant 1972 dollars. My conversion formula were:

$$1972 \; \$U.S. = (1960 \; \$U.S.) * 1.45$$
$$1972 \; \$U.S. = (1977 \; \$U.S.) * 0.71$$

It must be realized that the constant dollar figures are

only estimates, and should only be used for comparison purposes. See, <u>Statistical Abstract of the United States 1980</u>, (Washington DC: U.S. Bureau of the Census, 1980), p. 439. After 1977, the Korean per capita national income surged past Japan's 1960 value. For original figures, see, <u>Yearbook of National Accounts Statistics 1980</u> Vol II, (New York: United Nations, 1982), p. 14.

[88]Macrae, p. 45.

[89]The two value theory is from, Silvanandan, p. 39.

[90]Frank suggests that Third World exploitation of industrial workers takes three forms: "Increase in the intensity of work, extension of the working day, [and] payment of labor power below its value. See, Frank, <u>Third World</u>, p. 161.

[91]Frank, <u>Third World</u>, p. 83.

[92]Rachael Grossman, "Woman's Place in the Integrated Circut," <u>South East Asia Chronicle</u>, cited in Silvanandan p. 38.

[93]Frank, <u>Third World</u>, p. 163-164.

[94]Silvanandan, p. 38.

[95]Frank, <u>Third World</u>, p. 175.

[96]Ibid.

[97]Silvanandan, p. 38.

[98]<u>Yearbook of Labour Statistics</u>, p. 522-586.

[99]All of the working hour statistics in this paragraph are from, Frank, <u>Third World</u>, p 168-170.

[100]LaPalombera, p. 133.

[101]Grossman, cited in , Silvanandan, p. 38.

[102]All of the wage rates in the table were calculated from, <u>Yearbook of Labour Statistics 1977</u>. The wage tables with the country averages were on pages 640-649. However,

all of the figures were in native currencies, so I had to use the dollar conversion tables on pages 856-861. Also, the wages were often expressed in weekly and monthly units rather than the desired hourly ones. When this discrepancy occurred, I consulted the section on the average hours worked per week to determine the hourly rate (pages 519-521). When there were no values for the hours, I estimated 45 hours per week. My formula to convert the wages from the monthly to the hourly wage rates was: Hourly rate = (Monthly wage) / (4.3 weeks per month * $ conversion rate * hours worked per week).

[103]Morris David Morris, <u>Measuring the Condition of the World's Poor The Physical Quality of Life Index</u>, (New York: Pergamon Press, 1979), p. 30-34.

[104]Frank, <u>Third World</u>, p. 166.

[105]<u>Far Eastern Economic Review</u>, May 14 1976, cited in Frank, <u>Third World</u>, p. 193.

[106]Warren F. Ilchman, Alice Stone, and Philip K. Hastings, <u>The New Men of Knowledge and the Developing Nations Planners and the Polity: A Preliminary Survey</u>, (Berkley: Institute of Governmental Studies, 1968), p. 18.

[107]Benjamin, p. 18.

[108]Macrae, p. 46.

[109]Carl H. Lande, "Technocrats in Southeast Asia: A Symposium--Introduction," <u>Asian Survey</u> 16 (December 1976), p. 1153.

[110]Gregors S. Treverton, "Latin America in World Politics: The Next Decade," cited in Frank, <u>Third World</u>, p. 244.

[111]<u>Amnesty International Report 1976</u>, quoted in, Frank, <u>Third World</u>, p. 223. In 1977, all of the Asian governments, except for Japan and sometimes India, were authoritarian. Macrae, p. 56.

[112]<u>Amnesty International Report 1976</u>, quoted in, Frank, <u>Third World</u>, p. 223.

[113]Ibid.

[114]Ibid, p. 225.

[115]Frank, Third World, p. 190.

[116]Ibid.

[117]John James MacDougall, "The Technocratic Model of Modernization: The Case of Indonesia's New Order," Asian Survey 16 (December 1976), p. 1150.

[118]Lande, p. 1151.

[119]Richard Hooley, "The Contribution of Technocrats to Development in Southeast Asia," Asian Survey 16 (December 1976) p. 1156.

[120]A look at some of the heads of the national development agencies from various countries reveals the transnational nature of these men:
    Philippines--Head of the National Economic Development Agency
        Gerardo Sicat--MIT economics PhD.
    Thailand--Director of the National Economic and Social Development
        Board
        Sanoh Unakul--Columbia University economics PhD.
    Laos--Director of the Plan of the Lao Government
        Pane Rassavons--University of Paris PhD. See, Hooley, p. 1159.

[121]Bell, Post-Industrial Society, p. 392.

[122]Ibid, p. 403.

[123]The history of the ascendency of the technocrats in Indonesia is from, MacDougall, p. 1166-1167.

[124]MacDougall, p. 1167.

[125]Lande, p. 1152. Ilychman, p. 69. One Middle Eastern technocrat explititly expressed his feelings for the politicians with the statement, "I don't like politicians." Ilchman, p. 26. Another technocrat described politicians

as "economically _irrational_ and. . . lacking economic
knowledge, scientific learning, expertise and education."
They argue that the political figures cannot have the
national interest in mind when they must appeal to local
interests to win an election. MacDougall, p. 1179.

[126]Frank, _Third World_, p. 243.

[127]Ilchman, p. 29.

[128]Ibid, p. 23.

[129]Lande, p. 1152.

[130]LaPalombra, p. 132.

[131]Ilchman, p. 21.

[132]Ilchman, p. 19.

[133]Ibid, p. 22.

[134]MacDougall, p. 1171.

[135]Hooley, p. 1164.

[136]Lande, p. 1154.

[137]Hooley, p. 1165.

[138]Timothy M. Shaw, "The Semiperiphery in Africa and
Latin America: Subimperialsim and Semiindustrialism,"
Review of Black Political Economy 9 No. 4 (Summer 1979),
p. 341.

[139]Fernandez, p. 17.

[140]Shaw, p. 344.

[141]Moira Johnston, "High Tech, High Risk, and High
Life in Silicon Valley," _National Geographic_ 162 No. 4
(October 1982), p. 472.

[142]Frank, _Third World_, p. 98.

[143]LaPalombra, p. 122.

[144]Ibid.

[145]Anne O. Krueger, "Alternative Trade Strategies and Employment in LDCs," The American Economic Review--Papers and Procedings of the American Economic Association meeting in New York, December 28-30, 1977, 68 No. 2 (May 1978), p. 273.

[146]Silvanandan, p. 28.

[147]Carlos F. Diaz-Alejandro, "International Markets for LDCs -- The Old and the New," The American Economic Review -- Papers and Procedings of the American Economic Association meeting in New York, December 28-30, 1977, 68 No. 2 (May 1978), p. 268.

[148]For a look at the computer usage in developing countries, see, Svein Erik Nilsen, "The Use of Computer Technology in Some Developing Countries," International Social Science Journal 31 No. 3 (1979), p. 513-528. For an description of the development of the computer industry in a Third World country, see, B. Bowonder and Sumit K. Pal, "High Technology and Development in India," Futures 10 No. 4 (August 1978), p. 337-341.

[149]Clyde H. Farnsworth, "New Trade Struggle: Services -- GATT Talk on Barriers is Set in Geneva," New York Times November 23, 1982, p. 1-D.

Bibliography

Bell, Daniel. The Coming of Post-Industrial Society A Venture in Social Forcasting New York: Basic Books, Inc., Publishers, 1973.

-----"The Future World Disorder: The Structural Context of Crises." Foreign Policy 27 (Summer 1977): 109:135.

Benjamin, Roger. "The Political Economy of Korea." Paper--Department of Political Science, University of Minnesota: n.d.

Bowonder, B, and Pal, Sumit K. "High Technology and Development in India." Futures 10 No. 4 (August 1978): 337-341.

Bugler,Jeremy. "Towards the No Growth Society." New Statesman, March 11, 1977, p. 312-316.

Danzin,Andre. "Are Science and Technology Leading to a New Pattern of Developmetn?" Impact of Science on Society 29 No. 3 (1979): 201-210.

Diaz-Alejandro, Carlos F. "International Markets for LDCs -- The Old and the New." The American Economic Review Papers and Procedings of the American Economic Association meeting, New York, December 28-30, 1977, 68 No. 2 (May 1978): 264-269.

Ernst, Martin L. "The Mechanization of Commerce." Scientific American 247 No. 3 (September 1982), p. 132-145.

Farnsworth, Clyde H. "New Trade Struggle: Services -- GATT Talk on Barriers is Set in Geneva." New York Times, November 23, 1982, p. 1-D.

Fernandez, Raul. "Third World Industrialization: A New Panacea?" Monthly Review 32 (May 1980): 10-18.

Frank, Andre Gunder. Crisis: In the Third World. New York: Holmes and Meier Publishers, 1981.

------Crisis: In the World Economy. New York: Holmes and
     Meier Publishers, 1980.

Frobel, Folker;   Heinrichs, Jurgen;   and   Kreye, Otto.
     "Export Oriented Industrialization of Underdeveloped
     Countries." Monthly Review 30 (November   1978):
     22-27.

Geographical Distribution of Financial Flows to Developing
     Countries. Paris: OECD, 1981.

"Hooked on Growth." Time, November 27, 1982, p. 58-66.

Hooley, Richard.   "The Contirbution of   Technocrats to
     Development in South East Asia." Asian Survey   16
     (December 1976): 1156-1165.

Hoon, Shim Jae. "South Korea Pays a Price For Success." New
     York Times International Economic Survey, February 4,
     1979, p. 59.

"Hurt the Poor and You Will Hurt   Yourselves."   The
     Economist 268 (September 9, 1978):   87-88.

Ilchman, Warren F.; Ilchman, Alice Stone;   and Hastings,
     Philip K. The New Men of Knowledge and the Developing
     Nations- Planners and the Polity: A Preliminary
     Survey. Berkley: Institute of Governmental Studies,
     1968.

Industrial Develpoment in the Far East Vol. II.   Selected
     documents   presented to   the Asian   Conference on
     Industrialization in Manila, December 6-20, 1965, New
     York: United Nations, 1966.

Johnston, Moria.   "High Tech, High Risk, and High Life in
     Silicon Valley." National Geographic 162 No. 4
     (October 1982): 459-477.

Krueger, Anne O.   "Alternative Trade   Strategies and
     Employment in LDCs." The American Economic Review
     Papers and Procedings of the American Economic
     Association meeting, New York, December 28-30, 1977 68
     No. 2 (May 1978): 270-274.

Lande, Carl H. "Technocrats in Southeast Asia: A Symposium - Introduction." _Asian Survey_ 16 (December 1976): 1151-1155.

LaPalombara, Joseph and Blank, Stephen. "Multinational Corporations and Developing Countries." _Journal of International Affairs_ 34 (Spring 1980): 119-136.

MacDougall, John James. "The Technocratic Model of Modernization: The Case of Indonesia's New Order." _Asian Survey_ 16 (December 1976): 1166-1183.

Macrae, Norman. "Two Billion People." _The Economist_ 263 (May 7, 1977): 7-67.

Morrie, David Morris. _Measuring the Condition of the World's Poor - The Physical Quality of Life Index._ New York: Pergamon Press, 1979.

Nilsen, Sven Erik. "The Use of Computer Technology in Some Developing Countries." _International Social Science Journal_ 31 No. 3 (1979): 513-528.

Palmer, Jay. "The Debt Bomb Threat." _Time_, January 10, 1983, p. 42-51.

Rostow, W.W. _The Stages of Economic Growth - A Non-Communist Manifesto._ London: Cambridge University Press, 1960.

Schmid, Gregory. "Interdependence Has Its Limits." _Foreign Policy_ 21 (Winter 1975-1976): 188-197.

Scott-Stokes, Henry. "Third World Industrializes, Challanging the West. . . " _New York Times_ International Economic Survey, February 4, 1979, p. 13-14.

Shaw, Timothy M. "The Semiperiphy in Africa and Latin America: Subimperialism and Semiindustrialism." _Review of Black Political Economy_ 9 No. 4 (Summer 1979): 341-358.

Silvanandan, A. "Imperialism in the Silicon Age." _Monthly Review_ (July-August 1980): 24-42.

Statistical Abstract of the United States 1980. Washington DC: US Bureau of the Census, 1980.

Street, James H. and James, Dilmus D. "Closing the Technological Gap in Latin America." Journal of Economic Issues 12 No. 2 (June 1978): 477-496.

Tomassini, Luciano. "Industrialization, Trade, and the International Division of Labor." Journal of International Affairs 34 (Spring 1980): 137-152.

Vidal, David. ". . .But Brazil Finds It Costly." New York Times International Economic Survey, February 4, 1979, p. 13.

Yearbook of Labour Statistics 1977. Geneva: International Labor Office, 1977.

Yearbook of National Accounts Statistics 1980 Vol II. New York: United Nations, 1982. 22620

THE NEW INTERNATIONAL INFORMATION ORDER:

THIRD WORLD DEMANDS IN A POST-INDUSTRIAL ERA

Kathleen C. McDonald

Information is the dominant resource in post-industrial society. Accordingly, those who have wide access to this resource will have an opportunity to make important advances in the style and quality of their lives. But what does the growing importance of information mean for the information poor in today's increasingly troubled interdependent world? Developing countries,[1] the information poor of the world, have little control over information in the mass media.[2] The Thirld World must fight the industrialized nations' dominance over the means of production and distribution of information and communications resources to develop internally and to increase international understanding in this era of interdependence. Thus, the developing nations have launched a campaign for a new, more just and efficient world information and communication order.

The New International Information Order (NIIO) is more of a process than an event, document or declaration.[3] This process of global negotiations to correct systemic inequities is of vital interest in a society where information has become such an important resource. According to Daniel Bell, the new problems surrounding information pose "the most fascinating challenge" to this emerging post-industrial society.[4] These problems also are a great challenge to Third World nations who already face many obstacles in their quest to develop. Poverty, overpopulation, racism, famine, inflation, scarcity of natural resources, unemployment, and the arms race are international problems that no single nation can cope with effectively. However, these complex problems are most prevalent in precisely those nations, usually in the Third World, which lack the resources to combat them. Today's interdependence requires the advanced nations to help the poorer nations to some degree just to keep everyone afloat. But the Third World obviously does not feel the advanced nations of the North have paid enough attention to global inequities. One clear sign of neglect is in the realm of news and information dissemination and control.

There are reasons besides a sense of human fairness why persons in emerging post-industrial societies should be concerned with demands for more balanced news flows. Two major trends in our society lend primacy to Third World demands for an NIIO. One is the shrinking size of our globe; the growing inextricability of the domestic and international spheres. Roger Benjamin noted this trend,[5] as did Alvin Toffler who expresses it this way, "Third Wave culture emphasizes contexts, relationships, and wholes," as opposed to Second Wave Culture which studied issues in isolation.[6] Bell describes how the U.S. and other advanced nations depend increasingly on foreign business,[7] and recounts Paul Samuelson's notion of the U.S. as a "headquarters" economy for the rest of the world.[8] This internationalization of our lives has resulted in what Toffler calls an emerging "planetary consciousness."[9] He even went so far as to describe the nation-state as a "dangerous anachronism."[10] Hence, we are obviously moving closer to Marshall McLuhan's global village, a trend which requires us to increase our understanding of the entire globe.

The second trend which is relevant to understanding the importance of the NIIO is the increasing significance and volume of information in post-industrial society. As Bell explains, however, "more information is not complete information; if anything, it makes information more and more incomplete."[11] To be informed in international affairs today, one must understand a wide range of arcane political, social, and economic issues. The media could help provide people with the genuinely broad base of information they need today as a result of the new importance of international understanding. There is a "greater need for mediation, or journalistic translation" to deal with the expanding mass of information.[12] As Bell explained, "for the optimal social investment investment in knowledge, we have to follow a 'cooperative' strategy."[13] These principles are the basis of the demands for the NIIO.

The Old International Information Order: What's Wrong with it?

The old information order which the Third World is battling is based on the free flow of information. The problem with the principle of free flow is that, in reality, a free flow is a very uneven flow with the Third World almost entirely on the receiving end. The industrialized nations have an historical advantage in development that keeps the developing world dependent on foreign media when many nations need to strengthen their own national cultures. In addition, the world's media, which is dominated by the North, gives a rather bleak impression of events in the Third World at a time when these poorer nations need the support of the advanced nations desperately. This imbalance and foreign bias in information flows manifests itself in many ways and is the root of the call for the NIIO.

Mustapha Masmoudi, former Secretary of State for Information in Tunisia, is an articulate critic of the current order who is often regarded as a spokesman for NIIO. Masmoudi outlined his perspective in a paper he prepared for UNESCO (United Nations Education, Science, and Cultural Organization) in 1978). Masmoudi put forth several major complaints against the present system that are widely accepted in the developing world. He started by describing "a flagrant quantitative imbalance" regarding the amount of information flowing between the North and the South.[14] For example, United Press International (UPI), Associated Press (AP), and Agence France Presse (AFP) together send three times as much information to Asia as they receive from that continent on an average day. Another example is that on a particular day in 1977, Venezuela dispatched only seven news items via UPI and AP for every one hundred items it received from the U.S.[15] The Western news agencies bear the brunt of many accusations from the Third World because these agencies--AP, UPI, AFP, and Reuters (based in England)--are practically the only sources of international and regional news to which many nations have access.

Masmoudi then points out great inequalities in information resources, both material and human, that exist between the North and the South.[16] The five transnational news giants, the four Western agencies plus the Soviet agency Tass, own the bulk of the world's information resources which distribute most of our international news. The four Western agencies claim to distribute 90% of the free world's foreign news.[17] Meanwhile, a third of the developing nations do not even have their own national agencies.[18]

Furthermore Masmoudi says there is a "de facto

hegemony" of the North over the South. This hegemony is shown in the indifference of the foreign media to "problems, concerns, and aspirations of developing countries," he explains. He goes on to say that Northern dominance is based upon financial, industrial, technological and cultural power which essentially forces much of the developing world to be mere consumers of information. This complaint is indicative of the general sense of "impotence and exploitation" felt throughout the Third World.[19]

Another complaint Masmoudi has is that there is a lack of information on developing countries in the North.[20] There are not a great many correspondents covering the developing world so there is little detail or depth of coverage on it in Northern media. Much of the world relies in large part on the international agencies for their Third World news, because few other organizations find it profitable to have correspondents throughout the developing world.[21] But even the five giants distribute their correspondents quite unevenly; together, they have 34% in the U.S., 28% in East and West Europe, 17% in Asia and Australia, 11% in Latin America, 6% in the Middle East, and 4% in Africa.[22]

Moreover, much of the news from the Third World is crisis reporting. Thus, many people are only acquainted with disasters and problems in developing nations, such as coups, famines, and fighting. Even crisis reporting has the potential to be serious journalism, but it often lacks the depth that would put the event in a cultural or historical perspective. Also, there is rarely much follow-up reporting to keep the public informed after the issue has become routine news.[23] This problem is due, in part, to insufficient expertise in covering Third World issues and concerns. Reporters often work with little previous knowledge of a developing nation, and a single press bureau may have to spread its limited resources over several diverse countries.[24]

In addition, Masmoudi perceives the survival of the colonial era in biased reporting and interpretation of news.[25] Western media claim to be objective, but many critics detect ethnocentric and racist slants to what is and is not reported. The press may distort reality by labeling national liberation movements and freedom fighters as guerillas, terrorists, and extremists or vice versa. For instance, the Iranian story as told by Western reporters "naturally" led Western publics to side with the deposed Shah instead of explaining the basis for the Ayatollah Khomeini's popularity and ways in which the Shah's regime had been oppressive.[26] Reporting like the following lent credence to the Shah's position:

> Much of the recent rioting has grown from demonstrations called by religious extremists opposed to the Shah's attempt to Westernize this oil-rich, anti-Communist nation and to loosen the traditional firm grip of the Moslem clergy.[27]

Also, when there is a coup, as in Afghanistan, Western reporters often inquire whether the new regime is pro-Soviet or pro-West rather than look at what the new government means for the people and conditions in that country. An example of subtly racist reporting is that when there are rebellions or coups in Africa, Northern papers often focus more on the slaughter of whites than the murders of blacks.[28] Thus, it is with sound reason that the Third World asks for better and fairer reporting of issues concerning them.

There is also a neo-colonialist bent in the way information is disseminated; messages are often ill-suited to areas where they are distributed. Since the news coverage is usually designed to meet the needs of the country in which the agency is based, international agencies often disregard the impact of news beyond their own borders. All the news in the above examples (often including items like Farrah Fawcett's divorce), written for Western audiences, also is sent from international agencies back to the Third World. There it only reinforces local feelings of inferiority and dependence rather than fostering national unity and cultural pride. Furthermore, nations geographically close to each other receive much news of their neighbors only through the transnational media. Thus, fellow developing countries get the Western-colored impression of these countries constantly having riots and floods.[29]

Lastly, Masmoudi complained that there are many alienating influences in economic, cultural, and social spheres which endanger the integrity of traditional cultures. The one-way information flow promotes alien,

225

irreverent values and lifestyles of the affluent West. Many feel this cultural invasion has resulted in a practical, if not intentional, cultural imperialism.[30] The negative influence of multi-national advertising is often especially emphasized. Advertising usually exalts Western materialistic values; but such values are expensive and the Third World is poor. Multinational corporations often succeed in playing on emotions and anxieties, manipulating consumers worldwide to increase the demand for their products. However, most people in the Third World do not need the advertised products from the West: Coca-Cola is often an example. But advertising can encourage people to spend precious money on such non-essential goods or it could frustrate those who cannot afford them.[31] Western marketing practices are a threat to traditional values as they promote "a worldwide cultural homogenization based on Western materialistic values."[32]

The development of communications satellites, comsats, has increased the complexity of the information debate by multiplying the amount of information that can be quickly and easily disseminated. The advanced nations who developed comsat technology control this increased information, thereby giving them even more power and leverage in global society. The Third World is justifiably unhappy with this widened gap between themselves and the advanced nations.

A particularly thorny issue in this field deals with direct broadcast satellites which allow a television set to receive foreign programming directly from a comsat. This technology bypasses the earth station so that a nation can no longer screen out undesirable material. The developing world and many socialist nations claim this infiltration of foreign influence threatens their traditional culture and values (Apparently television broadcasts are felt to be more influential than radio messages which have been crossing national borders with short wave radio for years.).[33]

Proposals for a New International Information Order

As a result of the described imbalances in information flows and subsequent dangers to non-Western cultures, the developing world has demanded a new information order. Third World spokesmen insist the developing world must establish its right to information, the right to communicate in the midst of an international system which furthers

226

individualist concerns at the expense of the entire
community. The Third World is asking for the NIIO for their
own benefit and for the good of the global village as well.
Details may differ in the various proposals for the NIIO,
but the basic components are as follows:

1. World media should devote more time and space to
   issues concerning developing areas.

2. Reports of Third World events should be put in
   proper perspective and expand beyond crisis
   reporting to include a wider range and more depth
   of information, from a variety of sources.

3. Nations should be able to reasonably protect
   their traditional cultures from alienating
   influences.

4. Freedom to communicate must be accorded to all
   nations, not just an international elite.

5. The international flow of information must become
   balanced and equitable to be truly free.

6. The international community must prevent abuses
   by those with current access to information.

7. States should be permitted to have published a
   communique rectifying and supplementing the false
   or incomplete information already disseminated.

8. A new supranational agency should promote the
   professional integrity of journalists and protect
   them against human rights violations. The
   question of domestic control of this area should
   be settled in accordance with national laws.[34]

All the proposals for the NIIO are based on the hope that
the international flow of information will become more
equitable, and increase international cooperation and
understanding as a step towards world peace.

History of the NIIO

This is not the first time nations have complained
about the flow of information and tried to change the
situation. In fact, around the turn of the century,

Americans accused the major European news agencies in operation at that time of dominating the communications scene. They claimed the agencies opposed their American counterparts and favored European points of view. Eighty years later, the U.S. has obviously reversed its position as the international information underdog. Today it is developing countries and socialist nations who are complaining that the largest agencies oppose newer national and regional agencies in the Third World, and that they ignore issues that are relevant to their non-western neighbors.[35]

However, it is not likely that the Third World will reverse its position as quickly as the Americans did. The world situation was quite different in 1900; the gap between European and American information resources was not so pronounced and the entire world was at a very different stage of historical development. Among today's developed countries, many years passed between their industrial revolutions and the wide application of technology to everyday life, which has only occurred recently. But in developing countries, both the information revolution and the industrial revolution are taking place at once. Moreover, this change is happening when many of these nations are trying to form a national identity among disparate groups within arbitrary borders left over from colonial days. Thus, one sees a different concept of the press in the Third World than in the advanced nations. These newer countries feel their arduous process of development requires an "exceptional reliance" on national mass media.[36]

This is the case in much of the South where the press has evolved in a similar way in many nations. Under colonial rule, the press was essentially an arm of the colonial government and it did not tolerate dissent. All forms of expression were strictly controlled. Then, during their struggles for independence, many leaders, authors, and journalists defied colonial regulations and expressed their hope for liberation. Finally, these nations won their independence and everyone was optimistic. The leaders truly spoke for the people and the people gave them their full confidence.

Unfortunately, the optimism faded as these young societies were flooded with the tensions of development.

The nations' meager resources could not meet the demands of competing forces in their societies; disillusionment was everywhere. In order to maintain the current power structure, most Third World governments aligned themselves with established powerful actors in society: the military and the business elite. To control criticism of the fragile governments, many leaders revived colonial regulations of expression. While governments encouraged solidarity and patience, the people remained poor and hungry and complained about the wealthy few.[37]

This is the general state of affairs in many developing countries today: the government sits on a delicate power structure and the nation faces a long hard road of development, so the press is employed to facilitate the development process. One author suggests that a nation can risk a free press only if it has developed a "broad national consensus based on tradition and a vision of the future acceptable to all interests, and minorities."[38] Few developing countries have reached that point.

In most of the Third World, the press is geared to fostering a sense of national unity and "mobilizing the people's will to work more effectively by stimulating progress and knowledge, as well as raising social expectations and demands." With such a monumental task before them, Third World media cannot play the passive role that the Western media play in regard to processes of national change.[39] The plight of the developing world requires a socially responsible press, and, many feel, a new information order.

The debate surrounding the NIIO has always been centered in the U.N. and its agencies. The concept of a genuine free flow of information found its way into the U.N.'s Universal Declaration of Human Rights in 1948. Article 19 says "Everyone has the right to freedom of opinion and expression; this right includes the freedom to hold opinions without interference and to <u>seek</u>, <u>receive</u> and <u>impart</u> information and ideas through any medium regardless of frontiers (emphasis added)."[40] The issue found a home in UNESCO, which has become the international forum for the discussion of information and communication issues. UNESCO's basic purpose is to foster international understanding and cooperation so that nations can better work for world peace. The information debate fits well into that scheme.

From the start, UNESCO stood behind the free flow principle. Its constitution promotes "the unrestricted pursuit of the truth and . . . the free exchange of ideas and knowledge," as well as "the mutual knowledge and understanding of all peoples, through all means of mass communications."[41] Despite its lofty ideals, UNESCO officials were quick to recognize the imbalance in international information flows. It was rather obvious that most media were Western-owned and controlled. However, an early UNESCO study also found that international news agencies operated according to the profit motive, which led them to neglect the unprofitable areas of the globe, namely the Third World.[42] To help correct this imbalance, a proposal was made in UNESCO to create and improve the competitive capacity of national and regional news agencies. Unfortunately, neither the developing world nor the international community paid a great deal of attention to this proposal. It did not occur to the Western powers which then dominated the U.N. that the Third World's dependence on foreign media and the foreign bias in news could be an important abrogation of the free flow principle.[43]

In the 1960s, UNESCO began to take the developing countries seriously and look at the information problem in depth. They worked on projects to set up national new agencies and regional cooperation for the exchange of news, films, and satellite broadcasting for education and development.[44] With the growing participation of newly independent Third World countries in international conferences, the idea of not just a free, but a more balanced flow of information emerged.[45]

Representatives of several Third World nations stressed this idea for the first time at an inter-governmental meeting in 1970 at the sixteenth session of UNESCO's General Conference. These delegations spoke of the unequal distribution of the media and asked for a more balanced flow of information. They emphasized their right to preserve their cultures and indigeneous values, which they felt were threatened by the dominance of foreign influences in the media. A majority of member states at the General Conference in 1972 recognized the danger to cultural integrity posed by a one way flow of news from advanced to developing nations. Then in 1974, the eighteenth session of the General Conference incorporated the concept of a "free and balanced flow of information" into UNESCO's program. The 1974 conference recommended that an international meeting dealing specifically with communications policies be held to encourage better international communication, and to work towards an understanding of the possible roles the media could play in development plans.[46]

The first Conference on Communications Policies, held in San Jose de Costa Rica in 1976, unanimously recommended the formulation of new national and international communication and information policies. In particular, it urged the set up of national communications councils, the development of scientific research in the communications field, and the development of more national and regional news angencies. The San Jose Declaration was indicative of the fact that the international community fully recognized the importance of information in national development and international relations.[47] The more international organizations investigated the subject of information, the more importance they have accorded it throughout the years.

The concept of a New International Information Order emerged first at the Symposium of Non-Aligned Countries on Communication in 1976.[48] The notion of any sort of new international order was first used in the U.N. General Assembly in 1974. At that time, Third World nations demanded a full program of new opportunities in the international marketplace and a large transfer of resources to aid them in the development process. They were demanding a New International Economic Order to replace the old one which they believe discriminates against the poor nations of the world.[49] Accordingly, as Third World countries realized the increasing importance of information, they demanded a NIIO to accompany their demands for the NIEO. But as with demands for the new economic order, advanced Northern nations have responded to demands for a new information order with marked hesitancy for fear the new order would weaken their strength and seriously threaten what they feel are valuable civil rights.

Northern worry over NIIO demands displayed themselves at the UNESCO General Conference in Nairobi in 1976. There some member states made an effort to adopt a set of principles addressing the "intellectual and moral unity" of the international community on the information issue.

However, there was so much opposition to each of several drafts that the Conference simply decided to postpone actual decision on the issue. The opposition came from many sides, including those who feared that the principles as stated would justify censorship or the expulsion of foreign correspondents from certain nations. Others objected to statements regarding the responsibilities of journalists, which the critics felt could enable governments to arbitrarily discriminate against "irresponsible" reporters. Critics also felt the draft lacked a positive reference to human rights, did not guarantee access to a diversity of sources, and possibly aimed to control the international flow of news.[50] As a result of the confusion surrounding the issues, the Conference recommended further study of contemporary communications in light of recent technological advances and the full complexity of developments in international relations.[51] The result was the formation of the International Commission for the Study of Communications Problems, chaired by Sean MacBride, then President of the Republic of Ireland.

In 1978, as the MacBride Commission worked, UNESCO's General Conference adopted by consensus the Declaration on Fundamental Principles Concerning the Contribution of the Mass Media to Strengthening Peace and International Understanding, and to Countering Racialism, Apartheid and Incitement to War. That lengthy title is indicative of UNESCO's broad recognition of the powerful roles that news and information play in world affairs today. The Declaration refers to many concepts regarded as essential to the NIIO, such as a free and balanced flow of information, freedom of journalists, social responsibility of the media, and increasing the sources of information used in news reporting.[52] Nevertheless, the MacBride Commission stresses that such an "agreement by governments does not automatically bring the support of all the professional and other interests concerned, nor of all sections of opinion."[53]

Differing Views of the Media: A Sources of Conflict

There are some people who question the entire concept of developing the NIIO in our complex world. These people are not belittling the importance of information and the media; few will do that anymore. In fact, the control of or alliance with the media is generally recognized as essential to maintaining the status quo or to changing the current power structure. One commentator noted that "on a scale of importance, [the media] are ranked alongside such uncontested elements of state power as political, economic, and military institutions."[54] Those who question the principle behind the NIIO may simply recognize that, on a global level, there are strongly diverging concepts of how the media and information services should be developed and run.

Western democracies hold a view of the press as a "critical appraiser" of contemporary political, economic, and social developments.[55] In these nations, the public has in effect given its trust to the private media organizations that disseminate information into the marketplace of ideas. Western nations usually concern themselves with technical problems of communication rather than with information content.

Some Western critics are afraid the NIIO solutions could be taken further than originally intended and may go so far as to legitimize censorship. Some nations might claim to protect their people from dangerous ideas, which would in effect deny them the right to choose and select ideas to form their own opinions. Many Westerners will agree the Third World needs more of its own publishing firms and news agencies, but are afraid these may be developed to the exclusion of foreign, especially Northern, correspondents.[56] These fears are still very much a part of the Western, particularly the American, attitude toward the NIIO. A 1982 report of U.S. policy toward the U.N. deplored "the efforts by the Soviet Union and some Third World governments to legitimize, under the rubric of a 'NIIO,' government controls over the collection, transmission, and publication of news by the press and other mass media."[57]

Extended into the international sphere, the Western approach basically asserts that the free flow of information is a "primary prerequisite" for healthy international relations, peace, and security. The West generally believes that a free flow of information automatically increases understanding; the rest--development, national pride and unity--can only follow it. President Ford described the Western philosophy in a statement directed to the Eastern bloc nations at the Helsinki Conference in 1975:

But it is important that you recognize the deep devotion of the American people and their government to human rights and fundamental freedoms and thus to the pledges this conference has made regarding the freer movement of people, ideas, and information.[58]

Those words are typical of the Western tendency "to make the operative level absolute in international relations."[59]

On the other hand, developing and socialist countries have quite a different view of the media's role in both the domestic and international realms. Internally, these countries have been organized in such a way that the public must grant its complete trust to the government-run media; there communication is a political issue concerned more with content than with technology. As a result, these nations believe the media should serve goals of national unity and development rather than perform a critical role as they do in the West.[60]

At the international level, many developing and socialist countries believe they should be selective in choosing the means for cultural and informational cooperation to improve security and create a relaxed atmosphere in international relations. A Bulgarian official clearly expressed this philosophy at Helsinki:

The People's Republic of Bulgaria also attaches great importance to international cooperation in the field of education and culture, of information and human contacts. Opened doors are a symbol of trust and hospitality. Our doors will be open to all people with open hearts, with good and honest intentions, who observe the laws, traditions and customs of their hosts.[61]

This kind of thought more or less underlies the NIIO; such an approach seems more realistic in today's world of cultural diversity.

As the above indicates, socialist nations of the North support the concepts behind the NIIO. According to a Polish author who cites several official and non-official sources, socialist states "believe [the NIIO's] implementation would serve to improve international relations, leading to greater mutual understanding and increased national independence and sovereignty for all states." Since the socialists' basic concept of the press if different than the West's, the principles of the NIIO are not so alien. Furthermore, the socialists feel the NIIO is necessary not only to close the information gap between developed and developing nations, but also to deal with the perceived interference of what they see as Western propaganda in internal affairs of foreign states. From their point of view, the West uses the concept of a free exchange of ideas as a "weapon in the psychological war against the socialist states."[62]

Representatives in U.N. bodies have proposed the NIIO as a solution to perceived imbalances in information flows and to propaganda tactics. But even though they have finally begun to realize the tremendous obstacles the developing world faces, Western observers simply feel development should not come at the cost of freedom. They emphasize that, once lost, individual liberties are hard to regain. Westerners' concept of the press makes them rather skeptical regarding the amount of reliable information that comes from government sources.[63]

However, proponents of the NIIO argue that a totally free press is too expensive for the Third World; "freedom from want must come before freedom of expression."[64] With this different concept of the press in mind, perhaps a Westerner can more readily understand Third World skepticism about the possibility of the media playing a socially responsible role without public direction. NIIO supporters insist they do not wish to restrict the flow of information; they only wish to correct systemic imbalances so the arguments for restricting the free flow will disappear.[65] Only then may the world have an international information order based on a genuinely free and balanced flow of information. As Elie Abel, a member of the MacBride Commission from the U.S. said, "we are dealing with a genuine Third World movement to gain control of its future."[66]

Towards a New International Information Order

As previously mentioned, the NIIO may be better described as a process rather than a set of conditions and practices. In the preface to his commission's report on communication, Sean MacBride explained that

> the particulars of the process will continually
> alter, yet its goals will be constant--more
> justice, more equity, more reciprocity in
> information exchange, less dependence in
> communication flows, less downward diffusion of
> messages, more self-reliance and cultural
> identity, more benefits for all mankind.[67]

Thus far there have been many encouraging initiatives in the
process of developing the NIIO. For instance, arrangements
have been made for more national and regional news agencies
in non-aligned nations which would disseminate news from and
about the Third World more widely. The best known attempt
may be the Non-Aligned News Agencies Pool (NANAP), led by
the Yugoslav agency Tanjug. Started in 1973, NANAP's
moderate success has been limited because of a conflict of
interests among the diverse member nations (including India,
Cuba, Cyprus, and Sri Lanka) and a lack of resources and
trained personnel.[68]

International and professional organizations have
continued to increase their efforts to help with technical
and educational cooperation for communications development
in the Third World. There have been a growing number of
conferences, seminars, and symposiums relating to the NIIO,
often depicting it as a necessary part of a new economic
order. Also, work has begun on the creation of resource
centers in Asia and Africa for the exchange of news, films,
television programs, and scholarly publications.[69]

Individual nations have even taken initiatives to
encourage communication development. The French under
Mitterand, for instance, have made a major commitment by
opening a "World Center in Computers and Social
Development." France is funding the project at $20,000,000
a year, which has attracted leading technical experts.[70]

Furthermore, there has been increased interest on the
part of major newspapers, publishers, and news agencies of
the North in devoting more of their resources to Third World
issues; the international media have finally made a general
commitment to provide more and better information on the
developing world. Significantly, sixteen major newspapers
worldwide have begun to publish a quarterly supplement

dealing with the debate surrounding the New International Economic Order.[71]

Although they may lack depth of coverage, the international news agencies have long claimed their services gather ample amounts of foreign, including Third World, news, but say that their newspaper and broadcast clients do not use a great deal of it.[72] The clients in turn maintain that their readers, viewers, and listeners simply are not that interested in such news. As William Randolph Hearst once said, "News is what is interesting, not necessarily what is important."[73]

Critics say the public in advanced nations is relatively uninformed about world affairs. This is especially so in America even though the American public has access to the world's most pervasive media. Perhaps because they are not visibly threatened by the rest of the world, Americans have retained a continental outlook and are often slow to realize their actual dependence on other parts of the world. They often fail to understand that domestic problems have roots in events across the globe. Recent events, however, such as the oil embargo, the Iranian hostage crisis, and the Societ invasion of Afghanistan, have taught more of the world, particularly Americans, that internal problems may be transnational in origin and that an individual country often cannot solve the problems alone.[74] This may mean there will be an increasing interest in good, sensitive news coverage of the developing world.

Despite these heartening developments and elegant Third World explanations, some Western nations still criticize the basic concept of the NIIO. They misunderstand the developing nations to mean they want some sort of "Orwellian design"--legitimate government control of communication, licensing of journalists, a mandatory code of ethics policed by governments. Instead, the Third World only advocates a system of professional self-regulation and social responsibility of the press.[75]

To cope with these problems, in 1978 UNESCO created the International Program for the Development of Communication (IPDC). The IPDC was created with hopes that, if the Third World would moderate its rhetoric, the industrialized world would increase assistance for developing communications capabilities in the developing countries.[76] The IPDC is supposed to be non-ideological and simply facilitate the provision of practical assistance for building up the communications skills and structures which are a precondition to the exercise of freedom of information. The free flow of information has little meaning to those who do not have access to radio, newspapers, or television.[77]

Much of the important recent discussion relating to the NIIO has taken place within the forum of the IPDC. One commentator noted that at the January 1982 meeting in Acapulco, there was a welcome reduction in the polemics and hostility that have marked the information debate thus far. Clifford Block, who directs the U.S. Agency for International Development's Research and Development program in development communications, explained that "the shared perception is that the developing world has been accommodating; the industrialized world must now keep its bargain if the compact is to last."[78]

While many nations in all parts of the world contributed relatively generously towards the solutions to information imbalances, the U.S.--the most advanced nation in information resources--still holds back. In Acapulco, the U.S. promised only "aid in the form of expert services" of $100,000, a contribution equal to that of East Germany and barely ten percent of the Soviet contribution in that category alone.[79] Even poor nations like Bangladesh and Benin gave a symbolic few thousand dollars in direct funding; but the U.S. did not give even one dollar in that category where the funds would be used according to the decisions of the IPDC council.[80]

It is understandable that some nations may be reluctant to support the NIIO. Some groups, like the whites in South Africa of power-hungry Argentine generals, could misuse the faith underlying the concept of the NIIO and impose a type of censorship or outlaw foreign journalists. However, the West, especially the U.S., should realize that the NIIO was not developed for the benefit of such prejudiced, intolerant groups; the concept emerged to correct imbalances in information flows which have traditionally discriminated against the poorer developing nations. In the process of negotiating for the NIIO, many involved began to realize the benefits of increased international understanding and respect the new order could bring to everyone in the world. Certainly, some groups could abuse their privileges; but these groups are probably already maligning the international community's trust in them.

Regardless of such explanations and reassurances, the U.S. and a few other advanced nations appear unwilling to enter a new information order very quickly. Perhaps they are unsure of how to deal with developments that tie them more closely to the rest of the world, while at the same time changes lead these nations into post-industrial society. No matter how uncertain and hesitant any nation might be, the entire world is moving closer and closer to actually becoming a global village. But, as the MacBride Commission cautioned, only if the media emphasize more of what joins people together rather than what divides them will all the peoples of the earth be able to aid one another through peaceful exchange and mutual understanding.[81]

ENDNOTES

[1]In this paper, the terms Third World, developing countries/world, and the South, i.e. the southern hemisphere, will all be used to refer to those nations in Central and South America, Asia, and Africa which have not gone through industrial revolutions entirely or at all and are underdeveloped. The terms the developed, advanced, industrialized world/countries and the North will refer to nations of the northern hemisphere which have industrialized already. The West refers specifically to the U.S. and Western Europe.

[2]As we move into post-industrial society, all kinds of information take on an increasing importance. However, this paper will consider only information as it is used in the mass media.

[3]Sean MacBride, Pres., International Commission for the Study of Communications Problems, <u>Many Voices, One World</u>, (New York: UNESCO, 1980), p. xviii.

[4]Daniel Bell, <u>The Coming of Post-Industrial Society: A Venture in Social Forecasting</u>, (New York: Basic Books, 1973), p. xix.

[5]Roger Benjamin, "The Political Economy of Korea," (Paper presented to the Department of Political Science at the University of Minnesota, 1982), p. 21.

[6]Alvin Toffler, <u>The Third Wave</u>, (New York: Bantam Books, 1980), p. 300.

[7]Bell, p. 484.

[8]Ibid., p. 485.

[9]Toffler, p. 324.

[10]Ibid., p. 327.

[11]Bell, p. 467.

[12]Ibid., p. 468.

[13]Ibid., p. xviii.

[14]Moustapha Masmoudi, "The New World Information Order," (Paper presented at the third session of the International Commission for the Study of Communications Problems, July 1978), cited in Hatchen, p. 95.

[15]MacBride, p. 146n.

[16]Masmoudi, p. 101.

[17]Anthony Smith, The Geopolitics of Information: How Western Culture Dominates the World, (New York: Oxford University Press, 1980), p. 73.

[18]Masmoudi, p. 101.

[19]Ibid., p. 101.

[20]Ibid., p. 102.

[21]MacBride, p. 157.

[22]Smith, p. 73.

[23]William A. Hatchen, The World News Prism: Changing Media, Clashing Ideologies, (Ames, Iowa: Iowa State University Press, 1981), p. 95.

[24]Oliver Boyd Barret, The International News Agencies, (Beverly Hills, California: Sage Publications, 1980), p. 88.

[25]Masmoudi, p. 102.

[26]Smith, p. 99.

[27]Los Angeles Times, 10 September 1978, cited in Smith, p. 99.

[28]Smith, p. 98.

[29]Hatchen, p. 102.

[30]Masmoudi, p. 103.

[31]MacBride, p. 110.

[32]Hatchen, p. 106.

[33]Ibid., p. 100.

[34]Masmoudi, p. 107.

[35]Jerzy Oledzki, "Polish Perspectives on the New Information Order," Journal of International Affairs 35 (Fall/Winter 1981/2): 157.

[36]Ibid., p. 159.

[37]Altaf Gauhar, "Third World: An Alternative Press," Journal of International Affairs 35 (Fall/Winter 1981/2): 166.

[38]Ibid., p. 170.

[39]Oledzki, p. 159.

[40]Doudou Diene, "UNESCO and Communications in the Modern World," Journal of International Affairs 35 (Fall/Winter 1981/2): 219.

[41]Ibid., p. 219.

[42]Reports on the Facilities of Mass Communications: Press, Film, Radio (Paris: UNESCO, 1947-1951), 5 vols., cited in Oledzki, p. 156.

[43]Oledzki, p. 156.

[44]MacBride, p. 40.

[45]Diene, p. 220.

[46]MacBride, p. 40.

[47]Ibid., p. 41.

[48]Diene, p. 220.

[49]Joan Edelman Spero, The Politics of International Economic Relations, (New York: St. Martin's Press, 1981),

p. 200-208.

[50]MacBride, p. 41.

[51]Ibid., p. xiv.

[52]Diene, p. 221.

[53]MacBride, p. 42.

[54]Oledzki, p. 155.

[55]Hatchen, p. 89.

[56]Ibid., p. 107.

[57]United States Policy Toward the United Nations, ad hoc group, "The United States and the United Nations . . . A Policy for Today," October 1981, mimeo released on March 16, 1962, p. 16, cited in Kaarle Nordenstreng, "U.S. Policy and the Third World: A Critique," Journal of Communication 32, (Summer 1982): 54.

[58]Kaarle Nordenstreng and Herbert I Schiller, "Helsinki: The New Equation," in National Sovereignty and Internation Communication, edited by Nordenstreng and Schiller, (Norwood, New Jersey: Ablex Publishing Corp., 1979), p. 239.

[59]Ibid., p. 239.

[60]Oledzki, p. 160.

[61]Nordenstreng and Schiller, p. 239.

[62]Oledski, p. 158.

[63]Ibid., p. 160.

[64]Hatchen, p. 103.

[65]MacBride, p. 144.

[66]Joseph A. Mehan, "UNESCO and the U.S.: Action and Reaction," Journal of Communications 3 (Autumn 1981): 163.

[67]MacBride, p. xviii.

[68]Hatchen, p. 108.

[69]MacBride, p. 142.

[70]Clifford H. Block, "Promising Step at Acapulco: A U.S. View," _Journal of Communications_ 32 (Summer 1982): 66.

[71]MacBride, p. 142.

[72]Hatchen, p. 83.

[73]MacBride, p. 157.

[74]Hatchen, p. 5.

[75]Kaarle Nordenstreng, "U.S. Policy and the Third World: A Critique," _Journal of Communication_ 32 (Summer 1982): 55-56.

[76]Block, p. 60.

[77]"Mexico Meeting on Development of Communications Approves Fifty-Four Projects," _U.N. Monthly Chronicle_, March 1982, p. 53.

[78]Block, p. 69.

[79]Nordenstreng, "U.S. Policy," p. 58.

[80]"Mexico Meeting," p. 53.

[81]MacBride, p. 35.

BIBLIOGRAPHY

Argumedo, Alcira. "The New World Information Order and International Power," Journal of International Affairs 35 (Fall/Winter 1981/2): 179-188.

Barton, Richard L. and Gregg, Richard B. "Middle East Conflict as a TV News Scenario: A Formal Analysis," Journal of Communication 32 (Spring 1982): 172-185.

Bell, Daniel. The Coming of Post-Industrial Society: A Venture in Social Forecasting. New York: Basic Books, 1973.

Benjamin, Roger. "The Political Economy of Korea." University of Minnesota Department of Political Science, 1982, 21 pp.

Block, Clifford H. "Promising step at Acapulco: A U.S. View," Journal of Communication 32 (Summer 1982): 60-70.

Boyd-Barret, Oliver. The International News Agencies. Beverly Hills, California: Sage Publications, 1980.

Diene, Doudou. "UNESCO and Communications in the Modern World," Journal of International Affairs 35 (Fall/Winter 1981/2): 217-224.

Gauhar, Altaf. "Third World: An Alternative Press," Journal of International Affairs 35 (Fall/Winter 1981/2): 165-178.

Hatchen, William A. The World News Prism: Changing Media, Clashing Ideologies. Ames, Iowa: Iowa State University Press, 1981.

Harley, William G. "The U.S., Stake in the IPDC," Journal of Communication 31 (Autumn 1981): 150-158.

Hill, Martin. The United Nations System. Cambridge: Cambridge University Press, 1978.

Hopple, Gerald W. "International News Coverage in Two Elite Newspapers," Journal of Communication 32 (Winter

1982): 61-73.

Knight, Grahm and Dean, Tony. "Myth and Structure of News," Journal of Communication 32 (Spring 1982): 144-161.

Mehan, Joseph A. "UNESCO and the U.S.: Action and Reaction," Journal of Communication 31 (Autumn 1981): 159-163.

"Mexico Meeting on Development of Communication Approves Fifty-Four Projects." U.N. Monthly Chronicle, March 1982, p. 53.

Nordenstreng, Kaarle. "U.S. Policy and the Third World: A Critique," Journal of Communication 32 (Summer 1982): 54-59.

Nordenstreng, Kaarle and Schiller, Herbert I., eds. National Sovereignty and International Communication. Norwood, New Jersey: Ablex Publishing Corporation, 1979.

Oledzki, Jerzy. "Polish Perspectives on the New Information Order," Journal of International Affairs 35 (Fall/Winter 1981/2): 155-164.

Ravault, Jean. "Information Flow: Which Way is the Wrong Way," Journal of Communication 31 (Autumn 1981): 129-134.

Report by the International Commission for the Study of Communication Problems, Sean MacBride, President. Paris: UNESCO, 1981.

Roach, Colleen. "Mexican and U.S. News Coverage of the IPDC at Acapulco," Journal of Communication 32 (Summer 1982): 71-85.

Sussman, Leonard. "Independent News Media: The People's Press Cannot Be Run by Government," Journal of International Affairs 35 (Fall/Winter 1981/2): 199-216.

Smith, Anthony. The Geopolitics of Information: How Western Culture Dominates the World. New York: Oxford University Press, 1980.

Spero, Joan Edelman. <u>The</u> <u>Politics</u> <u>of</u> <u>International</u>
<u>Economic</u> <u>Relations</u>. New York: St. Martin's Press,
1981.

Toffler, Alvin. <u>The</u> <u>Third</u> <u>Wave</u>. New York: Bantam Books,
1980.

UPLINK, Newsletter of the Rural Satellite Program, US AID,
July 1982 and June 1981.

Weaver, David W. and Wilhoit, G. Cleveland. "Foreign News
Coverage in Two U.S. Wire Services," <u>Journal</u> <u>of</u>
<u>Communication</u> 31 (Spring 1981): 55-64.

POVERTY AND DEMOCRACY IN AMERICA

Carl Cordova

> In the optimistic theory, technology is an
> undisguised blessing. A general increase in
> productivity, the argument goes, generates a
> higher standard of living for the whole people.
> And indeed, this has been true for the middle and
> upper thirds of American society, the people who
> made such striking gains in the last two decades.
> It tends to overstate the automatic character of
> the process, to omit the role of human
> struggle....Yet it states a certain truth--for
> those who are lucky enough to participate in it.

> Michael Harrington
> The Other America

> Free, dost thou call thyself? Thy ruling thought
> would I hear of, and not that thou hast escaped
> from a yoke.

> Friedrich Nietzsche
> Thus Spake Zarathustra

As our economically mature nation enters the
Post-Industrial Age, many Americans anticipate widespread
improvements in their standard of living. One might assume
that technological progress refines society to the extent
that philosophical and artistic concerns prevail in an
unprecedented fashion: the efficiency of the country's
economic machinery increases; people have to work less to
accomplish the same tasks--yielding greater leisure time;
and the general quality of life elevates our consciousness
for the sublime. However, such expectations betray a
disregard of the plight facing the lowest echelons of the
nation. Those Americans struggling against poverty do not
look forward to progress with hope but see it only as
increasing their potential for misery. In a nation imbued
with all of the opportunities afforded and equality
guaranteed by "democracy," can this outlook be called
"excessively pessimistic?" Evaluating this issue requires
an examination of poverty, an inquiry into the efforts made

to alleviate this problem, and a determination whether democracy in the United States promotes an egalitarian society.

The millions of poor in America not only face a standard of living below decency; they fight the despair that comes from being ignored by the upper two-thirds of the nation. As Michael Harrington points out in The Other America, misunderstanding poverty in the United States results from the way it remains hidden as "the other America." He explains how it exists as invisible:

> Poverty is often off the beaten track. It always has been. The ordinary tourist never left the main highway, and today he rides interstate turnpikes. He does not go into the valleys of Pennsylvania where the towns look like movie sets of Wales in the thirties. He does not see the company houses in rows, the rutted roads (the poor always have had bad roads whether they live in the city, in towns, or on farms), and everything is black and dirty. And even if he were to pass through such a place by accident, the tourist would not meet the unemployed men in the bar or the women coming home from a runaway sweatshop.[1]

The majority of Americans have no idea of the extent that the "invisible land" suffers. Some naive people look at rural poverty with romantic illusions. For example, a drive through the Appalachians in the summer season might give the uninformed an impression of a pastoral setting. An Upper or Middle-class person might look at the "quaint" surroundings and decide that the local inhabitants have an idyllic existence; instead of recognizing the lack of prosperity, education, medical care and technological progress, the well-off observer could think that a life without worrying about car payments, mortgages, higher tax brackets, politics at the country club, the kids' braces, problems with "the help," or maintaining a good credit rating would be "fortunate." For what more could one ask than to be "exempt from the strains and tensions of the middle class?"[2]

One cannot mask urban poverty in the same manner. However, it remains easy to avoid confronting the misery:

249

workers travelling to their jobs in the city go to the business districts, and if they do happen to catch a glimpse of the slums they usually ignore the brief and removed encounter; shoppers from the suburbs go about their business in a self-absorption that blocks out most of the visible signs of the poor; and the children from the suburbs go to economically "segregated" schools, associating with other children from their own area and not the inner city. As the affluent Americans become increasingly removed from the poor, they fail to recognize the problem as well as a sufficient solution.[3] As poverty remains easy to overlook, ineffective and quick "remedies" are simplified to give the rest of the society a false sense of egalitarianism:

> This new segragation of poverty is compounded by a well-meaning ignorance. A good many concerned and sympathetic Americans are aware that there is much discussion of urban renewal. Suddenly, driving through the city, they notice that a familiar slum has been torn down and that there are towering, modern buildings where once there had been tenements or hovels. There is a warm feeling of satisfaction, of pride in the way things are working out: the poor, it is obvious, are being taken care of.[4]

Manifestations of Harrington's observation can be witnessed throughout the United States in most of the major cities. For example, concern grew in the late 1960's and early 1970's in Philadelphia over the decay of buildings around the city's waterfront area. The older warehouses and rowhomes crumbled from neglect and developed into a great eyesore for the historic city as the country's bicentennial rapidly approached. Politicians, planners, and builders undertook a massive project to clean up the area around Delaware Avenue to improve the city's image and "relieve" the plight of the economically depressed. A good part of the docks were given a modern "facelift," including the building of a small shopping area with several restaurants, and many square blocks of houses and apartments were rebuilt in a colonial motif replacing the squalid wrecks that were either condemned or close to collapsing on their own. The repaired homes were later sold for over $100,000 each, and the area named "Society Hill" became an exclusive place to live on the waterfront.

The metamorphosis that occurred in Society Hill did not help the former urban dwellers who were either driven out by the decay or displaced by the construction. The price for the new real estate became much too high even for most middle class citizens, let alone the poor ones who used to surround and pervade the same area before rebuilding occurred. Similiar rebuilding took place in slums nearby, only the impoverished inhabitants were displaced in a different manner. The planners incorrectly computed the number of square blocks to be destroyed in order to make way for new highway and building construction. After promising all of the displaced people from the slum area that there would be plenty of new housing to accommodate them when they returned, local leaders discovered that too many square blocks had been destroyed--needlessly uprooting hundreds of families. Someone on the planning commission had drawn a line on a map in a wrong place, but the city would not increase the funds allocated for the project to correct their error. The razed city blocks remained a surplus wasteland, and the displaced people were left on their own. Lacking effective lobbying influence, their story quickly died in the press and passed away from the public's consciousness.

One usually thinks of a lack of food, money, and shelter when thinking of poverty. The poor suffer greater misery than an absence of necessities--many social critics argue that they are most susceptible to physical illness directly resulting from their living conditions. Harrington calls this inevitability "one of the most familiar forms of the vicious cycle of poverty":

> The poor get sick more than anyone else in the society. That is because they live in slums, jammed together under unhygenic conditions; they have inadequate diets, and cannot get decent medical care. When they become sick, they are sick longer than any other group in the society. Because they are sick more often and longer than anyone else, they lose wages and work, and find it difficult to hold a steady job. And because of this, they cannot pay for good housing, for a nutritious diet, for doctors. At any given point in the circle, particularly when there is a major illness, their prospect is to move to an even

lower level and to begin the cycle, round and
round, toward even more suffering.[5]

Another "invisible" misery (to the upper two-thirds of
society) that the poor face cannot be avoided or ignored by
anyone in all levels of income; death strikes everyone
regardless of their income, but the poor face mortality
sooner and more often than anyone else. An article
discussing mortality rates in a July 1976 issue of the
American Journal of Epidemiology concludes that a "vast body
of evidence has shown consistently that those in the lower
classes have higher mortality, morbidity, and disability
rates ... in part due to inadequate medical care services as
well as to the impact of a toxic and hazardous physical
environment."[6]

In the mid-1970's about one of every three blacks lived
below the poverty level, while the figure for whites was one
of every eleven people. The median income for black
families at this time was a little under three-fifths of the
median income for white families. These figures showed an
increasing gap of median incomes between the two groups
since the late 1960's, and the trend has continued since
that time. Because of this racial disparity in economic
levels and the large percentage of blacks living in urban
slums, a comparison of health statistics for blacks and
whites can lend evidence to the view that "poverty
kills".[7]

Statistics from 1973 exemplify health trends from the
last two decades: black mothers died in childbirth three
times more often than white mothers; the black infant
mortality rate was two-thirds higher, and black mothers lost
their infants more than twice as often in the age range of
one to eleven months (postneonatal mortality rate); and life
expectancies for black males and females were 61.9 and 70.1
years respectively, compared to those figures for their
white counterparts--68.4 and 76.1 years. There exists no
evidence that indicates genetic causes for the racial
disparity of life expectancies. Consequently, the
environment must account for the difference: the economic
statistics (poverty levels, median incomes) tend to favor
the notion that poverty has a direct and hazardous influence
on human life.[8]

Life in the "other America" affects mental health as

much as it influences physical well-being. Freud made this observation a half a century ago:

> We shall probably discover that the poor are even less ready to part with their neuroses than the rich, because the hard life that awaits them when they recover has no attraction, and illness in them gives them more claim to the help of others.[9]

Much more has been made of the trials and tribulations of upper and middle-class life. Society seems to sympathize with the pressures one faces from the "rat-race." Many people assume that the lower down on the economic ladder one lies, the less upsetting and traumatic day-to-day living becomes. "Breakdowns" and other nervous disorders are mostly limited to the executives and other people in important positions, but the poor lead a relatively peaceful existence free from the pressures to succeed. However, the truth of this matter contradicts this common myth completely. Although mental illnesses are harder to document than physical sichnesses, psychologists universally acknowledge the vulnerability of the poverty-stricken Americans to mental disease. Some of their illnesses are different, and many are more serious than those from other classes. Harrington discusses the duress suffered by the poor:

> Indeed, if there is any point in American society where one can see poverty as a culture, as a way of life, it is (in the area of mental illness). There is, in a sense, a personality of poverty, a type of human being produced by the grinding, wearing life of the slums. The other Americans feel differently than the rest of the nation. They tend to be hopeless and passive, yet prone to bursts of violence; they are lonely and isolated, often rigid and hostile. To be poor is not simply to be deprived of the material things of this world. It is to enter a fatal, futile universe, an America within America with a twisted spirit.[10]

In order to understand poverty one must realize the emotional consequences as well as the economic hardship; inequality is a human story, not merely a statistical disparity. Herbert Gans asserts in his More Equality that inequality "results in feelings of inferiority, and these in turn generate inadequacy and self-hate or anger."[11] Gans maintains that these feelings of inadequacy and self-hatred contribute greatly to the large incidence of mental illness among the poor. The anger elicited from a poverty situation leads to frustration, and ultimately, senseless violence. Instead of political protest through lobbyists, the pent-up outrage has other manifestations: vandalism, crime, and delinquency. However, there are less obvious and radical effects of being poor that leave indelible marks on a person's character, influencing people in a manner that cannot be counteracted by any "generous" programs from the rest of society:

> ... some poor mothers refuse to send their children to school because they cannot afford to dress properly; shabby clothes may protect a youngster from the elements--a flour sack made into a suit or dress would do that--but shabby clothes also mark the child as unequal, and mothers want to protect their children from this feeling even at the cost of depriving them of schooling. And if a father cannot find work, not only will his child's feelings of inferiority be more intense, but the child may justifiably conclude that it is doomed to the same fate, and that there is no reason for staying in school or continuing to hope for a fair shake from "the system".[12]

One cannot understand the true misery in the "other America" without considering such less visible circumstances. Elliot Liebow wrote a sensitive account in the early 1960's of the lives of young black men in a blighted section of Washington, D.C. The Streetcornermen met frequently at a place called Tally's Corner, from which Liebow's book acquired its name. Originally written as a Ph.D. dissertation in anthropology, Tally's Corner creates lucid portraits of several of society's "losers." Lacking education and income, they exist "in a social milieu between that of the relatively stable and upwardly mobile lower-class workingmen, and that of the derelict and bum."[13]

Liebow spent about two years with the men at Tally's

Corner, gathering information that insightfully shattered many myths surrounding the poor, especially blacks from ghettoes. One important area of concern that he discusses centers on the relationship between men and jobs.[14] A common misconception is that most of these "losers" would not take a job if it were "handed to them on a platter." Liebow describes a scene on a street where a white man drives around the neighborhood, stopping abreast of idling men and asking them if they want a day's work. Most of the men turn him down and the driver leaves the area with only a few out of the forty of fifty men he approached. One might get the immediate impression that most of these men do not want to support themselves or make something of their lives, but Liebow explains this misconception. Many of the men have full or part-time jobs, but they might happen not to be working at the moment a would-be employer approaches them. In their off-hours they gather with their friends on the street or sit on their steps alone because they are too depressed to stay at home by themselves, or they need to get away from their wives and families. Some men work at nights, or some have jobs that do not begin until later in the day. Sometimes more personal reasons cause them to turn down temporary employment; one man may have to go to a funeral, another might have to answer a court subpeona, while someone else might be having problems with his wife and not want to leave home for the whole day. Liebow also points out that many of the men one sees on the streets have physical disabilities that a causual observer might not notice; some men have arthritic hands, polio-withered legs, or a habit of coughing up blood if they bend or move suddenly. One man, "who hangs out in front of the Saratoga apartments has a steel hook strapped onto his left elbow."[15] These are some of the possible circumstances in a slum area that lead outsiders to create myths and stereotypes about the poor.

The association of crime and poverty contributes to the misconception of the "other Americans." Jeffrey Reiman describes how the criminal justice system often affects egalitarian views within society:

My view is that since the criminal justice system--in fact and fiction--deals with individual legal and moral guilt, the association of crime with poverty does not mitigate the image of individual moral responsibility for crime, the image that crime is the result of an individual's poor character. My suspicion is that it does the reverse: it generates the association of poverty and individual moral failing and thus <u>the belief that poverty itself is a sign of poor or weak character</u>. The clearest evidence that Americans hold this belief is to be found in the fact that attempts to aid the poor are regarded as acts of charity rather than acts of justice. Our welfare system has all of the demeaning attributes of an institution designed to give handouts to the undeserving and none of the dignity of an institution designed to make good our responsibilities to our fellow human beings. If we acknowledged the degree to which our economic and social institutions themselves breed poverty, we would have to recognize our own responsibilities toward the poor.[16]

As the United States enters the Post-Industrial Age and one wonders if the poor can transcend their plight and share the fruits of advancing society, the manner in which Americans have dealt with the poor must be examined and corrected as well as the symptoms and manifestations of poverty ifself. We cannot continue to "avoid asking the question of why the richest nation in the world continues to produce massive poverty."[17] To eradicate the miseries in the other America, the nation must become more informed of the true nature of the problem at hand in order for a change in attitude to occur. The government and citizens must approach the situation with the thought in mind that helping the poor will benefit the whole country; Harrington reminds us that "the spirit of a campaign against poverty does not cost a single cent." This sensitivity and vision should characterize all democracies.[18]

The question of whether or not the poor can struggle free from their bonds and enjoy a better place within the anticipated "leisure society" of the Post-Industrial Age seems to pose a challenge for American democracy. Does our nation possess a sufficient egalitarian consciousness to be sensitive to the needs of the less fortunate members of society? Democratic nations should realize that most social handicaps originate from circumstances of birth. Herbert Gans offers a way to judge the workings of democracy in America:

America can be described as an unequal society that would like to think of itself as egalitarian. While officially dedicated to equality of opportunity, to enabling the disadvantaged to succeed on the basis of their individual ambition and talent, America has not acted to remove the group handicaps--of class, race, and sex, among others--which prevent many people from actually realizing that opportunity. Rich and poor, for example, have an equal opportunity to work as common laborers, but the poor rarely obtain the education and social contacts that provide access to the executive positions. Equality, therefore, cannot be defined solely in terms of opportunity, it must also be judged by _results_, by whether current inequalities of income, and wealth, occupation, political power, and the like are being reduced.[19]

How does our democracy work? The people making decisions in our country hold the key to whether certain groups in society will be helped or neglected. Does the majority and their decision-making yield results which benefit the whole of society or just themselves? This question was one which deeply concerned Alexis de Tocqueville in his _Democracy in America_.

In 1831, Alexis de Tocqueville began to travel for nine months throughout America in order to examine democracy in the working. His two volumes which praise and criticize the socio-political organization of the United States comprise a comprehensive analysis that has transcended time and the modernization of the world that accompanies it. The questions he asked people and the opinions and theories he posited are still pertinent today as one tries to evaluate our system of democracy. Tocqueville warned that precautions should be taken to guard against a "Tyranny of the Majority," the title of one of his better-known essays. Democracies should be careful not to allow the laws and decisions of the Majority to disregard the needs of the impotent in society:

A majority taken collectively is only an individual, whose opinions, and frequently whose

interests, are opposed to those of another
individual, who is styled a minority. If it be
admitted that a man possessing absolute power may
misuse that power by wronging his adversaries, why
should not a majority be liable to the same
reproach? Men do not change their characters by
uniting with one another; nor does their patience
in the presence of obstacles increase with their
strength. For my part, I cannot believe it; the
power to do everything, which I should refuse to
one of my equals, I will never grant to any number
of them.[20]

Tocqueville foresaw the inevitable rise and dominance
of a particular kind of Majority in America; since the
United States at that time was a fast-growing manufacturing
nation, the French visitor recognized the probability of an
"industrial aristocracy." In explaining how an aristocracy
might be created by manufacturers, Tocqueville outlined a
process the worker experiences which Karl Marx labeled later
in the nineteenth century as "alienation of labor." As
manufacturing becomes more specialized, the division of
labor becomes greater and the worker gets "more weak, more
narrow-minded, and more dependent." The manufacturer,
however, increases in wealth and independence in direct
proportion to the worker's degradation. This type of
aristocracy does not resemble ones that have preceded it,
for the rich businessmen remain independent of one another
and the poor. Tocqueville maintains that this new type of
society can be more oppresive in the end:

The territorial aristocracy of former ages
was either bound by law, or thought itself bound
by usage, to come to the relief of its serving-man
and to believe their distresses. But the
manufacturing aristocracy of our age first
impoverishes and debases the men who serve it and
then abandons them to be supported by the charity
of the public. . . . Between the workman and the
master there are frequent relations, but no real
association. . . . I am of the opinion, on the
whole, that the manufacturing aristocracy which is
growing up under our eyes is one of the harshest
that ever existed in the world.[21]

This industrial aristocracy still exists today, only it

has widened the scope of its influence to include the state and federal governments. Richard Reeves retraced Tocqueville's travels 150 years later, and he also observed an aristocracy the same way the Frenchman had. When Reeves travelled to Albany he met with New York State's chief legislator, Speaker Stanley Fink. Discussing the influence of the state government, Fink mentioned the state's efforts to "democratize the balance of power between the state and the banking and insurance industries." He also admitted that the government cannot control these industries: "We regulate them, but we really don't always know what they are doing." Reeves noticed from this interview and other discussions in Albany that the regulators are often dependent on those they regulate, "for information, for services, for taxes, and, often, under the American system of campaign financing, for the money that fuels the regulators' political careers." Once, Speaker Fink and other legislators were threatened by Mobil and Texaco that a mass relocation of the two international oil companies' offices and employees out of state would occur if the New York Assembly approved a gross receipts tax on their companies. Newspapers also reported that the Comptroller, Edward Regan, was receiving contributions from 21 state banks, unions, and organizations of state employees to help pay off $842,000 in loans (some from the same banks) that financed his campaign and election. Since Regan is officially supposed to oversee the operations of banks and state employees, this situation seemed very murky indeed.[22]

The industrial aristocracy that Tocqueville saw growing and Reeves reaffirmed can be seen as one of the major obstacles to fulfilling egalitarian ideals within a democracy. Even though democratic systems supposedly have built-in mechanisms that protect the interests of all citizens, executives within the government are still human and inevitably weaken under the influence of the industrial aristocracy. This political reality moved former President Nixon to remark to Reeves that, "The secret of American leadership is convincing Americans that what you want to do is in their self-interest."[23] Instead of Tocqueville's "Tyranny of the Majority," New York City's Mayor Koch feels that "If there's a tyranny right now . . . it's a tyranny of a minority. The elitists." Koch told Reeves how the powerful groups use information as a means to retain control and exert influence in government. The information of witnesses, professors, and "experts" offered simplistic "evidence" to influence significant groups, like Congress, to make certain decisions. Such empirical data accessible only to a few becomes used as ammunition to get important bills passed.[24] Thus, American government has two types of "tyrannies" at work: the tyranny of the Majority that Tocqueville described; and the Tyranny of the Elitists, a select and powerful "Minority." The poor in society are left out in the cold in each case; they are neither a majority, nor part of the elitists.

In the past couple of decades, Americans have increasingly turned to the courts, demanding "enforced fairness." Professor Laurence Tribe of Harvard Law School says that the trend of citizens using the courts in the process of decision-making instead of the representative government is growing:

> The miracle, mystery, and authority of the courts in basically political situations is growing because it does not depend on the notions of judges as super-human. . . . The authority is rooted in the common and earthy idea that everyone can get in . . . the conception of "your day in court."[25]

Judges have assumed much of the power and responsibility of elected officials. Many feel that this trend results from the inferior caliber of most current politicians. The lawyer seems more and more like the Egyptian priest -- an interpreter of an occult science. The lawyers in our nation have accumulated enormous power due to their special, ritualized knowledge and the current trend of "legal egalitarianism."[26] However, legal egalitarianism remains a limited force for changing society and helping the powerless; the courts' influence usually has "negative" manifestations, or the ability to stop injustice without creating solutions. Hence, the poor cannot rely on the courts to lift them from their socio-economic inferiority; they can only have hope that legal egalitarianism will keep their situation from getting worse.

As we enter the Post-Industrial Age and wonder if the poor can advance with the rest of society, it becomes obvious that outside help is needed. If left alone, the poor will only suffer greater misery:

As the society became more technological, more skilled, those who learn to work the machines, who get the expanding education, move up. Those who miss out at the very start find themselves at a new disadvantage. . . . The good jobs require much more academic preparation, much more skill from the very outset. Those who lack a high-school education tend to be condemned to the economic underworld -- to the low-paying service industries, to backward factories, to sweeping and janitorial duties.[27]

Hence, the post-industrial society will make it even more imperative for society to take positive measures against poverty. Any success in such egalitarian programs will largely depend on the modern machinery of our working democracy. Can post-industrial democracy correct inequality in America? One obstacle to meeting this challenge remains the selection of officials for national and local offices. Television and other modern means of publicity allow single names and faces to become instantly familiar in an unprecedented fashion. Reeves noted this phenomenon after completing American Journey:

With great competition for the public's attention, name recognition is valuable and celebrity seems to have become the coin of the realm of American politics. The U.S. Senate was becoming a reward for people who won fame in or had famous relatives in other fields -- astronauts, professional athletes, television personalities, the sons and daughters of tycoons and politicians -- and, at the end of my journey, a movie actor was elected President.[28]

The prevalence of signs saying "So-and-so Cares" or "So-and-so Listens" does seem to indicate a decline in the selection process of American politics. The fact that an increasing number of candidates rely on such publicity reflects poorly on those who make the signs as well as those who are influenced by them. Such modern campaigning by a particular candidate almost suggests that that person will only "listen" and "care" in office, not make important decisions:

Many professional So-and-sos, in fact, don't intend to do anything -- at least anything that might make voters mad. Politicians don't so much try to make friends as they try to avoid making enemies. Making decisions can make enemies. If you "listen" and "care" you may go unnoticed forever, you may be reelected forever. Decisions are for judges.[29]

Some politicians deny Tocqueville's observation that American politicians rank second or third in society's hierarchy of intellect. Still more object to the observation that elected officials are declining rapidly in quality. What type of "quality" denotes excellence in a politician of the post-industrial era? At the local level, one must certainly have knowledge and compassion for one's constituency in order to ensure socio-economic stability, political freedom and justice. On national levels, politicians should be (as they always should have been) sensitive to particular problems (such as unemployment, inflation) that seem to plague one group of society more than others, asking why this pattern of degradation remains concentrated in some areas and how it can be eradicated.

Even the least cynical observer must note that many leaders do not consider themselves servants of the masses who are part of a democracy that looks after the public. Stanley Fink of the New York State Assembly clearly states his responsibilities as Speaker:

I was elected by my colleagues to be their leader because they thought I was the best. . . . That's where my loyalty is, to them, not to the Governor, not to the leaders of my party, not to "the people." "The people" didn't elect me leader.[30]

What about legislating, lobbying, and making decisions to benefit one constituency? William Hoyt of Buffalo, an assemblyman who used to teach American History, responds to those concerns, saying, "I don't worry much about that. There are only eight or ten votes a year that could possibly affect my reelection. I'm very careful about those. The rest I don't worry about too much."[31] Hoyt also explains his motives for becoming an elected official: "Because it's exciting. I'd like to be a congressman next. It's just a

game, but it's fun. It's better than teaching school."[32]
Hoyt's sentiments and aspirations cannot be considered
isolated within the political machinery of America. Perhaps
the dearth of sympathy for the poor, resulting from
ambition-minded politicians whose priorities rest on
self-perpetuation, remains the most colossal obstacle to
arresting the plague of poverty in the United States.

Understanding and recognizing the poor should be the
first step, for we can no longer ignore or hide the fact
that the wealthiest country in the world has a substantial
part of its population living below levels of decency. The
greatest challenge facing American democracy remains the
necessity to transcend the "tyrannies" of the Majority and
Elitists in favor of directing egalitarian movements for the
sake of the "other America." However, there must be a
restoration of confidence in elected officials, and only
superior minds in these positions can accomplish this
important task. "Legal egalitarianism" has only limited
potential, so our nation must improve the machinery of its
government if the poor are to look forward to the emergence
of the post-industrial state with any shred of hope.

ENDNOTES

[1]Michael Harrington, The Other America: Poverty in the United States, (N.Y.: Penguin Books, 1981), p. 3.

[2]Ibid., p. 4.

[3]Ibid., pp. 4-5.

[4]Ibid., p. 5.

[5]Ibid., p. 16.

[6]Jeffrey Reiman, The Rich Get Richer and the Poor Get Prison, (N.Y.: John Wiley and Sons, 1979), p. 84.

[7]Ibid., pp. 82-85.

[8]Ibid., pp. 85-86.

[9]Harrington, p. 128.

[10]Ibid., p. 129.

[11]Herbert Gans, More Equality, (N.Y.: Vintage Books, 1973), p. 21.

[12]Ibid., p. 21.

[13]Elliot Liebow, Tally's Corner: A Study of Streetcorner Men, (Boston: Little, Brown, and Company, 1967), p. xii.

[14]Ibid., pp. 29-71.

[15]Ibid., p. 32.

[16]Reiman, p. 154.

[17]Ibid., pp. 154-155.

[18]Harrington, p. 177.

[19]Gans, p. xi.

[20]Alexis de Tocqueville, <u>Democracy in America</u>, (N.Y.: Vintage Books, 1945), I, 269-270.

[21]Tocqueville, II, p. 170-171.

[22]Richard Reeves, <u>American Journey</u>, (N.Y.: Simon and Shuster, 1982), pp. 48-49.

[23]Ibid., p. 69.

[24]Ibid., pp. 321-322.

[25]Ibid., p. 100.

[26]Ibid., p. 106.

[27]Harrington, p. 13.

[28]Reeves, p. 91.

[29]Ibid., p. 102.

[30]Ibid., p. 40.

[31]Ibid., pp. 49-50.

[32]Ibid., p. 50.

Bibliography

Harrington, Michael. The Other America: Poverty in the United States. N.Y.: Penguin Books, 1981.

Harris, Senator Fred R. and Mayor John V. Lindsay. The State of the Cities. N.Y.: Praeger Publishers, 1972.

Gans, Herbert J. More Equality. N.Y.: Vintage Books, 1973.

Liebow, Elliot. Tally's Corner: A Study of Streetcorner Men. Boston: Little, Brown, and Company, 1967.

Miller, Zane L. The Urbanization of Modern America A Brief History. N.Y.: Harcourt Brace Jovanovich, Inc., 1973.

Reeves, Richard. American Journey. N.Y.: Simon & Schuster, 1982.

Reiman, Jeffrey H. The Rich Get Richer and the Poor Get Prison. Ideology, Class and Criminal Justice. N.Y.: John Wiley and Sons, 1979.

Tocqueville, Alexis. Democracy in America. Volumes I and II. N.Y.: Vintage Books, 1945.

THE AMERICAN FAMILY LIVES ON

Lynne Royer

I. Introduction

Each generation of the twentieth century has predicted the inevitable doom of the American family, and yet each generation carries on the same as before.[1] Pessimists, with regard to the future of the family, have pointed to macrofunctional as well as microfunctional issues in predicting the fate of the family. Macrofunctionalists speak of the loss of functions the family has experienced since the onset of industrialism. Microfunctionalists emphasize the divorce rate, declining fertility, and the increase in female labor force participation, among other things. These pessimists do not give the American family enough credit though. The institution of the family is an adaptive one. Just as the family transformed to meet the changing needs of industrial society, so will it transform to meet the new requirements of post-industrial society.

The present paper will give attention to both macro- and microfunctional issues, but will emphasize the microfunctional matter of changing sex roles. My argument is that our society still needs the functions of the family, and that current instabilities such as divorce are not signs of its inevitable extinction but rather are indicators of change occurring within the family. The sexual division of labor in the family, which used to be a source of solidarity, has partially broken down and is no longer functional. In order for the nuclear family to remain prevalent as a social system, it will have to adopt a new structure. The family of post-industrial society will be founded on egalitarian marriage where the traditional male/female sex roles will have no place.

II. Methodology

Before embarking on a speculative look at the controversial future of the American nuclear family, I will outline a conceptual framework. Sociologists employ such perspectives as the interactional, developmental, and structural-functional when analyzing the family.[2] The perspective adopted depends on the nature of the problem under study. Many sociologists of the twentieth century, including William Ogburn, Talcott Parsons, and William Goode, have examined the various functions performed by the family. Whether these functions are still necessary or not will determine its future existence. So the structural-functional approach will be utilized here. This approach is mainly concerned with "the place of the family in the society and of the individual family as a social system".[3]

Key concepts of the structural-functional approach are social system, subsystem, equilibrium, role differentiation, structure, status, role, function and dysfunction. A social system is "a system of two or more interdependent units which are at the same time actors and social objects to each other". The family then is a social system and the husband-wife dyad is a subsystem within the family. There are four requirements for a social system. They are differentiation, organization (established rights and obligations among members), boundary maintenance (more cohesive integration between members than between members and outsiders), and equilibrium tendency.[4] Differentiation in the family refers to the different roles performed by the members. Historically this role differentiation has been based on sex. Males performed instrumental functions, such as providing economic security, while females provided the expressive functions such as child care and emotional support. The family is boundary maintaining in that the members see themselves as a special group, and more intimate ties exist among themselves than with outsiders. Finally, there is equilibrium tendency because ideally conflicts are worked out and the unit endures over time.

In the structural-functional perspective, "The family may be viewed as one of several subsystems in the society, with the relationships between family and society or family and other subsystems as the focus of investigation. The individual nuclear family may also be analyzed as a system in its own right - a boundary-maintaining system which is under various internal and external pressures toward boundary dissolution or maintenance. The first type of analysis has been referred to as macrofunctionalism and the latter as microfunctionalism."[5]

The first section of this paper will be a macrofunctional analysis consisting of two sections: one is the family's loss of functions and the other is the present functions still performed by the family. The second section is a microfunctional analysis of three interrelated issues. Female labor force participation is the first concern from which follows a discussion of the changing form of solidarity within the family, which has been affected by decreasing role differentiation. Then, the third issue is make-female sex roles which need to become completely interchangeable so that the new structure of the family may emerge. The final section will propose a new structure for the post-industrial family.

III. Loss of Functions

In pre-industrial times, the family and community were tightly integrated. The family was a subsystem of the community, and as such there was a continual reciprocity between the two systems. For an example of this integration, just think of the community barn-building sessions of early New England. The family at this time had a wide range of functions, including work, education, worship, and leisure.[6] Division of labor was based on sex. For the most part, the men worked outside the home at various crafts while the women performed multiple tasks in the home. Women were responsible for many economic functions, such as gardening, spinning cloth, raising chickens, making medicines and candles, and brewing beer.[7] Of course, they were simultaneously chief guardians of the children.

With the onset of industrialism in the nineteenth century, many economic functions were taken over by the factories. Some unmarried women continued their economic function by working in the textile mills, but married women could not manage outside employment along with child-rearing duties.[8] Besides, female employment outside the home was not accepted by society at that time. Eventually industrial production took over virtually all manufacturing which used to be a function of the family.

The economic function was not the only one relinquished by the family. Protective, educational, religious, and leisure functions were also handed over to industrial society. Protection was transferred from the man of the family with his gun to police and fire departments. Education became more formal and specialized in outside

269

institutions. Both leisure and worship, formerly concentrated in the home, were transferred outside the home. Structures were designed specifically for leisure activities, such as swimming pools, tennis courts, and race tracks. Worship was no longer a private family affair, but occurred outside the home in a special place at a specific time.

Sociologists began to voice concern over the loss of functions the family was experiencing in the 1930's. William F. Ogburn noted that the family was left to, "Provide affection and understanding for its members but little else. . ."[9] In 1937, Pitirim Sorokin made the statement:

The family as a sacred union of husband and wife, of parents and children will continue to disintegrate - the main socio-cultural functions of the family will further decrease until the family becomes a mere overnight parking place for sex relationships.[10]

Pessimists like Sorokin undervalued the importance of the remaining functions performed by the family - reproduction and child bearing, sexual relations, and the provision of intimacy and a sense of belonging. The family specialized in these personal functions.

In the marketplace, specialization leads to more efficient production and a generally raised standard of living. The same can be said of the family. Devoid of most of its former functions, the family was enabled to increase its efficiency in personal functions. Talcott Parsons held this same view.

When two functions, previously imbedded in the same structure, are subsequently performed by two newly differentiated structures, they can both be fulfilled more incisively and with a greater degree of freedom.[11]

The family is more specialized than before, but not in any general sense less important, because the society is dependent more exclusively on it for the performance of its various functions.[12]

The family's loss of institutional functions then does not imply its eventual extinction.

Burgess and Locke, writing in the late 1940's, also took a non-alarmist stand. They pointed out that the family had moved from an institutional orientation to a companionship orientation.[13] In other words, the family's primary concern was no longer practical matters such as production but was now more personal matters. The home had become an escape from the stress of the outside world where one could unwind, talk, love, and receive moral support.

All of the above writers seemed confident that indeed the family had lost functions. But it may be argued that actually there was not a loss, but a "change in the content and form" of functions. While the family was no longer an economic producing unit, it did perform a vital function as a consuming unit. Though religion and education were carried on outside the home, parents continued to influence behavior patterns and value systems concerning these institutions. Considering education, "Did the pioneer parents who withdrew their children from school to work on the farm perform more of an educational function than today's parents who save, borrow, and mortgage to provide 16-plus years of schooling for their children?"[14] Thus, the family did not experience an absolute loss of functions, and the ones it still dominated could be performed more efficiently. It is to these remaining vital functions we now turn.

IV.   Present Functions Of The Family

The functions the family still controls are the restriction of sexual relations, reproduction and child bearing, and perhaps most importantly the function of supplying the love and sense of belonging of a primary group. Each of these functions is still vital to society for reasons to be explained. First I will discuss the personal needs met by the family as a primary group. Then a discussion of the family's control of sexual relations in a mature industrial society and of the reproductive function will follow.

The family is the closest form to an ideal primary group in society.[15] It is termed the "molecule of society".[16] The primary group is a common sociological

271

concept originated by Robert K. Merton. Five conditions are required to meet the recognition of an association of individuals as such a social system. Face-to-face relations, permanence, affectivity, noninstrumentality, and diffuseness. The first three conditions should be clear. Noninstrumentality infers that the primary group is in itself a goal, and not merely a means to a goal. Instrumental association on the other hand, would be fraternizing with one's boss under the assumption it will beget a promotion or raise. Diffuseness implies that relations are not focused around any single activity. The opposite of diffuseness is specificity wherein individuals meet only under special conditions - such as a labor union. The nuclear family meets all five of these requirements. Though the family has always functioned as a primary group, this function has become increasingly important with industrialization.

In pre-industrial times, there was little differentiation between family and community. Everybody knew everybody. Industrialization and urbanization destroyed this integration. The geographic and social mobility that industrialism required of the family, made communities less permanent and less unified.[17] Individuals came to depend more and more on the nuclear family to fill personal needs. These personal needs include love, self-esteem, and a sense of belonging. As society, the work place especially, became more impersonal, higher paced, and more demanding, the fulfillment of personal needs in the family seemed all the more important. The family became the eye of the storm.

Inglehart classifies the personal needs met by the family as "self-actualization needs".[18] These needs are the highest on Maslow's hierarchy of human needs. Self-actualization needs become important only when material well-being and physical security can be taken for granted. Love and sense of belonging provided by the family, Inglehart claims, are first level self-actualization needs. Once they are fulfilled, individuals then strive to satisfy second level actualization needs such as intellectual and esthetic propensities.[19] From this it may be stated that the family is a prerequisite for meeting esthetic values (second level self-actualization needs) in post-industrial society. Only when individuals have satisfied their need for love and other intrinsic personality needs, will they be prepared to move to the top of Maslow's hierarchy. Kahn and Pepper hold this same view. They state that increased emphasis on personal growth and self-expression can evolve only when the "well-being of society, community, and family" are taken "for granted".[20]

The family, as primary group, is the best means towards satisfying first level self-actualization needs. Because of the high degree of mutual rights and obligations between family members, the family can fulfill long-term commitments which could not be expected of friends or neighbors.[21] Having a permanent, obligatory association within a family is like money in the bank. It is insurance against future, unforseen losses. Problems are "considerably softened" if an "individual has a close personal relationship".[22] The permanent affectivity provided by the family allows the individual to depend on the stability of love and sense of belonging which are so important. Research has established that "maintenance of a stable intimate relationship is...closely associated with good mental health and high morale...".[23] Thus, one of the functions the family still holds is supplying the primary self-actualization needs of love and belonging. The family's function as a primary group is likely to extend far into post-industrial society. The remaining functions - control of sexual relations, and reproduction and child bearing are likely to endure as well.

The control of sexual relations is still a necessary function that only the family can fill. In the transitional period between industrial and post-industrial society, the familial control of sexual relations is particularly important for discouraging a leisure-oriented ideology.

Such a leisure-oriented ideology would result from two facts. One is that sex is a leisure activity, and if it became a preoccupation then so would leisure at the same time. The second reason a leisure-oriented ideology would result is that sex would indeed become a preoccupation in the absence of a restrictive norm. This is because sexual appetites are ever-present and limitless.

Durkheim say bodily desires as infinite and saw nothing in the body or mind of man to limit such desires.

The operation of the individual's life process

does not require that those wants stop at one
point rather than another...Therefore, insofar as
those desires are boundless...But, then if no
external force limits our feeling, it can be by
itself nothing but a source of pain. For
unlimited desires are insatiable by
definition..."[24]

In order to diminish the pain of unfulfilled feelings,
people would increasingly search for sexual gratification.
Since those feelings are limitless, a preoccupation with sex
would emerge along with an emphasis on leisure which is not
congruent with a premature post-industrial society such as
ours.

Kahn and Pepper warn that premature post-industrialism
is apt to occur if a society adopts a post-industrial
lifestyle without obtaining the necessary level of economic
wealth.[25] In a mature post-industrial society, the level
of material well-being is so high that it can be taken for
granted, and then leisure can become increasingly
important. Highly industrialized countries, on the other
hand, must not take material well-being for granted but must
continue to strive to increase their real GNP. If such
countries did take their material well-being for granted,
then they risk an indefinite delay of the arrival of
post-industrialism. Therefore to avoid risking this
indefinite delay, the familial control of sexual relations
must remain to offset a possible premature leisure emphasis.
This brings us to the third function of the family, the
reproductive function.

Replacement of members is necessary for the
continuation of society. Thus the reproductive function of
the family has and will remain intact. Though fertility
levels in the United States are considerably lower than
those a century ago, this is not due to a decreased
interest in children.[26] It is not the case that fewer
couples are having children, rather more couples are having
fewer children, i.e. family size has decreased.

The total fertility rate declined twenty-three percent
between 1940 and 1979 in the United States. In 1940, the
rate of natural population growth (births - deaths,
excluding immigration) was 4.6.[27] A growth rate of 4.6
means it will take the population fifteen years to double

itself. In 1979, the rate of natural increase had declined to -6.0[28] The negative figure seems to imply the population is decreasing, but due to immigration the population is still increasing slightly. Fewer children are being born, but in order to interpret the significance of this fact the distribution of births has to be considered. In other words, are many couples having no children while some couples have many children, or is everybody having few children? The answer is that most people are having fewer children. Few couples are choosing to remain childless, but the norm for family size is at an all-time low of 1-2 children.

Statistics on current childlessness and fertility in the United States provide solid evidence that the reproductive function of the family is alive and well despite any ongoing changes. Voluntary childlessness has actually decreased to around five percent since its rate of seven percent in the 1930's.[29] It is not clear why this is so, but nevertheless it is a significant trend.

As already stated, fertility levels have decreased throughout this century, but in order to evaluate our society's true attitude toward child bearing we must examine the change in family size norms. The method used for estimating family size norms is to compare the ratio of low-order births to high-order births across time. For clarity, a 1st-order birth is a birth where the child is the woman's first child. A 3rd-order birth is where the woman has had two previous children. In any particular year, if there is a high ratio of low-order births (1st, 2nd, 3rd) to high-order births (4th, 5th,...8th or more) then the family size norm is that of less children.

In 1950, there were 13.32 1st-order births for every 8th or more-order birth. By 1979, there were 93 1st-order births for every 8th or more-order birth. This means that for every woman giving birth to her eighth child, there were 93 women giving birth to their first child. For 6th and 7th-order births, the ratio was 9 1st-order births to every 6th or 7th-order birth in 1950, and in 1979 this ratio increased to 35 1st-order births to every 6th or 7th-order birth.[30] The significance of all this is that few people are having 5-6 child families, but most people are still having 1-2 child families.

The 1-2 child family size norm is likely to remain relatively stable as women continue to take on more roles outside the home. Women today have more means by which to find personal fulfillment whereas children and a happy home used to be the only means. However this low family size norm does not imply that the desire to bear a child and socialize it into a lovable, functioning human being will dissipate. Most career women still express an interest in having children. In one recent survey, 350 liberated college women - liberated in that they wanted full-time careers and expected an egalitarian division of labor in their homes - expressed the desire both to marry and have children.[31] The collection of findings reported above all indicate that the reproductive function of the family will continue into the Post-Industrial Age.

Before turning to the microfunctional analysis of changes occurring within the structure of the family, I will evaluate the significance of the divorce rate, for no discussion of the future of the American family would be complete without considering this social phenomenon.

V. Divorce: Is Marriage Going Out Of Style?

The statistics on divorce are less than encouraging. The divorce rate has been steadily increasing in this country since 1930.[32] There was an upsurge in divorces in the late 1960's. The divorce rate doubled between 1966 and 1976, and in certain states there are more than fifty divorces to one hundred marriages.[33] It is understandable that one could become pessimistic when surveying these statistics. However, two phenomena in society at the present support the belief that divorce is not a sign of the decline of the family. One is the remarriage rate. The second stems from two social changes, the increased acceptance of divorce and the higher expectation of happiness in marriage. First we will look at the remarriage rate.

The majority of divorced persons do remarry. The remarriage rate in 1930 was 61 per thousand - but consisted primarily of widows remarrying. In 1977, the rate was 134 per thousand, and consisted mostly of divorced individuals remarrying.[34] Today four out of five divorced persons remarry.[35]

Of course, the critic may ask whether these remarriages are any more satisfactory than first marriages. Most research suggests that remarriage is a satisfactory institution. In one study, researchers interviewed first married and remarried couples and calculated a "mean marital happiness" score. There was no significant difference reported between the marital happiness of the two groups. The researchers, Glenn and Weaver, concluded that divorce and remarriage are not signs of marital decline, but rather they are institutions for upgrading the quality of married life.[36] Another study by Albrecht was even more encouraging. He found that remarried persons rated their current marriage as better than their first marriage.[37] These findings relate to remarital success, but there is a distinction between marital success and marital stability. A marital couple may be unhappy together but still decide to stay together, or conversely a couple that has a happy marriage but cannot overcome certain obstacles may break up.

While many remarriages seem to be more successful in terms of happiness, they are not on the whole more successful in terms of stability. In general, remarriages of divorced persons are just as likely to end up in divorce as first marriages.[38] This instability may stem from society's lesser approval of remarriages than first marriages and not from inherent flaws in remarriage as an institution. Remarriage takes second seat to first marriages and this differential approval is displayed at the very outset in the belief that one's second marriage ceremony should be plain and simple. Andrew Cherlin, a noted Johns Hopkins sociologist, has proposed that part of the difficulty in obtaining stability in a remarriage results from the incomplete institutionalization of remarriage.[39]

Remarriage is not a complete institution in American society because it does not have a standard set of norms to govern it. The absence of norms ranges from technicalities, such as what term the children from a former marriage should call their step-mother, to issues of etiquette such as whether a new spouse should socialize with the former spouse when the latter comes to the house to pick up the children for a weekend. In addition to this lack of appropriate norms, often the extended family of the divorced person does not lovingly accept the remarriage partner, or even the remarriage itself, thus creating more conflict. Because remarriage is not fully institutionalized, strain is placed upon the second marriage at the outset. Remarriage must become socially accepted as a valid institution with a fixed set of norms before such marriages can hope for stability as well as success. The fact remains that divorced persons are remarrying. It is not that people have given up on marriage as an institution; rather, they have given up on their particular mate.

So far I have shown that divorce is not a threat to the family, because people are remarrying. Analysis of two social changes of the last few decades provides further understanding of the true significance of the high rate of divorce. One is the increased acceptance of divorce as a solution to an unhappy marriage, and the second is the higher expectation of happiness in marriage.[40]

The introduction of No-fault divorce in 1970 was a hallmark in the increasing acceptance of divorce as a justified solution to an unhappy marriage. By making divorce less of a personal stigma, which was achieved through reducing the harshness of the conditions for divorce, divorce entered people's minds as a realistic option. More people did choose divorce as evidenced by the sharp increase in the divorce rate in the 1970's. In the early part of the twentieth century when divorce was not such an easy solution to marital tension, people had to remain married and try their best to cope with their differences. Families were not in any sense healthier simply because there was a relatively low divorce rate. Coupled with the increased acceptance of divorce is the higher expectation of happiness in marriage which has evolved over the last half a century. Divorce was not as prevalent in the early twentieth century because happiness was not an assumed condition of marriage. If it was there, that was a blessing, but if not, the marriage was not doomed to fail.[41]

Mate selection in former times was based on practical qualities. A woman looked for a man who would be a good provider, and a man looked for a woman who was a good cook, housekeeper, and showed potential for child rearing. Emphasis in mate selection today is on love and companionship. Just as the family has shifted from institutional functions (economic, educational, etc.) to personal functions, the basis of marriage has shifted from practical qualities to personal qualities. There is research evidence to support this statement. A questionnaire was administered to six hundred college students asking what they considered the most important determinant in the selection of a marriage partner. The majority said "dependable character" and "emotional stability".[42] A divorced man gave the following explanation as to why his marriage had ended: "We're still very compatible; we just started growing in different directions. Our marriage was constraining. The romanticism dissolved. There was resentment that we were not understanding each others' needs...".[43]

The greater the emphasis on marital happiness the less will be the overall marital stability. Lee compares the "love theme" in America to the "success theme". The drive to succeed economically has been with us for some time. It is so strongly ingrained in our culture that many individuals do not feel satisfied with themselves unless they have acquired economic success. Finding love in a long-term intimate relationship has become another success drive in our culture over the past half century. As with economic success, virtually all people strive for it, but many fail.

The possible dissociation between goals and means (the structural cause of anomie) need not be restricted to the economic sphere. One might argue that the goal of marital success is also strongly emphasized by American culture, and is, like economic success, "conveniently indefinite and relative". The means by which marital success may be attained are also highly ambiguous.[44]

As long as the goal of marital happiness remains of high precedence, people will continue striving for it. Divorce is no longer a quirk in the system but a possible means to achieving marital happiness. From this view, divorce can finally be seen as functional in that it allows individuals a second chance.

Divorce then becomes the safety valve that makes the system workable. Those who are frustrated or oppressed can escape their families, and those who fail at what is regarded the most important human

> activity can gain a second chance. Divorce is,
> therefore, not an anomaly or flaw in the system,
> but an essential feature of it.[45]

The true significance of the divorce rate is not in showing
the decline of the family, but rather in signaling a major
structural change within the family. This structural change
is the decrease in role differentiation between husband and
wife.

As women work outside the home they assume an
additional role of the traditional husband-provider.
However, men have not yet adapted their own roles. Instead
of becoming both husband-provider and wife-caretaker, most
men have stayed exclusively within the husband-provider
role. This imbalance has created disequilibrium within the
family.

The next section will examine the rise in female labor
force participation, and how it has contributed to the
decrease in role differentiation. I propose that both men
and women need to share egalitarian roles in order to
restore equilibrium to the family system. After considering
the female work force and the resulting implications for
marital solidarity, I will discuss sex role ideology which
must become egalitarian before the needed completion of the
structural change toward an absence of role differentiation
can occur.

VI. Women At Work

Women are gaining more strength in the work force every
year. The 1980 Census revealed that over fifty percent of
all women now work outside the home.[46] Women are
acquiring roles traditionally associated exclusively with
the husband's role set, such as that of economic provider.
This has brought about a slackening of the sharp role
differentiation of husband-provider and wife-caretaker.
Women's new economic security gives them additional rights
within the family. Since they have contributed to the
traditional husband-provider obligations, women have earned
the right to expect their husbands to contribute more to the
traditional wife-caretaker role. The new power of women is
a relatively recent phenomenon. It has emerged only with
the dramatic rise in women working outside the home in the
past twenty to thirty years.

Females working outside the home was an impossibility in early industrial society.[47] Life expectancy was only around fifty years, and women spent most of their adult life in child rearing. Also, in less technological society, there were few labor-saving devices and women had to spend most of their physical strength in domestic activities. With increased industrialization, many single women became employed in factories, but married women remained at home for the most part until World War I. Even after World War I, the ideology "A woman's place is in the home" remained predominant.[48]

Whereas childless married women increasingly joined the labor force between the two world wars, married women with children did not do so in large numbers until after World War II. In 1950, less than twenty-five percent of wives worked. By 1976, that rate had doubled, and during the same period the rate for mothers with pre-school children tripled. In 1978, over half of the married women were employed, and forty percent of mothers with pre-schoolers and with husbands were employed.[49]

A number of factors lead to increased female employment. Mass education for women, the desire for economic security, the deflated status of the housewife role, and increased public acceptance of working wives - have all influenced female labor force participation positively.[50] It is not surprising that women should want to have careers of their own when our society appraises status and prestige by type of employment. Since the majority of married women now work, the housewife role has been considerably deflated. Non-working wives have been made to feel insecure because they are "just a housewife". This insecurity is embodied in soapbox speakers such as Phyllis Schlafly who galavant around the country telling women to stay at home. However, by asking women to stay at home, Schlafly is asking women to relinquish the personal fulfillment they stand to gain from outside employment.

Greater economic independence and higher prestige can make women happier and healthier. Researchers report better physical health and fewer symptoms of psychosomatic illness among working wives.[51] Of course, it may just be that healthier women are more likely to go looking for a job, but then many unhealthy women must work out of sheer economic necessity. Whether or not women are healthy from working, they definitely gain in self-esteem.

Many women report they have more confidence in themselves. However, those who are mothers show signs of role conflict as well. In situations involving a conflict between working and mother roles, women express feelings of inadequacy.[52] Guilt emotions are also expressed by working mothers. The most recent research indicates that such guilt is not externally imposed by others' explicit action, instead it originates from an internalized traditional value system.[53] Cognitively these women may feel justified in combining the two roles, but at a deeper emotional level they encounter dissonance.

Increased female labor force participation generated social problem-oriented research in the 1950's and early 1960's. Social scientists were concerned about effects of mothers' working on marriage stability, on children, and on the mothers' overall well-being. Research findings failed to support the validity of these concerns, and attention turned to other more theoretical questions such as the new bargaining position of women.[54]

As women were employed, they became "independent members of their families". Their day-to-day living no longer centered around the needs of their families. Identities became more completely crystallized. Also their new economic strength enabled them to demand more from other family members.[55]

Women hope for their husbands to take on more household and child rearing responsibilities. Unfortunately, not many husbands come through. The husbands seem to have no problem in accepting their wives' employment. In a study of working wife and non-working wife families, the husbands had more favorable than unfavorable or neutral attitudes concerning their wife's employment.[56] However, the husbands did little to assist in domestic duties. The husbands of working wives did do an average of one hour's housework and child care more than did the husbands of non-working wives. But both sets of husbands contributed minimally compared to their wives. The non-working wives did 34 hours more housework and 11 hours more child care than their husbands, and the working wives did 21 hours more housework and 5 hours more child care than their husbands. Thus in the

homes with working wives, since they put in fewer hours
than non-working wives and their husbands did not make up
for the deficit, presumably certain things just went undone.
Probably these husbands could not accommodate doing
traditional female activities with their traditional male
sex roles.

There is a cultural lag between the social fact of
women entering the work force and society's conception of
what the female's role should entail. Even though most men
now recognize women's right to have full-time careers, they
have not altered their expectations for women to fill most
of the household responsibilities. The underlying
assumption seems to be that housework and child care are
inherently the women's responsibility, regardless of job
patterns. Now that many husbands and wives are committed to
full-time careers, why should the wife be the sole candidate
for the demanding role of domestic manager? Even if the man
does contribute to domestic activities, it is always her
housework and not their housework.[57] For men, fatherhood
and career have always been compatible roles, but for women
- motherhood and career are seen as conflicting roles.
"Many people still assume that if a woman wants a full-time
career, then children must be unimportant to her. But of
course no one makes this assumption about her husband."[58]

Much of this inegalitarian ideology is the result of
the traditional historic sexual division of labor which
lingers on. The traditional sexual division of labor is now
dysfunctional because it places excess duties and
obligations on the woman. She must either be a
"superwoman" in managing her career and simultaneously
taking care of the housework and children, or she can forgo
some of the preceding and incur feelings of guilt and
inadequacy because she cannot fulfill her wife-mother role.

Society's reluctance to adopt an updated attitude of
the woman's role has created disequilibrium within the
family. In order to restore equilibrium, people must alter
their sex role attitudes, and begin to value the idea that
both sexes should share the same roles. By being able to
take on any role, men and women adopt more varied behaviors
which were formerly "not appropriate" for their particular
sex. Undoubtedly this will increase personal fulfillment
for all involved.

The value change concerning appropriate sex roles cannot be expected to occur overnight. Through the socialization process, values become internalized and people come to cling to and accept them as their own. It is nearly impossible to convince an older person that the value of housework being a woman's role is an arbitrary one that can and should change. Inter-generational value changes occur when changes have persisted for a considerable time span.[59] So it is only with the generation growing up - with dual-career parents and more flexible sex roles, that a complete overhaul of the social structure can be expected to originate.

A reordering of the social structure is what is needed to fully institutionalize the increasingly similar roles of men and women. Goode recognized this a few decades ago, although he underestimated the speed with which this need would evolve.

> ...We do not believe that any family system now in operation, or likely to emerge in the next generation, will grant full equality to women... we believe that it is possible to develop a society in which this would happen, but not without a radical reorganization of the social structure.[60]

The needed structural change is for the present trend of the converging of traditional male and female roles to continue to the point where they are completely interchangeable, i.e. they are identical. This structural adjustment is necessary to restore equilibrium in the family since the distinct sexual division of labor no longer exists to create marital solidarity. Durkheim stated that, "...When the mode of solidarity becomes changed, the structure of societies cannot but change."[61] He was mainly speaking to the change from organic to mechanic solidarity in society, made possible by the division of labor. Since the family is a social system just as society, the concept of solidarity is also relevant to the family.

Durkheim coined the phrases mechanic solidarity and organic solidarity to describe the general forms of social cohesion operating in pre-industrial and industrial societies. In mechanic solidarity there is no role differentiation and unity stems from common characteristics and values. In organic solidarity, individuals are diverse and are united by the division of labor which makes them interdependent.[62]

In pre-industrial society, homogamy was the state of existence. All members of the community performed the same tasks and shared common characteristics and culture. These individuals were then united by their "collective conscience". In industrial society the process of industrialization necessitated labor specialization to increase efficiency of production. Members are then united by the division of labor characterized in organic solidarity. Since each individual performs only one specialized function, he/she depends on others in the society to provide the other needed functions. For a simple example, the baker depends on the tailor to provide clothes and the tailor depends on the baker to provide bread. The division of labor in society, which has been a necessary phenomenon of the industrial era, is not a concept belonging solely to economists.

While society has evolved from mechanic to organic solidarity, the family is in a state of transition from organic to mechanic solidarity. The family was united by organic solidarity in early industrial society when there was a sharp division of labor based on sex. The husband depended on the wife to perform necessary domestic functions and child rearing - so called "expressive functions". Conversely, the wife depended on the husband to carry out "instrumental functions" - the primary one that of economic provider. Husband and wife were bound together by their mutual dependence upon one another for their respective specialized functions. As women became more involved in the labor force, especially during World War II, and no longer accepted their sole role of housewife, the sexual division of labor began to loosen. Marital solidarity also loosened as a consequence of this decreased role differentiation.

Although organic solidarity within the nuclear family has partially declined, in that most women no longer depend on their husbands for economic security, most men still depend rather heavily on women to perform the domestic functions. The evolution from organic to mechanic solidarity in the family must be completed in order to restore equilibrium to the family.

285

The evolution to mechanic solidarity within the family depends on men's acceptance of more expressive activities traditionally carried out by women. Equilibrium is generated in the family when there is a balance of instrumental and expressive activities.[63] Historically, this equilibrium has been maintained by women controlling home-oriented expressive activities and men controlling work-world instrumental activities. Now that women perform less of the expressive roles and more instrumental ones, men are the only ones who can restore familial equilibrium.

The first step in accomplishing role sharing within the family is to change the out-dated sex role ideology that opposes the needed structural alterations. Once these attitude changes permeate a large extent of the population, the necessary "sociopolitical climate" for structural change is likely to formulate.[64] There has already been a significant shift toward egalitarian roles in recent years, however male attitudes have not modernized at the same rate as female attitudes have.

VII. Changing Sex Roles

It is an established fact that sex role attitudes have changed dramatically in the past two decades. A multitude of research studies have documented this. A longitudinal panel study of women found that in 1962, thirty-two to fifty-six percent of respondents gave egalitarian responses concerning sex role attitudes, and in the follow-up 1977 measurement, sixty to seventy percent gave egalitarian responses.[65] Another two-period cohort analysis of women, measured the degree of traditional to feminist attitudes of Douglas College students in 1969 and 1973. Significant shifts toward feminism were reported between the two measurements. Attitudes concerning employment, financial responsibilities, and the sexual division of labor changed the most. "By 1973, a majority believed their careers were of equal importance to their husbands' and that they should share equally in the financial support of their families. They also expected to work all of their adult lives and to have substantial help from their husbands with household chores."[66]

Men have also altered their sex roles attitudes, though not to the same extent as women. Five surveys of both men and women showed a trend toward egalitarian attitudes

286

towards women's roles between 1964 and 1974.[67] A three-period survey conducted in 1972, 1975, and 1979 also of men and women revealed differential increases in egalitarian responses (to married women earning money in industry if she had a husband capable of supporting her). Although feminist changes occurred for both groups by about ten percentage points, the women were about six percentage points more feminist in each year than the men.[68]

Further evidence that men are less willing to adopt egalitarian attitudes came from a study of sixty-two Ivy League male college students.[69] Four categories of responses materialized in response to statements such as, "It is appropriate for a mother of a preschool child to take a fulltime job...". A quarter of these males gave "traditionalist" responses which meant that "they intended to marry women who would find sufficient fulfillment in domestic...pursuits without ever seeking outside jobs". Sixteen percent gave "Pseudo-feminist" responses. At an abstract level, they agreed to working wives, but their feminism was loaded with numerous unrealistic qualifications. Only seven percent gave "feminist" responses, wherein they would be willing to make accommodations in their own careers to help advance their wives' careers. Half of the sample gave "modified traditionalist" responses which meant a sequential work pattern for their wives - i.e. quit working during the child bearing years. This study was conducted in 1971, so probably these response categories have shifted in the interim. But this study is significant in displaying the reluctance of men to change their attitudes to the degree that women have.

Possibly men have not changed their attitudes as easily because they feel threatened by the increasing dominance of women.

> Perhaps as a function of the conflict involved in the progressive emancipation of women in the last century...the ideology of women tends to minimize the difference between the sexes...The men, on the contrary, tend to exaggerate the cleavage, and even, ideologically, to regard it as an impassable gulf to be accepted with good-humored tolerance.[70]

Maybe if the movement in this country had focused on changing the roles of both sexes simultaneously, instead of stressing women's cause, men would not have been forced into a defensive position. In Sweden, notorious for progressivism, there was not a "women's liberation", but rather a "sex-role debate".[71] "The isolation of the women's movement in America feeds the male tendency to avoid introspection."[72]

It is unfortunate that men have been placed in this defensive position, because men have just as much to gain from a new sex role ideology as women do. Rigid sex roles, whether male or female, restrict healthy behavior flexibility. Goldberg identifies the masculinity blueprint as, "...the pressure and compulsion to perform, to prove himself, to dominate, to live up to the 'masculine ideal' - in short, to 'be a man'...".[73] These traditional components of masculinity, Goldberg terms "a blueprint for self-destruction". Traditional masculinity is self-destructive in that it is not the man's core identity, but a psychological defense mechanism - a defense against being "feminine, dependent, emotional, passive, afraid, or helpless". All men possess traits associated with traditional femininity inside themselves which they hide with a mask of masculinity. The reluctance to display femininity restricts the personal fulfillment of the "macho" male. High masculinity in adult males has been associated with "high anxiety, high neuroticism and low self-acceptance".[74]

> It is time for the man, therefore, to reject the role demands on him to play superman: the unneeding, fearless, unemotional, independent, all-around strong man. Liberation for him would mean a re-entry into the world of playfulness, intimacy, trusting relationships, emotions and caring, and a priority on fulfillment of his needs and his growth.[75]

Women have become as liberated as is possible without a large-scale reform of ideas on traditional masculinity. Now that women's attitudes about themselves have changed, it is time for men to change their attitudes about their sex roles. This is imperative in two respects - one is the role of fathers in socializing sex role attitudes in their children, and the second is the power men now hold in the business world to help or harm women.

Parents are role models for their children's behavior. Fathers can facilitate egalitarian sex roles in children by accepting and reacting positively toward the mother's new power. If fathers react negatively or defensively, the children may see the mother's new sex role as a source of conflict and deem it maladaptive. Also, the father can help by approving of cross-sex behavior in children when they exhibit it. The father's acceptance of cross-sex behavior is especially vital since they are the ones preoccupied with proper sex role behavior in children. Fathers "get even more upset than mothers when their children deviate from 'proper' sex-role behavior - especially boys."[76] Thus egalitarian sex-role beliefs are crucial in fathers' willingness to socialize their children for such values.

Since men still dominate many of the decision-making ranks in business and industry, their beliefs about the "proper" role of women in career paths is vital. "...Changing men is connected to women's real liberation as a result of the ability men have to prevent women from seeking a whole range of alternatives...".[77] In order for women to fully penetrate the job market, the men at the top must be willing to let them in.

An obstacle to women's progress is the belief that most women will eventually want to have children, and that this will interfere with their careers. The "motherhood myth" holds that there is a maternal instinct, and that women cannot obtain fulfillment in the absence of children.[78] Also belonging to this myth is the notion that there is no substitution for a mother's undivided attentions. However, scientists from sociologists to biologists have refuted the existence of a mother instinct. "Women don't need to be mothers any more than they need spaghetti."[79] In Sweden, the motherhood myth has been shattered.

The common consensus in Sweden is that men and women should participate equally in employment and parenting roles. The basic idea is that, "In the long run, men and women should be given the same rights, obligations and work assignments in society."[80] Swedish parents of small children both take a reduction in hours at work and take turns being the caretaker of the children. Hopefully, such an egalitarian division of labor in the family will crystallize in American families before too long.

An egalitarian division of labor is what is needed to restore equilibrium to the American family structure. Once men and women participate equally in all roles, traditional obligations cast aside, then the evolution of the family from organic to mechanic solidarity will be complete. Marital solidarity will no longer depend upon the sexual division of labor, as it did in the past, but will come from the shared characteristics and values of the husband-wife dyad.

Instead of recapitulating the details of the foundation which has been set, I will venture to describe the post-industrial family to which I have been hinting. I have established why the American family will survive, why it is currently experiencing disequilibrium, and the changes within the family that must occur to restore equilibrium. What the result of these changes will be remains to be stated.

VIII. Conclusion: The Post-Industrial Family

The basis of the post-industrial family will be the egalitarian marriage. In egalitarian marriage, wife and husband will think of and treat each other as equals in all respects other than physiological differences. Each will respect the other's goals both within and outside the family. These goals will be compatible between the two, because an underlying assumption of their commitment to one another is a common value system. This common value system is what unifies the family in mechanic solidarity.

The married couple in an egalitarian marriage will strive for personal fulfillment. Sometimes this will occur through shared endeavors, but when necessary to sustain individual identities, each partner may singly pursue personal fulfillment. Neither would stand in the way, nor even want to, when the other desired, for example, to take a week's vacation alone for introspection. Because of this mutual respect for each other's identities, marriage will never be stifling again.

It is vital that there are no restrictions on self-expression in order for partners to attain their full identities and find fulfillment. Freedom of self-expression will result from the abolition of traditional sex roles. As the cultural folkways that discourage the use of cross-sex behaviors disappear, men and women will no longer hold back from expressing inner tendencies. The extinction of sex roles is also a necessary condition for the implementation of an egalitarian division of labor which is characteristic of the future mechanic solidarity of the family.

As already stated, in mechanic solidarity individuals perform the same roles in society and are united by their shared value system. When the family completes the evolution to mechanic solidarity, there will be no differences in the roles performed by partners. Men and women will share equally in both the instrumental and expressive functions of the family. Both will work at their own careers; both will share in their housework; and, both will be involved in the care and rearing of their children. Other institutions in society will have to adjust in order for men and women to be free to fulfill their new roles, but this is an entirely separate subject yet to be dealt with.

This is the end then to which the family system is approaching. As society revises its out-dated conception of what "a male" and "a female" should be to what "an individual" should be, then the family will be reunited in mechanic solidarity. An individual of the post-industrial family will have a varied role set including that of spouse, worker, parent, housekeeper, caretaker, and the list goes on. The post-industrial family will not be without its occasional imperfections, such as divorce, but it will allow for the identity expression and personal fulfillment of all of its members. It is highly possible that our children will witness the "golden age of the family".

Endnotes

[1]William L. O'Neill, Divorce in the Progressive Era (New York: New Viewpoints, 1973), p. xiii.

[2]Bert N. Adams, The Family: A Sociological Interpretation (Chicago: Rand McNally College Publishing Company, 1975), p. 8.

[3]Jennie McIntyre, "The Structure-Functional Approach to Family Study," in Emerging Conceptual Frameworks in Family Analysis, eds. F. Ivan Nye and Felix M. Berardo (New York: The Macmillan Company, 1966), p. 57.

[4]Ibid., pp. 54-59.

[5]Ibid., p. 54.

[6]Ernest W. Burgess, Harvey J. Locke, and Mary Margaret Thomas, The Family: From Traditional to Companionship (New York: Van Nostrand Reinhold Company, 1971), pp. 435-438.

[7]Mary Jo Bane, Here To Stay: American Families in the Twentieth Century (New York: Basic Books, Inc., Publishers, 1976), p. 27.

[8]Ibid.

[9]Adams, p. 86.

[10]Clark E. Vincent, "Familia Spongia: The Adaptive Function," Journal of Marriage and the Family 28:31.

[11]Talcott Parsons, "The Point of View of the Author," in The Social Theories of Talcott Parsons, ed. Max Black (Englewood Cliffs, N.J.: Prentice-Hall, 1961), p. 129.

[12]Talcott Parsons and Robert Bales, Family, Socialization and Interaction Process (Glencoe, Ill.: Free Press, 1955), pp. 10-11.

[13]Vincent, p. 31.

[14]Ibid.

[15]E. Litwak, and I. Szeleny, "Primary Group Structures and their Functions: Kin, Neighbors, and Friends," American Sociological Review 34:469.

[16]George F. Vincent and Albion W. Small, "The Family as a Primary Group," in Sociological Theory: Present-Day Sociology from the Past, eds. Edgar F. Borgatta and Henry J. Meyer (New York: Alfred A. Knopf, 1975), p. 270.

[17]Dr. Brazleton, "Images of the Family Then and Now," Tape Recorded Lecture, Harvard University.

[18]Ronald Inglehart, The Silent Revolution (Princeton, N.J.: Princeton University Press, 1977), pp. 22-23.

[19]Ibid.

[20]Herman Kahn and Thomas Pepper, Will She Be Right? The Future of Australia (St. Lucia, Queensland: University of Queensland Press, 1980), p. 57.

[21]Litwak and Szeleny, p. 469.

[22]M. F. Lowenthal and C. Haven, "Interaction and adaptation: Intimacy as a critical variable," American Sociological Review 33:29.

[23]Ibid., p. 29.

[24]Emile Durkheim, "On Anomie," in Images of Man: The Classical Tradition in Sociological Thinking, ed. C. Wright Mills (New York: George Braziller, Inc., 1960), p. 450.

[25]Kahn and Pepper, p. 120.

[26]Mary Jo Bane, "Here To Stay: Parents and Children," in Family in Transition, eds. Arlene Skolnick and Jerome H. Skonick (Boston: Little, Brown and Company, 1980), p. 95.

[27]U.S. Bureau of the Census, Statistical Abstract of the United States: 1981 (Washington, D.C.: 1981), Table No. 84, p. 58.

[28]Ibid.

[29]Bane, "Here To Stay: Parents and Children," p. 96.

[30]U.S. Bureau of the Census, Table No. 86, p. 59.

[31]Ann P. Parelius, "Emerging Sex-Role Attitudes, Expectations, and Strains Among College Women," Journal of Marriage and the Family 37:150.

[32]U.S. Bureau of the Census, Table No. 125, p. 80.

[33]Wesley R. Burr et al. eds., Contemporary Theories About The Family, vol. 1 (New York: The Free Press, 1979), p. 203.

[34]U.S. Bureau of the Census, Table No. 125, p. 80.

[35]Alvin L. Schorr and Phyllis Moen, "The Single Parent and Public Policy," in Family in Transition, eds. Arlene Skolnick and Jerome H. Skolnick (Boston: Little, Brown and Company, 1980), p. 555.

[36]Norval D. Glenn and Charles N. Weaver, "The Marital Happiness of Remarried Divorced Persons," Journal of Marriage and the Family 39:335.

[37]Stan L. Albrecht, "Correlates of Marital Happiness Among the Remarried," Journal of Marriage and the Family 41:864.

[38]Glenn and Weaver, p. 331.

[39]Andrew Cherlin, "Remarriage as an Imcomplete Institution," in Family in Transition, eds. Arlene Skolnick and Jerome H. Skolnick (Boston: Little, Brown, and Company, 1980), p. 375.

[40]W.F. Ogburn and M.F. Nimkoff, Technology and the Changing Family (Boston: Houghton Mifflin Company, 1955), p. 8.

[41]Ibid., p. 9.

[42]Ibid., p. 42.

[43]Kelin E. Gersick, "Fathers by Choice: Divorced Men Who Receive Custody of Their Children," in Divorce and Separation: Context, Causes, and Consequences, eds. George Levinger and Oliver C. Moles (New York: Basic Books Inc., Publishers, 1979), p. 315.

[44]Gary R. Lee, "Marriage and Anomie: A Causal Argument," Journal of Marriage and the Family 36:525.

[45]Adams, p. 351.

[46]Merrill Sheils, ed., "A Portrait of America," Newsweek January 17, 1983, p. 29.

[47]Sandra L. Bem and Daryl J. Bem, "Training the Woman to Know her Place," in The Future of the Family, ed. Louise Kapp Howe (New York: Simon and Schuster, 1972), p. 216.

[48]Burr et al., p. 203.

[49]"Work and Family Roles: Egalitarian Marriage in Black and White Families," Social Service Review 55:315.

[50]Adams, p. 264.

[51]Bane, Here to Stay: American Families in the Twentieth Century, p. 28.

[52]Burr et al., p. 211.

[53]Lucia A. Gilbert, Carole Kovalic Holahan, and Linda Manning, "Coping with Conflict between Professional and Maternal Roles," Family Relations 30:419.

[54]Burr et al., p. 203.

[55]William J. Goode, "Industrial and Family Structure," in A Modern Introduction to the Family, eds. Norman W. Bell and Ezra F. Vogel (New York: The Free Press, 1968), p. 118.

[56]Social Service Review, p. 319.

[57]Bem and Bem, p. 219.

[58]Ibid., p. 223.

[59]Inglehart, p. 121.

[60]Goode, p. 119.

[61]Anthony Giddens, ed. Emile Durkheim: Selected Writings (Cambridge: Cambridge University Press, 1972), p. 141.

[62]Emile Durkheim, Division of Labor in Society trans. George Simpson (New York: The Macmillan Company, 1933), p. 56.

[63]McIntyre, p. 60.

[64]Karen Oppenheim Mason and John L. Lzajka, "Change in United States Women's Sex-Role Attitudes, 1964-74," American Sociological Review 41:574.

[65]Arland Thornton and Deborah Freedman, "Changes in the Sex Role Attitudes of Women, 1962-1977; Evidence from a Panel Study," American Sociological Review 44:839.

[66]Parelius, p. 151.

[67]Andrew Cherlin and Pamela B. Walters, "Trends in United States Men's and Women's Sex-Role Attitudes: 1972 to 1978," American Sociological Review 46:456.

[68]Ibid.

[69]Mirra Komarovsky, "Cultural Contradictions and Sex Roles: The Masculine Case," in Family in Transition, eds. Arlene Skolnick and Jerome H. Skolnick (Boston: Little, Brown and Company, 1980), p. 210.

[70]John R. Seeley, R. Alexander Sim, and Elizabeth W. Loosley, "Differentiation of Values in a Modern Community," in A Modern Introduction to the Family, eds. Norman W. Bell and Ezra F. Vogel (New York: The Free Press, 1968), pp. 497-98.

[71]Warren Farrell, The Liberated Man (New York: Random House, 1974), pp. 3-4.

[72]Ibid.

[73]Herb Goldberg, The New Male: From Macho to Sensitive But Still All Male (New York: The New American Library, Inc., 1980), p. 5.

[74]Ibid., p. 10.

[75]Ibid., p. 183.

[76]Farrell, p. 12.

[77]Ibid., p. 14.

[78]Ibid., p. 121.

[79]Betty Rollin, "Motherhood: Who Needs It?" in The Future of the Family, ed. Louise Kapp Howe (New York: Simon and Schuster, 1972), p. 69.

[80]Olof Palme, "Lesson From Sweden: The Emancipation of Man," in The Future of the Family, ed. Louise Kapp Howe (New York: Simon and Schuster, 1972), p. 247.

BIBLIOGRAPHY

Adams, Bert N. The Family: A Sociological Interpretation. Chicago: Rand McNally College Publishing Company, 1975.

Albrecht, Stan L. "Correlates of Marital Happiness Among the Remarried." Journal of Marriage and the Family 41:857-866.

Bane, Mary Jo. Here To Stay: American Families in the Twentieth Century. New York: Basic Books, Inc., Publishers, 1976.

_____. "Here To Stay: Parents and Children." In Family in Transition, pp. 91-105. Edited by Arlene Skolnick and Jerome H. Skolnick. Boston: Little, Brown and Company, 1980.

Bem, Sandra L. and Bem, Daryl J. "Training the Woman to Know her Place." In The Future of the Family, pp. 202-223. Edited by Louise Kapp Howe. New York: Simon and Schuster, 1972.

Brazleton, Dr. "Images of the Family Then and Now." Tape Recorded Lecture. Harvard University.

Burgess, Ernest W.; Locke, Harvey J., and Thomas, Mary Margaret. The Family: From Traditional to Companionship. New York: Van Nostrand Reinhold Company, 1971.

Burr, Wesley R.; Hill, Reuben,; Nye, F. Ivan, and Reiss, Ira L., eds. Contemporary Theories About The Family. Vol. 1. New York: The Free Press, 1979.

Cherlin, Andrew. "Remarriage as an Incomplete Institution." In Family in Transition, pp. 368-381. Edited by Arlene Skolnick and Jerome H. Skolnick. Boston: Little, Brown and Company, 1980.

Cherlin, Andrew, and Walters, Pamela B. "Trends in United States Men's and Women's Sex-Role Attitudes: 1972 to 1978." American Sociological Review 46:453-460.

Cherns, Albert. "Work and Values: Shifting Patterns in Industrial Society." International Social Science Journal 32: 427-441.

Durkheim, Emile. Division of Labor in Society. Translated by George Simpson. New York: The Macmillan Company, 1933.

_____. "On Anomie." In Images of Man: The Classic Tradition in Sociological Thinking, pp. 449-485. Edited by C. Wright Mills. New York: George Braziller, Inc., 1960.

Farrell, Warren. The Liberated Man. New York: Random House, 1974.

Gersick, Kelin E. "Divorced Men Who Receive Custody of Their Children." In Divorce and Separation: Context, Causes, and Consequences, pp. 307-323. Edited by George Levinger and Oliver C. Moles. New York: Basic Books Inc., Publishers, 1979.

Giddens, Anthony, ed. Emile Durkheim: Selected Writings. Cambridge: Cambridge University Press, 1972.

Gilbert, Lucia A.; Holahan, Carole Kovalic, and Manning, Linda. "Coping with Conflict between Professional and Maternal Roles." Family Relations 30: 419-426.

Glenn, Norval D., and Weaver, Charles N. "The Marital Happiness of Remarried Divorced Persons." Journal of Marriage and the Family 39: 331-336.

Goldberg, Herb. The New Male: From Macho to Sensitive but Still All Male. New York: The New American Library, Inc., 1980.

Goode, William J. "Industrialization and Family Structure." In A Modern Introduction to the Family, pp. 113-120. Edited by Norman W. Bell and Ezra F. Vogel. New York: The Free Press, 1968.

Hick, M.W., and Platt M. "Marital Happiness and Stability: A Review of Research in the Sixties." Journal of Marriage and the Family 32:553-574.

Inglehart, Ronald. <u>The</u> <u>Silent</u> <u>Revolution</u>. Princeton, N.J.: Princeton University Press, 1977.

Kahn, Herman, and Pepper Thomas. <u>Will</u> <u>She</u> <u>Be</u> <u>Right?</u> <u>The</u> <u>Future</u> <u>of</u> <u>Australia</u>. St. Lucia, Queensland: University of Queensland Press, 1980.

Komarovsky, Mirra. "Cultural Contradictions and Sex Roles: The Masculine Case." In <u>Family</u> <u>in</u> <u>Transition</u>, pp. 205-216. Edited by Arlene Skolnick and Jerome H. Skolnick. Boston: Little, Brown and Company, 1980.

Lee, Gary R. "Marriage and Anomie: A Causal Argument." <u>Journal</u> <u>of</u> <u>Marriage</u> <u>and</u> <u>the</u> <u>Family</u> 36: 523-532.

Litwak, E., and Szeleny, I. "Primary Group Structures and their Functions: Kin, Neighbors, and Friends." <u>American</u> <u>Sociological</u> <u>Review</u> 34: 465-481.

Lowenthal, M.F., and Haven, C. "Interaction and Adaptation: Intimacy as a Critical Variable." <u>American</u> <u>Sociological</u> <u>Review</u> 33: 20-30.

Mason, Karen Oppenheim, and Lzajka, John L. "Change in United States Women's Sex-Role Attitudes, 1964-74." <u>American</u> <u>Sociological</u> <u>Review</u> 41: 573-596.

McIntyre, Jennie. "The Structure-Functional Approach to Family Study." In <u>Emerging</u> <u>Conceptual</u> <u>Frameworks</u> <u>in</u> <u>Family</u> <u>Analysis</u>, pp. 52-77. Edited by F. Ivan Nye and Felix M. Berardo. New York: The Macmillan Company, 1966.

Ogburn, W.F., and Nimkoff, M.F. <u>Technology</u> <u>and</u> <u>the</u> <u>Changing</u> <u>Family</u>. Boston: Houghton Mifflin Company, 1955.

O'Neill, William L. <u>Divorce</u> <u>in</u> <u>the</u> <u>Progressive</u> <u>Era</u>. New York: New Viewpoints, 1973.

Palme, Olof. "Lesson From Sweden: The Emancipation of Man." In <u>The</u> <u>Future</u> <u>of</u> <u>the</u> <u>Family</u>, pp. 247-258. Edited by Louise Kapp Howe. New York: Simon and Schuster, 1972.

Parelius, Ann P. "Emerging Sex-Role Attitudes,

Expectations, and Strains Among College Women."
Journal of Marriage and the Family 37: 146-153.

Parsons, Talcott. "The Point of View of the Author." In
The Social Theories of Talcott Parsons, pp. 311-363.
Edited by Max Black. Englewood Cliffs, N.J.:
Prentice-Hall, 1961.

Parsons, Talcott, and Bales, Robert. Family, Socialization
and Interaction Process. Glencoe, Ill.: Free Press,
1955.

Rollin, Betty. "Motherhood: Who Needs It?" In The Future
of the Family, pp. 69-82. Edited by Louise Kapp Howe.
New York: Simon and Schuster, 1972.

Seeley, John R.; Sim, R. Alexander, and Loosley, Elizabeth
W. "Differentiation of Values in a Modern Community."
In A Modern Introduction to the Family, pp. 497-508.
Edited by Norman W. Bell and Ezra F. Vogel. New York:
The Free Press, 1968.

Sheils, Merrill, ed. "A Portrait of America." Newsweek,
January 17, 1983, pp. 20-33.

Smith, Richard M., and Smith, Craig W. "Child Rearing and
Single-Parent Fathers." Family Relations 30:
411-417.

"Work and Family Roles: Egalitarian Marriage in Black and
White Families." Social Service Review 55: 314-26.

Stuckert, R.P. "Role perception and marital satisfaction -
a configurational approach." Marriage and Family
Living 25: 415-419.

Thornton, Arland, and Freedman, Deborah. "Changes in the
Sex Role Attitudes of Women, 1962-1977; Evidence from a
Panel Study." American Sociological Review 44:
831-842.

U.S. Bureau of the Census. Statistical Abstract of the
United States: 1981. Washington, D.C., 1981.

Vincent, Clark E. "Familia Spongia: The Adaptive
Function." Journal of Marriage and the Family

28:29-36.

Vincent, George F., and Small, Albion W. "The Family as a Primary Group." In Sociological Theory: Present-Day Sociology from the Past, pp. 269-284. Edited by Edgar F. Borgatta and Henry J. Meyer. New York: Alfred A. Knopf, 1975.

Weller, Robert H., and Bouvier, Leon F. Population: Demography and Policy. New York: St. Martin's Press, 1981.

COMPUTERS, COMMUNICATIONS, AND PRIVACY

Gary Margiotta

The advent of computer technology has given rise to a myriad of problems throughout its development. Initially, conquering the consequences of the new computer technology was confined to the physical technology itself--the computer hardware and software. Man still faces the task of keeping pace with the physical technological advances. Understanding and using the new technology, the silicon chip, optical fibers, and communication satellites, to name but a few, must be attained. Yet, computers and computer technology through their pervasiveness in society have become commonplace, accepted, a given. Today, there is a shift in computer awareness away from the physical technology. A growing issue focuses on the more philosophical aspect of privacy in today's computer-based society. This paper will address the problem of privacy in today's computer-based information society.

In a post-industrial society,[1] emphasis is placed on information. Information becomes the "fuel" of society. In the United States today, over fifty percent of labor income derives from producing, processing, and distributing information, and not from the agricultural, manufacturing or service sectors of the economy.[2] The revolutionary advances in computer technology have greatly facilitated, if not given birth to, this information society. However, along with the benefits of today's advanced technological state comes concomitant costs. One such cost is the potential loss of personal privacy in a computer-based society.

My interest in this topic has been greatly fostered by the ideas espoused by John Wicklein, author of Electronic Nightmare and guest lecturer to the Senior Scholars' Seminar. Wicklein discusses the "marriage of the computer and electronics." He proposes the inevitability of a home communications set (HCS), a TV-like home computer terminal, controlling all communications between people (a central television, telephone, newspaper, mail service, banking service, entertainment center, ad infinitum). He further proposes that, left unchecked, this computer-based system could erode personal privacy due to the ease of accessing and storing private information through this central, universal computer communications system. Wicklein shows the need for laws, regulations, and computer systems to ensure the protection of one's personal privacy.

It is with these thoughts in mind that I have undertaken the task of examining how privacy can be protected in a computer-communications based information society. The role of the courts (especially the United States Supreme Court) and Congress (through federal legislation) in establishing and protecting privacy, as well as the media's relationship to privacy, will be examined. Various proposals for privacy protection in the near future are explored. Finally, the prospect of achieving these proposals and establishing the protection of personal privacy is evaluated.

DEFINITIONS

Before delving into the problems of privacy protection, it is necessary to understand exactly what it is that needs protection. There are various interpretations of "privacy" and "the right to privacy". However, a definition of privacy should be pertinent to the problems of an information society. Among the many available, I found the following definitions most helpful in guiding the study of privacy protection in today's computer-based information society.

David M. O'Brien tries to link the philosophical meaning of privacy with a legal connotation of "the right of privacy" when he defines privacy as "an existential condition of limited access to an individual's life experiences and engagements".[3] Arthur R. Miller defines privacy as "the individual's ability to control the circulation of information relating to him".[4]

However, the definition I found most germane to my study on privacy in our information society is offered by Alan F. Westin. Westin defines privacy as "the claim of an individual, group, or institution to determine for themselves when, how, and to what extent information about them is communicated to others.".[5]

In an information society fueled by the advanced

technological capabilities of the computer, Westin warns against the possible threat of someone, or some group, penetrating another's "inner zone" of privacy through the accessibility of computer store and transmitted data.[6] According to Westin, and numerous others, the computer threat to privacy is a real and present danger.

THE PROBLEM

The computer-communication system in our society is said to be a threat to our personal privacy. This problem has arisen through great technological advances, the costs, benefits, and subsequent threat of which must be examined to understand the scope of the problem.

It is difficult to argue against progress and growth. The growth in the computer industry over the past twenty to thirty years (with even more dramatic progress in recent years) has undoubtedly benefited mankind. The computer has opened up scientific, medical, and other social fields to the betterment of mankind. The computer probably "touches" all of us in some way, everyday. Computerization in education, banking (Electronic Funds Transfer), and especially communications has opened a new way of life. But with these advances come potential costs. Arguments about the computer "dehumanizing" man are often heard. The computer has made the production, processing, collection, and storage of data that is needed in society easier and more efficient. Paradoxically, the potential abuse of this power to control, collect, and use such information is also the problem.

The combination of the computer and communications industries will be the key to information and knowledge in the (near) future.[7] The computer will tie all communications together; technology will develop toward many specialized telecommunication services for specific purposes, but will be integrated by the computer and in-home microprocessors with various information and communication media.[8] With this premise or assumption that the union of computer-communications technology will integrate the flow and collection of information, specific questions can be raised in regard to the issue of privacy.

The first privacy issue raised is one based on our free, democratic socio-political system. In our free, open

305

society, the "right to privacy" comes into direct conflict with our "right to know" what is occurring in our government and our society. Privacy in conflict with other social values (good government, free press, crime prevention, social service provisions, etc.) is a key problem.[9]

Another problem concerns the role of the federal government in collecting and processing personal information. There has been a drastic increase in the amount of personal information collected and disseminated by the government on private citizens. The federal government uses 9,260 computers and has 6,723 records systems containing 3.8 billion records on individual citizens.[10] The threat of government power and control over such personal information is a potential danger to one's privacy. A corollary problem in this regard is the possibility of this vast amount of information collected by the government being used by third parties in both the public and private sectors.[11] Data collection services on individual citizens in the private sector is a growing industry. Equifax Services, Inc., a private firm that makes computer-based investigations of an individual's financial, medical, and other records, has 50 million personal computer dossiers it sells to 62,000 customers including government agencies.[12] Personal privacy protection of information in the private sector must be maintained. Yet, it will be shown how laws and regulations concerning the collection and dissemination of personal information in the private sector lag behind public sector (government) laws, which are only slightly better at that.

A final problem concerns the individual's ability effectively to control and protect his own privacy. In light of his definition mentioned above, Arthur R. Miller alludes to two key problems in privacy protection in our information society: first, the individual's loss of controlling the access to information about himself and, second, the loss of controlling the accuracy of information about himself.[13] If personal privacy is to be maintained, according to Miller, an individual must be able to effectively control information about himself. The possibility of harm to an individual by the new computer-communications technology is more likely if the individual loses this control.

As the availability of information through the new

306

computer- communications technology continues to grow, the threat of an individual's loss of privacy in his thoughts and actions grows simultaneously. If the computer-communications system becomes the fulcrum of informational transactions, the ease of accessing private data on private citizens will increase, unless sufficient laws, regulations, and safeguards are built into the system to assure privacy protection. The problems mentioned above are real, but solvable. With the problems in focus, it is now necessary to understand how privacy has been protected, and can be protected in an information society.

THE ORIGINS OF PRIVACY

The "right to privacy" is not a right recognized or guaranteed by the United Stated Constitution. Perhaps the right to privacy was taken for granted by the founding fathers. It is not known for sure. Nonetheless, the legal right to privacy was not an issue in the early years of our society. It was not until 1890 that privacy became a legal battleground.

In 1890, Samuel D. Warren and LouisD. Brandeis, two Harvard Law School professors, wrote an article in the Harvard Law Review entitled "The Right To Privacy."[14] This article became the basis of privacy law and social policy in the United States. Although the legal arguments of the article have been questioned,[15] the significance of the article was its recognition of privacy as a social and legal issue. Written originally to oppose the various practices of the press of gaining personal information, the Warren/Brandeis article began the legal ramifications of privacy versus the First Amendment guarantee of the free press. Many legal arguments and cases concerning privacy, especially information privacy, have come under the auspices of the press and the First Amendment. But, as Don R. Pember states in Privacy and the Press, the United States Supreme Court (and the courts in general) have placed freedom of the press above the right of privacy as a constitutional guarantee.[16] It is necessary to examine the Supreme Court's role in establishing and protecting personal privacy.

THE COURTS

As stated above, privacy is not mentioned in the U.S.

307

Constitution or Bill of Rights. However, a number of constitutional battles over the right to privacy have reached the Supreme Court through the judicial system. Privacy cases have arisen under claims of the First Amendment (freedom of speech and the press), Fourth Amendment (guarantee against unlawful searches and seizures), and Fifth Amendment (guarantee against self-incrimination). However, throughout the 1950's and 1960's, the Court failed to recognize the constitutionality of the claims for a right to privacy.

Throughout the 1950's and 1960's, federal and state courts upheld both the constitutionality of requiring certain information from citizens for government purposes (the census, for example--U.S. v. Rickenbacker, 197 F. Supp. 924 (1961)) and allowing wide-ranging personal information to be collected about them (as with law enforcement investigative files--Anderson v. Sills, 56 N.J. 210 (1970)).[17]

However, in 1965, the U.S. Supreme Court finally recognized the constitutional right to privacy in its decision in Griswold v. Connecticut. This decision struck down Connecticut's attempt to regulate the use of contraceptive devices.[18] The practical guarantee of the constitutional right to privacy can be found in the Court's decision. The Court noted that: "specific guarantees in the Bill of Rights have prenumbras, formed by emanations from those guarantees that help give them life and substance. . . (These) create zones of privacy."[19]

At the time, this decision seemed to instill new hopes for those seeking stricter privacy protection, especially in regard to informational privacy from increased public and private gathering of data. Yet, Griswold has only been limitedly applied to cases involving matters of family relations, contraception, and procreation (specifically the landmark Roe v. Wade abortion ruling in 1973).[20] The Supreme Court did not deal with the matter of informational privacy until 1977.

In 1977, the Supreme Court nearly gave more weight to the constitutional right of privacy established under Griswold beyond matters of procreation and family relations. The right to informational privacy was addressed in Whalen v. Roe. The case involved a New York state law requiring that the state be provided with information concerning prescriptions for certain dangerous drugs.[21] The state was concerned about controlling the distribution of various dangerous drugs. The contention in this case was that the information provided to the state would erode the confidential private doctor-patient relationship.

The Supreme Court recognized the privacy interests of "avoiding disclosure of personal matters" and "independence in making certain kinds of important decisions . . . (and) that information will potentially become publicly known and adversely affect their reputations."[22] However, the Court denied the contentions in Whalen by recognizing the broader interest of the state's, and the statute's, reasonable exercise of police power.[23] Here, the Court recognized the potential threat to informational privacy, but held that the potential invasion of privacy was sufficiently protected by the statute's security provisions.

It can be seen, then, that the courts, and Supreme Court specifically, have not guaranteed the constitutional right to privacy under Griswold to the fullest extent possible (i.e. toward informational privacy). As the new computer-communications technology advances, and information gathering, storage, and usage gets easier, it can be seen that the current legal framework may be inadequate to defend our information privacy.

MASS MEDIA AND PRIVACY

In the new computer-communications information society, the contemporary role of the mass media is likely to undergo some major changes. With the increased computerization of society, the familiar forms of media communication (newspaper, television, radio, movies, etc.) may be singularly phased out and integrated into one central home-computer terminal. Indeed, this is the view of many authors.[24] The home computer will deliver an electronic newspaper, and will become the central media provider.[25]

The media and privacy have come into conflict in the past primarily under the First Amendment charges of freedom of the press. A key problem mentioned above, the right to know versus the right to privacy, is the essential thrust of privacy concerns with the mass media. Privacy and the press, the communications industry (under the FCC), common

309

law remedies in cases involving the invasion of privacy and the future telecommunications systems will be examined here.

The print media (the press) have been the major disseminators of information to society since the founding of our country. The latitude given to the press under the First Amendment has been greatly exploited by the media. Indeed, people demand information and it has been the role of the press to serve this public interest. However, the press must temper its right to information by serving the public interest. The press is not totally at liberty to print everything. Common law has come to protect citizens against libelous and defamatory printed information. The subject of the press, the First Amendment and privacy protection, is excellently covered by Don R. Pember in Privacy and the Press. For this discussion, Pember neatly delineates the present common law tort of privacy pertaining to mass media action or conduct.

According to Pember, the tort of privacy can be raised when the media uses:

1. an individual's name or photograph in an advertisement without his consent.

2. the publication of private information about an individual (the original heart of the Warren/Brandeis article's argument).

3. the publication of nondefamatory falsehoods about a person (putting a person in a "false light").[26]

A similar set of guidelines is offered by Dean William Prosser, an eminent privacy scholar:

The common law right to privacy is protected against:

1. use of another's name of likeness for personal advantage

2. intrusion upon a person's seclusion or solitude

3. public disclosure of embarrassing private facts about a person

4. publicity that places a person in a false light in the public eye[27]

As Pember notes, there are two main defenses to the privacy tort. Publication of information in the public interest (highly arbitrary!) is one, while the individual's consent to publish the information is the other.[28]

The conflicting relationship between the press and privacy will continue as society debates the interests of the "right to know" versus the "right to privacy." The significance of the common law of privacy is whether or not it can be successfully adapted to the computerized information society. As more and more information is processed through and accessed from computer systems, the present legal recourse of privacy law is bound to take on new meanings. As Alan F. Westin and Michael A. Baker state, "the common law right to privacy does not play a role in the record keeping area. The courts have not held that individuals can invoke the common-law right to prevent the recording of personal information about themselves by a government agency or private organization."[29] If the home computer-communications system does become the new mode of media communication (Wicklein's HCS),[30] interesting questions can be raised about control and privacy matters.

The United States does not have a Federal Department of Communications that oversees the various communications industries. In the U.S., the Federal Communications Commission (FCC), an independent agency created in 1934 to regulate interstate and foreign commerce in communications by wire and radio, is the major authority in the communication field. The FCC is to serve the public interest by . . .

> making available, as far as possible, to all the people of the U.S. a rapid, efficient, nationwide and worldwide wire and radio communications service with adequate facilities at reasonable charges.[31]

The question becomes: How will (does) the FCC fit into the new computer- communications system and how will it protect information privacy?

The answer, to many, it seems is that the FCC is not

equipped to handle the new communications technology. Wicklein feels the FCC has let private business concerns precede the public interest in the new communications technology. By doing so, the U.S. has fallen behind many countries in developing efficient and less expensive communications systems and the FCC has lost much control over the communications industry.[32]

In protecting the privacy of communications, the FCC has limited power. The FCC says that, "no unauthorized person shall intercept any communication and divulge or publish the existence, contents, substance . . . of such intercepted communication to any person".[33] This narrowly relates to the transmission of the communication, and not the processing or storing of communication as common with computer usage today. With the new computer-communications technology, it is evident that new laws and/or extended FCC authority will be needed to meet this problem.

To close this section on media and privacy, an interesting idea of Arthur Miller's will be mentioned. Miller, like Wicklein, sees the home computer replacing the functions of the current press, radio, and television news media. With the increased computerization of personal information, there will be an increase in the amount and variety of information available to the mass media. If the home computer becomes the integrator of media information, Miller proposes that the computer networks will become part of the "marketplace of ideas" just like the contemporary media. To this extent, the "new" media should be entitled to First Amendment protection if their function is to disseminate information for non-commercial purposes.[34]

This scenario has not fully unfolded yet. But, with the development in the new computer-communications technology it may not be a totally unrealistic proposal. Until such time, informational privacy must be maintained. A large part of assuring privacy must come from governmental legislation. It is the role of Congress that must next be examined.

CONGRESS

As mentioned, the U.S. does not have a Federal Department of Communications. Along with the FCC, there are certain agencies such as the National Telecommunications and

Information Administration, under the Commerce Department, which attempt to guide federal communications policy.[35] Concerning information legislation in the new computer-communications field, especially the privacy issue, the major Congressional policy is governed under two acts--The Freedom of Information Act of 1966 and the Privacy Act of 1974. There have been a number of other legislative acts (The Fair Credit Reporting Act of 1970, Right to Financial Privacy Act of 1978) dealing with various issues of privacy in recording information. However, the Freedom of Information Act (FOIA) and the Privacy Act signify both the major strengths and weaknesses in the quest for informational privacy. Even as their names suggest, the two Acts underscore the contradictory nature of achieving a comprehensive information and privacy policy. Although the two laws have meant to supplement each other as a working policy, it can be shown that the key problem of this area--the "right to know" versus the "right to privacy"--is not adequately covered by this federal legislation.

The Freedom of Information Act 1970

The Freedom of Information Act (FOIA) is premised on the principle that the public has a right to know about the basic workings of its government. It established a policy of open government by providing the basic authority and procedure for individuals to petition the executive branch for unreleased documents and records. The purpose of the act was to establish a general philosophy of full government agency disclosure and to provide a court procedure by which citizens and the press may obtain information. Access to government information is available to "any person." The courts cannot consider the needs of the party seeking access unless the data sought falls within one of the nine statutory exemptions.[36]

The FOIA was an attempt to "open up" the files, so to speak, of the federal government. What must be remembered is that both the FOIA and Privacy Act of 1974 deal with the collection and access of governmental information. This is fine. The problem is the collection and access of private sector information. Private computer-based data services, such as Equifax Services Inc., are expanding rapidly. Also, the flow of information over the interactive home computer-communications system (as foreseen by Wicklein, Miller and others) in the private sector will need

313

regulating. The major legislation (FOIA and the Privacy Act) operable today only concern themselves with the government control of information. This must be kept in mind as the major problem facing privacy legislation.

The Privacy Act of 1974

The Privacy Act of 1974 embodies the principal legislative affirmation that the "right to privacy" is a personal and fundamental right protected by the Constitution of the United States."[37] The Act institutionalized individual interests in personal privacy by requiring safeguards on what information federal agencies may legitimately collect and how they may use the information maintained. The Act does not protect personal privacy in the "right to be let alone" sense. It does secure to individuals a legal right to exercise some measure of control over the information collected about them by the government. The Privacy Act only regulates the processes and procedures by which federal agencies acquire, store, and disseminate information.[38]

As previously noted, the Privacy Act fails to incorporate private information collection systems in its purview. The effectiveness of this legislation is confined solely to data collection and dissemination on a federal level. As part of the Privacy Act, a Privacy Protection Study Commission was formed to analyze the efficacy of the legislation on federal agencies. The Commission published its findings and recommendations in 1977. Personal Private in an Information Society discusses the problems and proposes solutions to attaining personal privacy under the 1974 Privacy Act.

Privacy Protection Study Commission

After three years of study, the Privacy Protection Study Commission (P.P.S.C.) published its findings concerning federal privacy legislation. The massive volume (654 pages with five separate appendices) covers the federal Privacy Act in its relation to computer-based record keeping in credit bureaus, employment records, education, mailing lists, and so forth. The main interest for this paper is the Commission's finding on the effects of new computer-communications technology and privacy. This topic was the subject of the fifth appendix, Technology and

Privacy.

The P.P.S.C. named the objective of having federal privacy legislation pertaining to the collection of information by the government as: first, to minimize intrusiveness; second, to maximize fairness; and third, to create legitimate, enforceable expectations of confidentiality.[39] The P.P.S.C. says:

> These three proposed objectives go beyond the openness and fairness concerns by specifically recognizing the occasional need for a priori determinations prohibiting the use, or collection and use, of certain types of information, and by calling for legal definitions of the individual's interest in controlling the disclosure of certain types of records about him.[40]

The major recommendations proposed by the P.P.S.C. in its report were:

1. The establishment of a Federal Privacy Board to oversee privacy matters and laws. 2. To allow the individual to serve as a check on the information about him by giving him greater access to records and the right to amend those records.

3. To limit the disclosure of personal information more than to

4. Greater regulation of the disclosure of personal information, especially by gaining the prior consent of the party involved, in some areas.[41]

These recommendations were the general overview of the many specific recommendations made by the Commission concerning the Privacy Act of 1974.

In its report, the Commission admitted that even areas subject to privacy legislation are subject to the danger that personal privacy will deteriorate due to new technology. The Commission noted a feature in America today that "neither law nor technology now gives an individual the tools he needs to protect his legitimate interests in the records organizations kept about him."[42] The report goes

315

on to say that, "any delay in addressing important privacy protection issues will narrow the options open to public policy-makers."[43]

The P.P.S.C. acknowledges the urgency of the problem which technology brings to privacy. The problem is somehow to gain control over the application of computer-communications technology to personal information record keeping. But as the Commission states, the problem is "the inability to anticipate and control the future uses of information (and information technology).[44]

As stated, the Privacy Act does not deal with the use and collection of private data systems. However, the P.P.S.C. in its report has recommended legislation relating to the collection, maintenance and use of private sector records. The P.P.S.C. hopes that the private sector organizations will become subject to statutes and regulations emanating from their recommendations.[45] However, the private sector has yet to be regulated by any such privacy legislation. This seems to be the main problem in assuring personal privacy protection in our information society today.

Privacy Protection Act of 1980

The most recent legislative action dealing with the issue of privacy is the 1980 Privacy Protection Act. This federal law concerns the relationship between Federal and State law enforcement officials and the news media. In this law, Congress has regulated the means by which Federal and State officers can obtain criminal evidence from the news media. Basically founded on Fourth Amendment principles, the 1980 Act further establishes the guarantee of "freedom of expression" by protecting the media from unlawful searches and seizures of their sources and materials. Stated briefly,

> the Act makes it unlawful for Federal, State, or local officers investigating a criminal offense to search for or seize "work product" or "documentary" materials possessed by a person in connection with, or with a purpose of, disseminating, a public communication in or affecting interstate or foreign commerce.[46]

316

The main objective of the 1980 Privacy Protection Act seems to be to protect individuals engaged in First Amendment pursuits who are not suspected of criminal activity (the news media) from illegal searches. The Act does not prevent the government from obtaining evidence from the media, but simply establishes the means by which that evidence may be obtained.[47]

The 1980 Privacy Protection Act shows how the United States government emphasizes First Amendment priorities. In our information society, the role of the media takes on greater importance. Indeed, with the increases in the computer-communications industry and the integration of the media via the computer, the entire concept of the media changes. The privacy afforded the media under the 1980 Privacy Protection Act seems necessary. However, as more personal information flows through this computer-communications system, it may be equally necessary to protect the privacy of this personal information from the media.

PROPOSALS TO ATTAIN PRIVACY

The threat of losing our personal privacy has increased in today's information society as the new computer-communications technology has increased. The technological advances in computers, communications, and information storage and access are unquestionably beneficial in and of themselves. The problem has been that privacy laws and regulations pertaining to the use of computerized information systems have not kept pace with the technology. This is undoubtedly a very difficult task to accomplish. Yet, the present laws and policies offer little in the way of protecting people against the threat of privacy invasion. Something must be done. Many proposals have arisen to guide public policy toward achieving personal privacy in this computer-based information society. The proposals of three major areas must be examined and perhaps synthesized into a more comprehensive privacy policy. The role of computer security systems, the courts, and privacy legislation through Congress in attaining a personal privacy policy must be explored.

Computer Security Systems

The foundation of our information society is the

computer. The interaction between computerized communication systems is the key to the transmission, collection, storage and use of knowledge and information in our society. To prevent computer failure and unauthorized access to information, the computer must be designed and programmed with highly sophisicated security systems. Illegal access to computer information must be prevented to assure the privacy of personal information stored in the computer.

Computer systems are designed with the goal of achieving high standards of security. Various security provisions can be built into the computer system depending on the level of security desired. Security measures to prevent illegal access to computer stored information can range from requiring terminal identification numbers to encoding data (cryptography). However, as computer expert Paul Armer has stated, "absolute security cannot be achieved."[48] To resort to an old adage, "rules were made to be broken." However, the rules that are broken in computer security systems are broken by people, not the computers. Computers can be designed to be secure until people violate that security.

James Martin, a computer systems expert, has written a number of books on computer systems and security.[49] Martin warns us that people are the preservers of privacy through programming, surveillance, and production of the computers.[50] But, Martin proposes that ultimately security of privacy will lie with the computer itself. To comply with and to carry out efficiently any privacy legislation will be the task of the computer--the computer will keep a check on other computers.[51] The computer will be the regulator of privacy laws concerning other computer-based information systems.

The way our technology and computerization of society is advancing, this proposal does not seem all that unlikely. As stated, computer systems are designed to be secure until humans violate that security. Skeptics of Martin's ultimate proposal will undoubtedly point this out. However, if computer scrutiny of computer security can limit the work and expense of human involvement, greater security might be achieved.

318

The Role of the Courts

The legal concept of a right to privacy has been argued in the courts since the 1890 Warren and Brandeis article on privacy. Throughout the history of this legal issue, the major emphasis has been on finding a balance between the right to privacy and the First Amendment guarantee of freedom of speech (most often expressed in litigation involving the press). The court system currently recognizes certain areas of civil invasion of privacy under common law. However, the court system, and especially the U.S. Supreme Court, has failed to fully extend the Constitutional guarantee of privacy declared in the Griswold decision. The Griswold decision has not led to a broader base for declaring grounds of personal privacy. This is an important point to remember as we enter the information society of today, and seek informational privacy.

Informational privacy has not been protected in the court system. The Supreme Court has affirmed in constitutional litigation (under the First, Fourth, Fifth, and Ninth Amendments) the broad political idea of privacy.[52] But at the same time, it has left issues in safeguarding informational privacy to the states (as seen in Whalen v. Roe), Congress, and generally political and administrative policy-making processes.[53] The Supreme Court has been less than receptive to recognizing and protecting the interests of personal information privacy. Said Justice Stewart in concurring in the Whalen decision, "Whatever the ratio decidendi of Griswold, it does not recognize a general interest in freedom from disclosure of private information."[54]

The Supreme Court, in establishing precedents for the entire judiciary system, must play an active role in promoting personal-informational privacy in our information society. According to Alan Westin, the judicial (Supreme Court) decisions are the key to privacy legislation and policy in the future.[55] Yet, it seems that the Court is slow to realize their importance and potential. In the past, the Court has based its privacy decisions (especially under Fourth Amendments claims) on strict, physical property concepts.[56] If the Court is to keep pace with the advancing technological impact of computer-communications and record keeping information systems, a new foundation for personal privacy decisions must be found. John Wicklein proposes the possibility of a constitutional amendment to guarantee personal privacy.[57] Alan Westin opposes this idea, stating that it would only lead to further confusion and argument in the courts.[58] In any event, the creation of personal privacy in our information society will undoubtedly involve the actions and decisions of the Supreme Court.

The Role of Congress and Legislation

The necessity of ensuring a coordinated national policy of protecting personal privacy in our information society must stem from federal legislation. The federal Privacy Act of 1974 was a good opening step in coordinating such a policy. The Act did have its faults, especially since it only regarded the public sector collection and dissemination of information. Also, discrepancy between the intentions of the Privacy Act and Freedom of Information Act tends to obfuscate a comprehensive federal information policy. Says David O'Brien:

> Only a comprehensive information policy establishing clear and consistent guidelines for (federal) agencies and appropriate mechanisms for implementing them will increase the practical feasibility of privacy safeguards, the reconciliation of interests in privacy and access, and promote agencies' accountability for their information practices.[59]

Combining public and private sector legislation and balancing the "right to know" with the "right to privacy" in any such legislation is the problem federal policy-makers face. The magnitude of this problem is large and achieving an operable solution will not be easy. As technology increases the amount and flow of information, the task will get increasingly difficult. Therefore, it is necessary to begin proposing new legislation now before the problem gets out of control.

The possibility of achieving a comprehensive federal legislative policy concerning privacy in our information society is questioned by Arthur Miller in Assault on Privacy. Miller stresses that the uncertain direction of the computer age, lack of an easy legislative solution, and ignorance of computer technology and privacy issues by lawmakers combine to make effective federal policy legislation unlikely in the near future.[60]

The most likely course of action in the future seems to be toward federal regulation of the information systems. The Privacy Protection Study Commission's major recommendation was for the creation of a Federal Privacy Board. The Privacy Board would receive complaints of citizens and challenge the appropriateness of information collected or used in the public and private sectors.[61] It is in the creation of this federal agency to oversee the protection of privacy in our information society that may see as the event most likely to happen.[62]

The creation of a Federal Privacy Board would be a welcome step in the development of privacy protection. An independent federal agency that would coordinate policy and regulations in both the public and private sectors would facilitate personal privacy protection. However, two points must be explored in this development. First, the formation of another federal agency may exacerbate the problems of our already inefficient bureaucratic nation. Second, and perhaps more important, the Privacy Board would need legislation and policy to regulate. Until more comprehensive information and privacy legislation is enacted, a Federal Privacy Board will be unnecessary.

CONCLUSION

Personal privacy protection is a growing concern of our technolog- ically-advanced information society. We live in an age where the union of the electronic computer and the communications industry emphasizes the primacy of information in our society. As computers continue to pervade society, more and more information concerning our personal thoughts and actions will be collected, processed, stored, and used. The protection of our personal privacy must be maintained in this information society.

The threat of losing control over the privacy of our personal information is not a futuristic problem. Interactive computer communications systems disseminate information throughout the public and private sectors. For example, someone's computerized banking transactions may be accessed by a private credit bureau who gives that person a bad credit rating. This data might, in turn, be accessed by the computer of an employment office who denies that person a job because of his negative credit rating. This scenario is not unrealistic. The threat of losing our informational privacy in our computer-based information society is a present danger.

A number of issues have been raised in this paper. The problems of the "right to privacy" versus the "right to know" and the lack of adequate privacy legislation in the public and, especially, private sectors of our society have been primarily emphasized. It has been shown that the courts and Congress must take a more active role to ensure personal privacy protection. However, as private citizens, each individual in this information society has the ultimate responsibility to elevate and maintain privacy as a value worthy of protection.

ENDNOTES

[1]See specifically the seminal book of this seminar, Daniel Bell's, The Coming of Post-Industrial Society: A Venture in Social Forecasting, (New York: Basic Books, Inc., 1973).

[2]Robert Ellis Smith, Privacy--How to Protect What's Left of It, (Garden City, N.Y.: Anchor Press/Doubleday, 1979), p. x.

[3]David M. O'Brien, Privacy, Law, and Public Policy, (New York: Praeger Publishers, 1979), p. 16.

[4]Arthur R. Miller, The Assault on Privacy, (Ann Arbor, Mich.: University of Michigan Press, 1971), p. 25.

[5]Alan F. Westin, Privacy and Freedom, (New York: Antheum, 1967), p. 7.

[6]Ibid., p. 33.

[7]See specifically Starr Roxanne Hiltz and Murray Turoff's, The Network Nation, (Reading, Mass.: Addison-Wesley Publishing Co., Inc., 1978); John Wicklein's Electronic Nightmare, (New York: Viking Press, 1981); and Miller's, The Assault on Privacy.

[8]Hiltz and Turoff, p. 399.

[9]James Martin, The Wired Society, (Englewood Cliff's, N.J.: Prentice-Hall Inc., 1978), p. 250.

[10]O'Brien, p. 205.

[11]Ibid., p. 203.

[12]Wicklein, p. 191.

[13]Miller, p. 26.

[14]Don R. Pember, Privacy and the Press, (Seattle: University of Washington Press, 1972), p. 33.

[15]See specifically Pember, p. 33, and O'Brien, p. 5.

[16]Pember, p. 226.

[17]Alan F. Westin and Michael A. Baker, Databanks in a Free Society, (New York: Quadrangle Books, 1972), p. 19.

[18]Miller, p. 204.

[19]Ibid.

[20]O'Brien, p. 192.

[21]Ibid., p. 193.

[22]Ibid.

[23]Ibid.

[24]See specifically Wicklein, Hiltz and Turoff, and Miller.

[25]Wicklein, p. 100.

[26]Pember, p. 241.

[27]Dean William Prosser, quoted in Miller, p. 173.

[28]Pember, p. 241.

[29]Westin and Baker, p. 19.

[30]Wicklein, p. 1.

[31]The Communications Act of 1934, quote in Stuart L. Mathison and Philip M. Walker, Computers and Telecommunications: Issues in Public Policy, (Englewood Cliffs, N.J.: Prentice-Hall, Inc., 1970), p. 3.

[32]Wicklein, p. 85.

[33]The Communications Act of 1934, quoted in Mathison and Walker, p. 214.

[34]Miller, p. 195.

[35]Wicklein, p. 85.

[36]O'Brien, p. 214-215.

[37]The Privacy Act (1974), Public Law 93-579. 5 U.S.C. 552(a), quoted in O'Brien, p. 207.

[38]O'Brien, p. 207.

[39]United States Privacy Protection Study Commission, Personal Privacy in an Information Society, (Washington, D.C.: U.S. Government Printing Office, 1977), p. 14-15.

[40]Ibid., p. 15.

[41]American Enterprise Institute, Privacy Protection Proposals, (Washington, D.C.: American Enterprise Institute, 1979), p. 8.

[42]United States Privacy Protection Study Commission, p. 8.

[43]United States Privacy Protection Study Commission, Technology and Privacy, appendix 5 to Personal Privacy in an Information Society, (Washington, D.C.: U.S. Government Printing Office, 1977), p. 3.

[44]Ibid., p. 26.

[45]Ibid., p. 54.

[46]Larry E. Rissler, "The Privacy Protection Act of 1980," FBI Law Enforcement Bulletin, February, 1981, p. 28.

[47]Ibid., p. 31.

[48]Paul Armer, "Social Implications of the Computer Utility," Computers and Communications: Toward a Computer Utility, Fred Gruenberger, ed. (Englewood Cliffs, N.J.: Prentice-Hall, Inc., 1968), p. 197.

[49]Martin excellently covers the area of computer systems in a number of books, including Computer Data-Base Organization, Systems Analysis for Data Transmission and Design of Real Time Computer Systems among others. For a discussion on security aspects of computer systems (too technical for this paper) see Security, Accuracy, and

<u>Privacy</u> <u>in</u> <u>Computer</u> <u>Systems</u>, (Englewood Cliffs, N.J.: Prentice-Hall, Inc., 1973), especially section two on computer design.

[50]Martin, p. 254.

[51]Ibid., p. 263.

[52]O'Brien, p. 194.

[53]Ibid.

[54]Ibid.

[55]Westin, p. 398.

[56]The development of privacy and the Fourth Amendment on property concepts dates back to the 1928 case of <u>Olmstead</u> <u>v.</u> <u>United</u> <u>States</u>. The Supreme Court ruled that no legitimate privacy claims existed with regard to messages passing over telephone lines outside a person's house or office. The Court held that the interception of such messages is carried out without entry into a person's premises, there is no "search" of a constitutionally protected area, and further that telephone messages are not things that can be "seized" under the Fourth Amendment. This basis was liberalized in the 1967 decision of <u>Katz</u> <u>v.</u> <u>United</u> <u>States</u> where the Court ruled that the Fourth Amendment "protects people, not places." However, property concerns still preclude protection of an individual's privacy interests in an information society as seen in the Court's ruling in <u>United</u> <u>States</u> <u>v.</u> <u>Miller</u> in 1976. The Court held that Miller's bank records and transactions were liable to subpoena and that they were not personal property.

This line of action prompts David O'Brien to state that "The contemporary Supreme Court fails to conceptualize Fourth Amendment interests more appropriate to the technological changes that threaten to diminish personal privacy in an increasingly information based society since it bases privacy on property concepts." See O'Brien, p. 233.

[57]Wicklein, p. 217.

[58]Westin, p. 395.

[59]O'Brien, p. 225.

[60]Miller, p. 220.

[61]American Enterprise Institute, p. 8.

[62]See United States Privacy Protection Study Commission; Wicklein; Miller; and O'Brien.

BIBLIOGRAPHY

American Enterprise Institute. Privacy Protection Proposals. Washington, D.C.: American Enterprise Institute, 1979.

Armer, Paul. "Social Implications of the Computer Utility." Computers and Communications--Toward a Computer Utility. Fred Gruenberger, ed. Englewood Cliffs, N.J.: Prentice-Hall, Inc., 1968.

Bell, Daniel. The Coming of Post-Industrial Society: A Venture in Social Forecasting. New York: Basic Books, Inc., 1973.

Hiltz, Starr Roxanne, and Turoff, Murray. The Network Nation. Reading, Mass.: Addison-Wesley Publishing Co., Inc., 1978.

Martin, James. The Wired Society. Englewood Cliffs, N.J.: Prentice-Hall, Inc., 1978.

Mathison, Stuart L., and Walker, Philip M. Computers and Telecommunications: Issues in Public Policy. Englewood Cliffs, N.J.: Prentice-Hall, Inc., 1970.

Miller, Arthur R. The Assault on Privacy. Ann Arbor, Mich.: University of Michigan Press, 1971.

O'Brien, David M. Privacy, Law, and Public Policy. New York: Praeger Publishers, 1979.

Pember, Don R. Privacy and the Press. Seattle: University of Washington Press, 1972.

Rissler, Larry E. "The Privacy Protection Act of 1980." FBI Law Enforcement Bulletin, February 1981, pp. 26-31.

Smith, Robert Ellis. Privacy: How to Protect What's Left of It. Garden City, N.Y.: Anchor Press/Doubleday, 1979.

United States Privacy Protection Study Commission. Personal Privacy in an Information Society. Washington, D.C.:

U.S. Government Printing Office, 1977.

United States Privacy Protection Study Commission. Technology and Privacy, appendix 5 to Personal Privacy in an Information Society. Washington, D.C.: U.S. Government Printing Office, 1977.

Westin, Alan F. Privacy and Freedom. New York: Antheum, 1967.

Westin, Alan F., and Baker, Michael A. Databanks in a Free Society. New York: Quadrangle Books, 1972.

Wicklein, John. Electronic Nightmare. New York: Viking Press, 1981.

A Case For Computer Literacy In Public Education

Marilyn Fricke

Technology has changed every aspect of American life, including lifestyle, politics, religion, and recreation. Its advancements and resultant changes based on electronics, automation, and computers can also change the quality and efficiency of our educational programs. Just as movable type, for example, expanded the impact of learning, so will the microcomputer affect the transition and handling of information. And, if this technology can help improve the quality and efficiency of our educational programs, American education should take advantage of that capability.

We live in the most highly technological society in history, where machines have already taken over much of the brute work in industry, heralding far-reaching changes in man's labor and leisure. However, in what way can we prepare those generations to follow in assimilating and developing skills to cope in such a time of rapid and fundamental change? A post-industrial society cannot afford to lag behind in an inefficient educational system still dependent upon a blackboard and a slide projector. Our society must be prepared to provide a system of instruction that will enable a student to make intelligent, productive, and creative decisions about computer technology, most notably microcomputers, during the early school years.

It is likely that the computer will force a radical reappraisal of educational content and method, just as the printed book did. The failure of public schools to make a major commitment in computer literacy now may have profound consequences for both the education of the American citizenry and the future of man in tomorrow's technological society.

Educators should not ignore the implications of the microcomputer revolution. Andrew Molnar, of the National Science Foundation sees computer literacy as the next crisis in American education. Only a few years ago most educators felt that the role of computers in education would be defined sometime in the future. Microcomputers have changed all that - the future is now.[1]

This paper endeavors to explore the instructional use of microcomputers in support of computer education. Careful consideration is given to the primary utilizations of the computer, as well as its systematic implementation in the school. Additionally, the consequences of this technology are recognized in terms of its social and economic costs. It is hoped that through this research, support for computer literacy will be built as a relevant program in public school curricula. But first, it may be beneficial to gather insight into the development and distinctive features of today's microcomputer.

The Computer: Its History and Hardware

The Mark 1, which was developed at Harvard University and became fully operational in 1945, was the first electro-mechanical digital computer.[2] Early computers like the Mark 1 occupied several rooms, equipped with elaborate wiring and environmental control. During the late fifties and early sixties, when the transister, the integrated circuit, and the printed board were developed, computers could be made considerably smaller than before.[3] In 1975, the microcomputer became available.

A computer is essentially a machine that receives, stores, manipulates, and communicates information.[4] What makes the microcomputer or micro exceptional is its memory. In a micro, the central processing unit is a microprocessor: a single integrated circuit on a chip of silicon. The advantage of the chip is its size, typically about a quarter of an inch per side. These microcomputers are scaled down versions of large computers with their own unique features. Their cost is low and steadily declining as new technologies increase the capacity and decrease the size of the equipment. For instance, since the microcomputer does not require a telephone connection to a large computer, line charges are eliminated. Additionally, they are portable and reliable. Microcomputers can be in a classroom where they are needed, and if one "crashes", it does not wipe out everyone's work. Availability is another characteristic that is frequently praised. While time sharing of computer use leads to competition, microcomputers allow for independence because they are self-contained units. Most

331

importantly, they are subject to the control of the student and the teacher, rather than controlling the student according to some remote plan. This control in the technology has aided educators during the last decade in their development of individualized learning techniques. With the advent of microcomputers, a whole realm of computing has been opened.

A typical microcomputer consists of five basic and fundamental electronic elements: (1) a central processing unit (CPU); a "processor", (2) single or multiple read-only memory (ROM) for preassembled control program storage, (3) one or more random access memory (RAM) for data and/or program storage, (4) an expandable universal input-output bus, (5) a computer clock.[5]

The processes of today's computer hardware are rather complex, but can be simplified for illustrative purposes. In basic terms, the clock sends out signals to initiate consecutive steps of the calculating cycle. Since most of today's computers work digitally (letters, numbers, and graphic symbols are represented by combinations of cues and zeros), this binary information is manipulated by the CPU. When the information is recalled, the information is converted back to letters and numbers. The storage of information is completed with ROM, which provides permanent storage and with RAM, if one prefers to store one's own programs and files. The bus is the connection path, which consists of a set of plugs or receptacles, for the peripheral equipment. Peripherals or Interfaces are such devices as disk drives, printers, and modems that connect to the computer. All these electronic devices constitute the computer's hardware. The hardware can do nothing by itself. It requires the array of programs or instructions called software, the merits of which will be discussed later.

During the past twenty-five years, spectacular developments have occurred in electronic technology. As current engineering and marketing trends suggest, the computer's capabilities will continue to increase, while costs will decrease. Given the new predictions by prognosticators, how will education make use of the advances? Additionally, even if the new methods are employed, will the new microcomputers be part of a fad or the beginning of a renaissance? This is a relevant question and the topic of further discussion.

Microcomputers in Schools; Fad or Function?

In the early part of this century, the son of a craftsman could enter the same trade as his father and expect to use the same tools and techniques. Today, however, the pace of technological development continues to accelerate and the span of time between significant changes in working practices continues to decrease. "The Malaise of our age is that our power increases faster than our ability to understand it and use it well."[6] In the words of Margaret Mead, "No one will die in the world in which he worked."[7] It could be safely stated that few of us would spend the whole of our working lives practicing a single set of skills. It would be detrimental then, to prepare students with a time-worn method from another era.

In the past, education existed in the form of mass schooling. The idea was borrowed virtually intact from industry, particularly from aspects of industry by Frederick Taylor.[8] Taylor, an American efficiency engineer devoted himself to expounding the best way to do a job, in the shortest time. Analyzing each operation in terms of its basic components became known as "Scientific Management", and this structuring became typical of schooling during the industrial age. The schools became professional, bureaucratic structures, defining the curriculum in terms of measured development and productivity. Society and the school it fostered adopted an "industrial consciousness", where students were the products of school factories. (This type of consciousness can still be found today in such areas as programmed instruction and behavioral objectives.) These ideals, by no means will be eradicated in the post-industrial era, but their importance will weigh less when adjusting the scales for a future orientation.

With innovations in education today, using new machines to retool our teaching techniques, we can anticipate change. The microcomputer revolution is a substantial part of instructional technology than can expose students to the type of technology they will need in a future-oriented curriculum. Introducing computers in the classroom is a natural progression in integrating the type of instruction necessary to aid those post-industrial pupils. For instance, the United States Department of Labor has estimated that by 1985, nearly half of 104.3 million

civilian workers will be employed in offices and the number
is expected to keep on growing.[9] Unquestionably, these
offices jobs will be influenced by systems combining word
and data processing, records management, telecommunications,
and electronic mail. If the office employees are expected
to have skills to utilize microcomputers and the development
of computers in business and industry continues to expand,
then this new technology will hold implications for the role
of the public school.

Schools of today need to integrate technology into the
public school curriculum. In the 1980's, the rapid infusion
of "high technology" devices into the workplace, home, and
marketplace will literally sweep schools up or away - ready
or not. The computer is one device that can improve the
quality of the future-oriented curriculum. And the students
have never been more prepared for this type of instructional
technology. Today's teenager has spent an estimated average
of 15,000 hours in front of the television set, as compared
to 11,000 hours in school (Kindergarten through twelfth
grade).[10] This generation grew up in a world dominated by
graphics and images on a screen. What better way to educate
our society than by a method more enthusiastically "viewed"
than the printed word?

The proper question to ask is not how computers can be
used in education, but how computers can help the quality of
education. One answer to this statement is that computers
should be used to their maximum advantage, leaving people
free for functions that are better performed by humans.
Using the computer for instructional support, for example,
frees teachers from the drill and practice work, allowing
them more time to search for new facts and implications
within their subjects. Teachers literally can become free
from roles as "judges" and expand their roles as mentors.
By liberating the teacher and student, technology should
allow for the long-sought goal of full individualization of
instruction. "Out of a total day, teachers average
twenty-five minutes on individual work with pupils."[11]
The ideals of individualization today exceed the capacity
of labor-intensive education as well as its human resources.
When the ratio of teachers to student is one per twenty or
thirty, the concept of individualization is thrown out of
the window and group-paced instruction becomes a virtual
necessity. Computers could be the answer.

We can envision the time of universal individualized education when every person will be educated and no two will be educated alike, when teachers deeply committed to the art of teaching and thoroughly versed in the science of learning will have at their disposal a full panoply of learning materials to which they will direct each individual student in accordance with his needs ... He will move smoothly and early from directed, highly structured learning situations to self-directed unprestructured activities, when he as a learner plays an active role in learning.[12]

With the entrance of the microcomputer, a technology can be offered that gears instruction to the specific abilities, needs, and progress of each individual. Today, students and teachers are increasingly exploiting the computer's power of computation, data processing, problem-solving, and simulation. Further, the computer's function can expand the standard Computer-Assisted Instruction (CAI) to new ways of exploring and manipulating the subject matter he is studying and the data at his disposal.

Additionally, microcomputers are likely to enhance learning by increasing the amount of information students have access to as well as the speed with which they can retrieve it. With a computer's electronic storage, retrieval, and presentation of information, any learning experience could benefit from immediate feedback. The average teacher, for instance, cannot correct each paper within seconds of its completion, while the computer can and does. With a drill and practice program, the computer even has the capability of immediate error detection. Later, at the end of a lesson, the computer gives the student a summary of his performance. Even when students are writing programs to solve a problem, they obtain feedback in the form of an unworkable response or unreasonable result. In all cases, this feedback is faster and more pertinent than the age-old standby, the red pencil.

We are now at the brink of a computer impact in the field of education that many believe will at least equal the impact of reading and writing on man's intellectual development. Computer power has the potential of providing

335

man with universal access and rapid synthesis of knowledge as well as providing a powerful creative tool for widening the intellectual horizon of man. In what is too often a boring environment, the computer is a novelty - yet this is one novelty that does not appear to wear thin. When the computer is used as a problem-solving tool, students work harder than ever to program a solution. In computer-assisted instruction, some instructors believe that the learning environment becomes unthreatening and noncompetitive to the underachievers as well as the above-average students. And when simulations and games are made available as a method of accessing key concepts, students will stop at nothing to "beat the machine." One fourth grade teacher and coordinator of the computer project at the Lamplighter School in Dallas commented, "I can't think of anything they (children) do elsewhere that makes their brains churn and work as much as when they are on the computer."[13] The computer can be an intense and challenging teacher. In fact, recent evidence indicates that students show increased motivation in classes where computers are utilized.[14] The computer impact can no longer be considered a fad.

On the contrary, the function of microcomputers is a preparation for the future. A knowledge of microcomputers is necessary, if only in preparation for a world beyond the classroom walls. The society into which today's kindergarten children will step after high school is only a well-defined prediction by futurists. Change is the only sure constant and the professional teacher can no longer be sure whether he 'knows' what is good for the child.

> Most of the information now being transmitted to them will be out of date within ten years. The rate of information change is actually accelerating, and the process will not stop... Even if the schools began to add subjects and subtract others, the problem would not go away. The proliferation of new subjects and the rot of old ones is so persistent that a curriculum will always be out of date as long as schools take as a major objective the transmission of a specific body of knowledge. Such an objective, however useful it may have been in the past, is now a formidable obstacle to any intelligent confrontation with the future. It creates the

illusion of knowledge stability. It fixes people to ideas, constructs, and information whose life expectancy is far shorter than peoples'. This is extremely dangerous. "What you don't know can kill you." is an important slogan, but even more important is Josh Billing's line, "that the world is plagued not so much by what we don't know but what we know that ain't so."[15]

With unprecedented changes occurring, the sensible use of technology in education will increase the alternative solutions to problems and will increase the student's flexibility in new directions. A richer store of experiences and realities surrounds the student using the computer and he can organize this wealth of information into units that may help in comprehending an environment of change.

Computer Literacy; What It Is and Whom It Affects

Computer Literacy is whatever a person needs to know and do with a computer in order to function in our information-based society.[16] The National Science Foundation has delineated four prerequisites to consider before a person is believed to be computer-literate.[17] Primarily, a person must have knowledge of how a computer works. The individual should know the five functional parts - the input, the memory, the central processing unit, the arithmetic and logical unit, and the output unit - as well as their capabilities. Second, the step-by-step technique for problem solving is important. While writing a program is not necessary, a basic understanding of how it can be derived is significant. The social implications of computer use, such as equity, is the third area an individual must be aware of to be considered computer-literate. Lastly, a student should be informed of the applications of computers and the extent to which we are affected. Knowledge in all four areas not only qualifies an individual as a computer-literate, but also prepares one for a practical future. According to a recent estimate, 80 percent of the jobs facing today's elementary students in 1990 do not even exist yet.[18] If this statistic is even partially true, the time to prepare is now. Learning to write a program may very well become one of the basic literary skills of the future. At any rate, the increased use of computers in society will force schools at all levels to provide training or at least some familiarity with computers.

The key to computer literacy is used basically to demystify the machine. Because the average worker comes in contact with at least fifty computerized effects by the time he or she arrives at the office, a person needs a functional knowledge of computers. Additionally, by starting early in elementary education, children can learn to control the computer, just as they learn to read and write in the earliest grades. In this way, a student's first encounter with a computer can put him in charge. Knowledge of how a program works, how information is represented, stored, and accessed, how bugs arise, etc., gives the user a far more sophisticated sense of the tool he is using. Learning to work with computers also teaches discipline, logic, and improves eye/hand coordination, as well as reading and math skills. According to Dr. Galanter at the Children's Computer School in Columbus, Ohio:

> We have seen that children take to this like ducks to water. They have no fear. They don't always understand the theory behind it, but they're open to learn. It is a natural thing like learning languages when they are small. They comprehend so quickly and it gets to be like second nature to them - just like using the telephone.[19]

Hands-on-experience is important; however using the computers solely as drill and practice devices (via Computer-Assisted Instruction), can apparently aggravate the problem of computer literacy. The computers are viewed as tools, which need not themselves be understood by the end-user. Using the tool without knowledge of how it works can literally put the user at the mercy of the machine. If the system fails, the user may feel lost.

Dr. Martin Ringle, from the Department of Computer Science at Vassar College, argues that the individual user of CAI drills builds up a blind trust in using microcomputers. Rather than coming to understand the computer, how specific programs work, and how information is represented, the individual builds up a subtle mystification rather than a true literacy. Learning involving both comprehension and production will enhance the student's use, while eliminating frustration, helplessness, ill-founded contempt, uncritical dependence and general technological alienation.[20] It seems logical that schools will offer a program to enhance the understanding of microcomputers.

Some might argue that computer literacy is the responsibility of the individual, and can be learned after one leaves school. However, there is a public cost of students who graduate without being exposed to computers, for they have not had a complete education. Underestimating the implications of technological illiteracy can only hinder the economy. To retrain after graduation is costly and an unnecessary waste. As past research has shown, failure in attempting to introduce computers in less-developed nations has been due to a lack of a significant number of computer-literates.[21] In order to educate our society to the highest degree, a familiarization with computer-based educational materials is our best prevention against modern-day illiteracy. In an information society such as ours, a populace of computer literates is as important as energy and raw materials are to an industrial society.[22]

Education is a process in which students, teachers, parents, administration, and environment are all integrated. Innovation, therefore, will not occur simply because of a program outline but will evolve out of the whole educational system. It does not work out of context. For there to be useful innovation, the educational community should have confidence and believe that there is a need to do something new. Therefore, to introduce computer literacy, the general community which supports education must recognize the need in order to overcome a reluctance to spend the necessary funds. To have a successful transformation within the school curriculum, the public must understand the implications of the computer revolution. They should also be introduced to computer literacy. Schools should also provide either preservice or inservice courses for teachers in such areas as computer-assisted instruction, computer-managed instruction, microcomputer programming, and computer applications. There is a growing trend toward involving instructional people in the purchasing decisions in the courseware, but their input is more valuable when they are familiar with the basic microcomputer operation. An educational tool is only valuable if it can help the teacher do a better job. With the introduction of the computer, teachers must be aware of the new avenues in approaching traditional methods to foster learning in their pupils.

Discontent by educators about their own computer illiteracy recently became more and more apparent. In response, the Congressional Office of Technology Assessment called for federal assistance in the computer-training programs for teachers just last month, (December of 1982).[23] Yet assistance is still unavailable and the problem of inadequate preparation of teachers in the use of microcomputers still exists.

Even members of the computer industry raise concerns about inadequate teacher training. As Lewis Branscomb, vice president and chief scientist of International Business Machines has contended, "People have been shoving things down the throats of schools. I see dangers in the industry if people don't take seriously the problems of providing the necessary support to the teachers and administrators who have to incorporate the computers into the curriculum."[24] Obviously, inadequate teacher preparation poses problems not only for the classroom, but for the computer industry itself.

Teachers should be aware that computers will drastically alter their role, but not replace him or her. As some teaching machine advocates put it, "any teacher who can be replaced by a machine deserves to be."[25] At this stage of technological development, the teacher can still play roles which a machine cannot perform: strengthening student-teacher relationships, changing perceptions towards those around us, as well as encouraging artistic expressions. Of course, there are those who believe that if the transition is to be made to a capital intensive education, the principal parties involved must be convinced that personal interests will be protected It may be true that many teachers do nothing about computers and computer-assisted instruction only because no one has really taken time to introduce them to teachers. Associated with the anti-technological bias of some teachers is not only a fear of the new situation in the classroom, but also the loss in job status with reference to job security. Certainly, the computer will greatly modify the role of the teacher. Some teachers will engage primarily in the preparation of the instructional system in cooperation with

340

other specialists. Others may need to prepare for different and more important roles, such as teaching new concepts and critical thinking skills. Generally, the new role of the teacher will be to work individually with all students on whatever problems and questions they may have in assessing and handling the basic ideas and procedures presented by a computer. The computer will change the role and function of the educator but not diminish his importance.

Teachers are aware of the possibilities offered by the microcomputer. It is up to the industry and the decision-makers in the educational community to make sure that teachers are given the key to unlock the capabilities of the computer. Sitting in America's classrooms are minds that can conquer the universe. The teachers in front of those classrooms guide those minds. Give the teachers the tools and they will help students to conquer.[26]

Utilization of the Microcomputer

The computer lends itself to both the instructional function in the classroom and the management of educational data in the form of personnel records, payrolls, and course scheduling. Focusing on the instructional uses of computers, there are three categories that seem to encompass the growing field.[27] The computer can be used as a teacher, as in Computer-Assisted Instruction. In this mode, the computer is preprogrammed to determine what a student will learn in the form of drill and practice, tutorial programs, curriculum-oriented games, simulations, and logic games. The second category includes the use of the microcomputer as a tool, in which the programs make available capabilities that the student did not have before, such as calculators and word processors. Third, the computer can serve as an object to be taught. Computers can be productively used as a tutee; the object to be taught, as in problem-solving or as a total learning environment. There are a variety of ways that computers can be implemented into instruction and alternatives are increasing as new software is developed.

Computer-Assisted Instruction - the computer as the teacher - has been the focus of most public schools' involvement with the computer revolution. This tendency prompts a word of caution. Dr. Patrick Suppes, a philosophy and mathematics professor at Stanford University and pioneer of Computer-Assisted Instruction, points out that children receiving CAI do achieve higher test scores but tend to make the same sort of errors, with the same frequency, as children who do not receive it. This suggests that test-taking skills, not understanding were improved.[28] Criticisms of CAI contend that this type of instruction may not expand one's reasoning.

When the computer is utilized as a tool, the programs typically give access to certain kinds of information not obtained before. The computer can be used as a laboratory instrument; to collect, analyze, and display data, and in the form of calculators and statistical packages. As a tool, the computer helps organize facts as well as increase the accuracy of the resultant information. The LOGO approach utilizes the computer as an object to be taught. The approach constitutes one of the practical departures from a more passive role in CAI, to a role which is designed to provide children with a problem-solving environment via computer programming. Seymour Papert, the developer of the LOGO language, feels that by using the computer as a tool for exploring mathematical and scientific concepts, children are learning how to think for themselves and to formulate problems on their own.[29] Using the language encourages children to write simple programs and manipulate a device called the LOGO turtle which generates graphics. Children quickly proceed to drawing more complex figures such as circles and flowers. Their process of experimenting; making mistakes and finally figuring out problems, gives them not only a successful feeling of accomplishment, but also a grasp of mathematical terms and concepts.

> It may also be of particular advantage for students to play the role of an "author" of instructional materials. The "author" will benefit in writing the machine directions by receiving feedback information from the system and other students. The total interaction will not only foster a self-directed inquiry and exploration of subject matter, but serve to motivate the student to improve the program's operation.[30]

342

The traditional role of computers in education was typically spoonfed knowledge administered to a child. But as small computers have grown more sophisticated, educators have begun to realize that the relationship need not be so one-sided. The new programs have the power to enhance greatly our "transfer" knowledge. The computer not only eases our ability to memorize, but allows us to apply the information to a simulated problem or situation. In transfer, a person must be able to recognize when a problem is present and be able to arrange problems in patterns to see elements that are common with other past problems.[31] Asking questions and generating hypotheses are a part of the skills needed for problem-solving. Given the advances made in computers, it may be correct to say that the practice gained in early LOGO experiences may be helpful in preparation for an environment characterized by change. In fact, Isaac Asimov, a futurist, recently summed up his view of the computer's impact on society: "We are reaching the stage where the problems that we must solve are going to become insolvable without computers. I do not fear computers. I fear the lack of them."[32]

The enormous growth in information has increased the diversity and complexity of education. The computer is changing the idea of competency from what one has stored in one's head, to what one can locate and the process of having access to it. Daniel Bell sees the character of knowledge changing. "If the atom bomb proved the power of pure physics, the combination of the computer and cybernetics has opened the way to a new "social physics" - a set of techniques through control and communications theory to construct a tableu entiere for the arrangement of decisions and choices."[33] Through the use of computers, simulation procedures will allow us to plot alternative futures in different courses, thus greatly increasing the extent to which we can choose and control matters that affect our lives. The school or some other form of "knowledge institute" will become the central institution of the future because it will be a new source of innovation.

What Bell finds central to his theory is the primacy of theoretical knowledge, a sort of logic combinational analysis. The access to this theoretical knowledge will produce an elite, not based on wealth and inherited social status but on intelligence and knowledge. It is not the accumulation of information that is important, but the understanding and critical judgment needed for a discipline of learning to adapt rapidly to new situations and problems.[34] The computer is the instrument by which it is possible to develop this methodology of organized complexity. Post-Industrial Society, with its principle of theoretical knowledge will present challenges in the character of knowledge. The shift of knowledge as something discovered and organized within itself, to something constructed by new phenomena will mean greater insight into individual thinking patterns. The computer will be liberating, offering the ideas of alternatives, helping people achieve certain kinds of control over their lives.[35]

If we place a value on new knowledge and development, allowing technologists to expand our capacity by several magnitudes, then our potential is one of sharing the theoretical knowledge with the whole society. If Bell is correct, schooling should concern itself not just with knowledge, but the structure of it, helping students to think continually and consciously about their unthinking, paying more attention to analytical skills. Computers used in schools can help accomplish the development of advanced problem-solving skills. Students would be able to structure problems in a logical form, construct graphs, simulate real systems as computer models, and to search data bases among other skills. The computer courses would not focus on the syntax of a common computer language but emphasize programming as a means for assimilating basic principles of computing.

With the computer, children can develop experiences with precision, repeatability, graphics, motion, and color. A computer-literate group actually sees how the presence of computers can enormously expand their freedom of action in creating learning paths into knowledge. Additionally, the students develop, as Papert describes, a social resonance.[36] Just as the typical child receives a "charge" from watching a TV screen, a child talking over a screen and making pictures receives a satisfaction of control. (This feeling is partially conveyed in our desire to own a calculator, our very own piece of technology.) The child feels that he is doing something meaningful, and it also provides a bridge between him and the world. The computer represents more than a basic skill. The microcomputer becomes a personal object a person can subtly identify with and 'master'.

Naturally, the worship of technology for the sake of technology is self-defeating. Technology is a means to an end in instruction and should be employed to benefit its society as an object to improve contributions to learning. The impact of the computer technology on future curricula will depend on the way in which it is integrated in an educational program. Successful improvement in education using the computer depends not solely on the technology itself, but on the planned, implementation of it into the school.

Implementation of the Computer Technology

The analogy between movable type, in the form of the printing press, and computers, can illustrate well the effective impact of new technologies. The press actually reduced the cost of delivering books, however the majority of the people were illiterate. The book was conceived as something that would take long hours of study to master. Computer technology today is seen in that same perspective. Technological devices already applied in schools have had only peripheral impact partly because educational technology is still quite primitive. Furthermore, the knowledge of the applications of the technology is even more primitive. In order for the computer to have a profound effect, computer literacy must be placed high on our educational priority list.

With all the alternative instructional approaches available with computer systems, the question might be asked - for what are we waiting? Some opponents of immediate implementation root their worries in concerns about obsolescence, incompatibility, and program availability. This conservative approach may well be justified, but waiting may also mean losing irreplaceable experience in the integration of various systems into instruction. There is always an economic opportunity cost of forgoing features that may greatly enhance future pioneering possibilities. A nation concerned with its social need and economic growth cannot be indifferent to the problems of illiteracy. If we are to capture the benefits of technology, we should certainly develop a computer-literate society. As Neil Postman made note in 1969,

The twenty-first century is only thirty-one years away and the schools are not as yet concerned to teach children how to think, how to master electronic media, how to deal with rapidly changing knowledge, how to produce knowledge, how to give direction to their own education, and how to understand themselves.[37]

Now, the twenty-first century is only seventeen years away. It is unfortunate that the same pronouncement is still applicable in many cases.

As a result of the microcomputer revolution, 50 percent of the students in the United States' schools and colleges will be receiving Computer-Assisted Instruction by 1987, as compared with the twenty-five percent today.[38] This statistic sounds impressive for those individuals exposed, but what about the remaining percentage? Moreover, who is to say that the instruction in the schools making use of computers will be constructive?

The National Center for Educational Statistics (NCES) of the Department of Education reported for 1982, that the availability of computers for instruction varies by grade level. About three-fifths of all secondary schools have at least one microcomputer or computer terminal, compared with one-fifth of all elementary schools. The NCES survey also showed the need for more microcomputers and suitable courseware. More than 60 percent of all the schools cited these needs as major. Although other needs were expressed, such as qualified teachers, start-up assistance and staff/community support, they were reported less frequently, cited as major problems by 41-50 percent of the schools surveyed.[39] With the introduction of micros, there must come an increase in the understanding of instructional applications and resources needed by schools, so to improve computer-based education.

Successful use of the computer depends on many variables, including the adequacy of the course objectives, the consistency of student interest, and most important, the software or courseware design itself. The software, the set of instructions that tells the computer how to handle a specific application, comes in many forms and is typically purchased on a "floppy" magnetic disc. Courseware, including concepts of instruction, sequence of presentation, and feedback construction, is at the heart of the computer's success. No matter how great the capability of a system, if the software is not of good quality, the instructional device will only be of marginal value. Often, the software fails to take into account knowledge about the learning process, because the product is often the brainchild of creative programmers and not of teachers.

For the most part, prepackaged experiences could be developed almost exclusively outside of the school. Textbook publishers, and teaching aid publishers could serve as a nucleus for an industry employing experts in learning theory, developmental psychology, and curriculum theory. However, the private business sector is not organized to make significant investments in the education of the public, except where there exists a distinct profit motive. While it is true that manufacturers have much to gain from computer literacy, it does not follow that they will invest a great deal of research into education.

The federal government should invest substantially in Research and Development.[40] Funds are required to carry out the educational task of reducing the inadequate quantity and quality of software, and the source should be federal. Although the ongoing cost of an educational program must be handled by the state and local school district, the impetus to initiate the program must have a source, one that would enhance the probability of curriculum developments. There is a national need to foster computer literacy and only some type of collective action between levels of government will ensure that all elementary and secondary students are receiving a fair understanding of the uses and applications of the computer. To be assured of equitable distribution and implementation of literacy programs in schools, a public policy, like that advocated by Roger Benjamin, might be a necessity.[41] To date, not even a national priority or funding program has been authorized in the United States.

Evidence of federal support of new national curriculum efforts tends to strengthen the idea that the federal government can prime the pump for commercial innovation. A study by the American Institute for Research (AIR) found that federal agencies supported 48 percent of all high impact educational products, while those products funded solely with private monies comprised only 24 percent.[42] They concluded that successful innovations of private

347

enterprise typically occur when they are built on prior development supported by federal sources. If new innovative products directly supported by government funds traditionally have been followed by private investment, the recognition of government funding is surely instrumental in computer exploitation. What is needed is a federal development effort that will reduce the uncertainty and risk related to new innovation.[43]

The economic feasibility of computers for every child is also a concern within the computer literacy argument. Seymour Papert in his text, Mindstorms, estimates that personal computers may stand to make education cheaper in the long run. He has calculated that "computer costs" for the class of 2000 would represent only about 5 percent of the total public expenditure on education, and this would be the case even if the remaining structure of education did not change.[44] Schools might even be able to increase class size by two or three students or reduce their school cycle from thirteen to twelve years, thus recouping the immediate cost of the computers. If productivity in education can improve with the use of such technology, society should move in that direction.

It is unlikely that computers will reduce the high cost of education in the short run. But then again, how cost effective is it to teach skills that are obsolete, for jobs that no longer exist, in a society that demands knowledge about computers? We should be asking what is basic in a predominantly information society. Computer education can be a most relevant part of the public school curriculum. Students should learn about the capabilities of computers if only to understand the role of this machine in modern society and to protect themselves against computer abuse - feeling inferior to the modern-day computer.

The Value of Computer Literacy

Schools may provide the only opportunity many students will have to learn computer skills. A tragedy it would be if schools of the inner-city and working class communities decide that they cannot afford to purchase computers or that the development of computer literacy is not a priority. If exposure to the computer only exists in the CAI mode, in which the computer tells the student what to do, the instruction may indirectly produce a "second class citizen" in a computer-based society. The deprivation effects of computer literacy forms a barrier that surrounds the ghetto child. These children who grow up in cities have a good reason to believe that the computer technology belongs to "others"; the education rich, who are eventually reduced to social enemies. If this trend continues, we could find ourselves with a generation of children divided between those who will feast in the huge range of possibilities that the computer can provide, and those who remain rooted in the ignorance of the past.

Although many educators take the proliferation of computers and their applications seriously and support some form of computer literacy, there is still no agreement as to which way is the best route in achieving that goal. However, this is no reason to divorce ourselves of the responsibility to introduce students to such a program early in their schooling. On the contrary, better training techniques or at least effective ways to reach computer literacy will be more readily at our grasp, as we determine what needs we must meet. It is essential that the growing disparity between the "haves" and the "have nots" of computing be addressed. For example, 44 percent of the current job vacancies in New York City are for white collar jobs, and only 13 percent of the unemployed in three major ghetto areas of New York City - Harlem, East Harlem, and Bedford Stuyvesant - have ever had a white collar job at any time.[45] What type of future do succeeding generations from this group have without access to knowledge of computers? Technological change is an especially relevant dilemma to 20 percent of our population, the poor. Andrew Hacker, associate professor of political science at Cornell University, explains the predicament of the under-educated or computer illiterate.

> Corporate America has escaped open attack because the victims of new technology do not yet outnumber its beneficiaries. But technology advances according to rules of its own and loyalty to new dominant institutions may diminish if accelerated automation or economic reverses reduce the corporate constituency. In this event, this second America, the society of losers, may grow in numbers and power with increasing rapidity. The resolution will not be a pleasant one.[46]

As a user of economic resources, public education reached the point where it must recognize some ultimate limitation of resource capacity to support education. Formal education will undoubtedly experience very severe financial strains in the next decade. The citizens of the United States seem to be approaching the maximum percentage of their income they are willing to spend for education (currently about 8.5 to 9 percent of GNP). "In round numbers, the per-pupil cost of public education in the United States has been rising at 10 percent per year compounded over the last ten years. Whereas the cost of computer technology has been dropping at the rate of about 25 percent per year for the last twenty years."[47] A prime reason why our educational share of the fiscal "pie" has continually increased is because education has been so highly labor-intensive rather than capital-intensive. Teaching has changed very little in the millennium, the only significant difference being the greater number of human brains that are subjected to the process.

The concern of education in the past was to provide a standard equitable education. But what we are finding today is that education should be individualized. Opportunities are offered to make schooling appear unequal, providing especially for the handicapped or the deprived in the best interests of all youth. Education will not serve the future well if the curriculum indoctrinates learners, confines them to past dogma, and refuses to take advantage of the systematic introduction of computers. Computers will not only personalize education for the diverse population in the public schools of the United States, but may make education more productive, and eventually less expensive.

With the almost infinite capacity for computers to catalog materials, to record pupil progress, and to deliver and receive prescribed educational stimuli to and from children, experiences themselves should be more interesting and certainly varied beyond the present teacher, text, and test. But with the new possibilities, the task of education is becoming more complex and difficult, obscuring the ultimate purposes and goals of the schools. With the approaching twenty-first century, one should be aware of the goals in education for a changing society. Primarily, education should provide for the well-being of the individual pupil and student, and his capacity to lead a productive and happy life. Because of the pace of transition in society, we should devote our attention to eliminating forms of schooling which lock young people into a vocational channel where they are mis- or underemployed. Education is also concerned with the strength of a nation, including industrial and commercial power, its economic integrity, political wisdom and military competence, and its total strength for security and survival.[48] The United States is the world leader in manufacturing microelectronic hardware, and education should concern itself with furthering this growth by educating the public in the use of the microcomputer. Mass computer literacy will not only increase the market for a leading United States industry, but more importantly, it will open gates to the public of both today and tomorrow, to the personal and occupational benefits of knowing how to use a computer.

Today, microcomputer innovations are fueled by a great deal of enthusiasm and by the conviction that they are a good thing. However, we need to acquire more direction in formulating a program to guide further developments. Several projects should be funded so that alternative models aimed at different age levels, using different content and methodology can be developed, evaluated, and offered to schools. Possibly what we need is a task force of teachers and administrators, to begin charting policies and goals toward a kindergarten through twelfth grade computer curriculum with a purpose. This curriculum should be implemented by teachers, teachers that are prepared by an inservice training to assure adequate staffing. The technological changes occurring in society are also affecting the schools and if not anticipated, may result in a failure to innovate in the valuable field of education. We must plan for the needs of society. Whether we will end up safe or sorry depends to a very great extent on the most practical educational aim for a post-industrial future, computer literacy.

End Notes

[1]Stuart D. Milner, "Teaching Teachers about Computers: A Necessity for Education," in Microcomputers in Schools, ed. James L. Thomas (Phoenix: The Oryx Press, 1981), p. 109.

[2]James L. Poirot, Computers and Education (Austin: Sterling Swift, 1980), p. 1.

[3]Christine Doerr, Microcomputers and the 3 R's (Rochelle Park: Hayden Book Co., 1979), p. 2.

[4]Hoo-min D. Toong and Amar Gupta, "Personal Computers," Scientific American 247 (December, 1982), 87.

[5]Robert Nomeland, "Some Considerations in Selecting a Microcomputer for School," in Microcomputers in the Schools, ed. James L. Thomas (Phoenix: The Oryx Press, 1981), p. 17-19.

[6]Jeanne Rieha, quoted in Chris A. DeYoung and Richard Wynn, American Education, (New York: McGraw Hill Book Company, 1972), p. 392.

[7]Murray Laver, Computers and Social Change (New York: Cambridge University Press 1980).

[8]Gary Benedict, "Educational Technology - Fad or Educational Renaissance?" NJEA Review 54 (1981), 36-37.

[9]William E. Doll, Schooling in a Post-Industrial Society (Educational Resources Information Center, 1977), p. 7.

[10]William Sharken and John E. Goodman, "Improving the Climate for Educational Technology," Instructional Innovator 27 (1982), p. 13.

[11]Edward J. Willett and Austin D. Swanson, Modernizing the Little Red Schoolhouse (Englewood Cliffs: Educational Technology Publications, 1979), p. 70.

[12]John Diebold, Man and the Computer (New York: Frederick A. Praeger Inc., 1969), p. 31.

[13]David E. Sanger, "The Computer Develops Some Glitches", New York Times,Education Winter Survey, 9 January 1983, p. 3.

[14]Doerr, p. 12.

[15]Neil Postman, "Curriculum Change and Technology," To Improve Learning: An Evaluation of Instructional Technology, ed. Sidney E. Tickton (New York: R.R. Bowker Co., 1971), p. 282.

[16]Rachelle S. Heller and C. Dianne Martin, Bits 'n Bytes About Computing (Rockville: Computer Science Press, 1982), p. 2.

[17]Carol Krucoff, "Education: The Computer Gap", Washington Post, 1 November 1982, sec. C, p.5.

[18]Introduction to James L. Thomas, Microcomputers in the Schools (Phoenix: The Oryx Press, 1981), p. XIV.

[19]Peggy Garvey, "School Offers Instruction in Computer Literacy," Today, (1982) p. 29.

[20]Martin Ringle, "Computer Literacy: New Directions and New Aspects," Computers and People, Nov.-Dec. 1981, p. 14.

[21]Andrew Molnar, The Next Crisis in American Education: Computer Literacy, (Arlington: Computer Microfilm International, 1981), p. 12.

[22]Molnar, p. 12.

[23]Sanger, "The Computer Develops Some Glitches," p. 3.

[24]Ibid.

[25]Charles E. Silberman, "Technology is Knocking at the Schoolhouse Door," The World of the Computer, ed. John Diebold (New York: 1980) p. 218.

[26]Chris Elliot, "The Latent Computer Literates," Media and Methods 19 (1982): 24.

[27]Adeline Naimen, Microcomputers in Education: An Introduction (Cambridge: Technical Education Research Centers, Inc., 1982), p. 17.

[28]Naimen, p. 19.

[29]Seymour Papert, Mindstorms: Children, Computers and Powerful Ideas (New York: Basic Books, 1980).

[30]Silberman, p. 220.

[31]John A. Starkweather, "Adaptive Machine Aids to Learning," in To Improve Learning: An Evaluation of Educational Technology, ed. by Sidney G. Tickton, II (New York: R.R. Bowker, 1971), p. 362.

[32]Thomas, p. XIV.

[33]Daniel Bell, The Coming of Post-Industrial Society, (New York: Basic Books, 1976), p. 347.

[34]Seymour Papert, "Personal Computing and Its Impact on Education," in The Computer in School: Tutor, Tool, Tutee, ed. by Robert Taylor (New York: Teachers College Press, 1980), p. 199.

[35]Thomas Dwyer, "Heuristic Strategies for Using Computers to Enrich Education," in The Computer in School: Tutor, Tool, Tutee ed. by Robert Taylor (New York: Teachers College Press, 1980), p. 88.

[36]Papert, "Personal Computing and Its Impact on Education," p. 200.

[37]Neil Postman, "Curriculum Change and Technology," in To Improve Learning: An Evaluation of Educational Technology ed. by Sidney G. Tickton, II (New York: R.R. Bowker, 1971), p. 284.

[38]"Microcomputers in Computer-Assisted Instruction," Computers and People, September-October 1982, p. 27.

[39]"Microcomputers Become Students' Tool," Guidepost; APGA, November 25, 1982, p. 12.

[40]Arthur Luehrmann, "Technology in Science Education," To Improve Learning: An Evaluation of Instructional Technology, ed. by Sidney E. Tickton, I (New York: R.R. Bowker Co., 1971), p. 284.

[41]Roger Benjamin, The Limits of Politics, (Chicago: The University of Chicago Press, 1980), See Chapter 9. Benjamin, professor of political science at the University of Minnesota, asserts that changes in the structure of political institutions will resolve some of the problems relating to the crisis of the Post-Industrial Society. The challenge of Post-Industrial Society brings challenges to the government, and the fundamental change of the political structure points toward collective action. A redesign of political institutions will undoubtedly hold implications for areas within education, especially new programs of computer literacy.

[42]Andrew Molnar, "National Policy Toward Technological Innovation and Academic Computing," The Journal; Technological Horizons in Education, 4(1977), 42.

[43]Molnar, p. 42.

[44]Papert, p. 17.

[45]John Diebold, Man and the Computer, (New York: Frederick A. Praeger Inc., 1969), p. 32.

[46]Andrew Hacker, The End of the American Era, (New York: Antheneum, 1980), p. 76.

[47]Arthur B. Shostak, "The Coming Systems Break: Technology and Schools of the Future," Phi Delta Kappan, (January, 1981), 357.

[48]Sterling M. McMurrin, "Technology and Education," To Improve Learning: An Evaluation of Instructional Technology, ed. by Sidney E. Tickton II (New York: R.R. Bowker Co., 1971) p. 284.

Bibliography

Bell, Daniel. The Coming of Post-Industrial Society. New
York: Basic Books, Inc., 1976.

Benedict, Gary. "Educational Technology - Fad or
Educational Renaissance?" in NJEA, 54 (May 1981),
36-37.

Benjamin, Roger. The Limits of Politics. Chicago:
University of Chicago Press, 1980.

Bitzer, Donald L. and Dominic Skaperdas. "The Design of an
Economically Viable Large-Scale Computer-Based
Education System," To Improve Learning: An Evaluation
of Educational Technology. Edited by Sidney Tickton.
Vol. II. New York: R.R. Bowker, Co., 1971.

Chi Chen, Tien. "Computing Power to the People - A
Conservative Ten-Year Projection," Computers and
Communications, edited by Robert J. Seidel and Martin
L. Rubin. New York: Academic Press Inc., 1977.

Dean, Jay W. "What's Holding Up the Show?" Today's
Education, April-May, 1981, p. 21-23.

Dede, Chris. The Next Ten Years in Education. Washington,
D.C.: National Institute of Education, 1980.

DeYoung, Chris A. and Richard Wynn. American Education.
New York: McGraw-Hill Book Company, 1972.

Diamantis, Rhea. "Micro Phobia." The Annual of the NEA:
Today's Education, 71 (1982), 91-93.

Diebold, John. Man and the Computer. New York: Frederick
A. Praeger, Inc., 1969.

Doerr, Christine. Microcomputers and the 3 R's. Rochelle
Park: Hayden Book Company, 1979.

Doll, William E., Jr. Schooling in a Post-Industrial
Society. Washington, D.C.: National Institute of
Education, 1980.

Dwyer, Thomas. "The Fundamental Problem of

Computer-Enhanced Education and Some Ideas about a Solution." The Computer in School: Tutor, Tool, Tutee. Edited by Robert Taylor. New York: Teachers College Press, 1980.

_____. "Heuristic Strategies for Using Computers to Enrich Education." The Computer in School: Tutor, Tool, Tutee. Edited by Robert Taylor. New York: Teachers College Press, 1980.

Elliot, Chris. "The Latent Computer Literates." Media and Methods, September, 1982, pp. 24-28.

Evans, Christopher. The Micro Millenium. New York: The Viking Press, 1979.

Fuller, R. Buckminster. "Education Automation: Freeing the Scholar to Return to his Studies." The World of the Computer, edited by John Diebold. New York: Random House, 1973.

Gallup, George. "Gallup Poll of the Public's Attitudes Toward the Public Schools." Phi Delta Kappan, 64 (September, 1982), 37-50.

Garvey, Peggy, "School Offers Instruction in Computer Literacy," Today, April 1982.

Goodlad, John L. "Education and Technology," To Improve Learning: An Evaluation of Educational Technology. Edited by Sidney G. Tickton. Vol. II. New York: R.R. Bowker Comp., 1971.

Hacker, Andrew. The End of the American Era. New York: Antheneum, 1980.

Hartman, Edward. "The Cost of CAI." The Computer and Education, Englewood Cliffs: Educational Technology Publications, 1973.

Hausmann, Kevin. "Statewide Educational Computer Systems: The Many Considerations." Microcomputers in the Schools, edited by James L. Thomas. Phoenix: The Oryx Press, 1981.

Heller, Rachelle S. and C. Dianne Martin. Bits 'N Bytes

About Computing. Rockville: Computer Science Press, 1982.

Johnson, Christopher. "Our Investment in Public Education." Today's Education, February-March 1982, pp. 15-17.

Joiner, Lee Marvin; Miller, Sidney R., and Burton J. Silverstein. "Potentials and Limits of Computers in Schools." Microcomputers in the Schools, edited by James L. Thomas. Phoenix: The Oryx Press, 1981.

Krucoff, Carol. "Education: The Computer Gap," The Washington Post, Nov. 1, 1982, C 5.

Kurland, Norman D. "Educational Technology in New York State," To Improve Learning: An Evaluation of Educational Technology. II. Edited by Sidney Tickton, New York: R.R. Bowker Company, 1971.

Laver, Murray. Computers and Social Change. New York: Cambridge University Press, 1980.

Levin-Epstein, Michael. "Plato, It's Not Greek in Baltimore." Media and Methods, November, 1981, pp. 7-8.

Luehrmann, Arthur. "Pre- and Post-College Computer Education," Computer in the School: Tutor, Tool, Tutee, edited by Robert Taylor. New York: Teachers College Press, 1980.

_____. "Technology in Science Education." The Computer in the School: Tutor, Tool, Tutee, edited by Robert Taylor. New York: Teachers College Press, 1980. pp. 149-157.

McMurrin, Sterling M. "Technology and Education." To Improve Learning: An Evaluation of Educational Technology. Edited by Sidney G. Tickton. Vol. II. New York: R.R. Bowker Co., 1971.

"Microcomputers Become Students' Tool." Guidepost; (APGA), November 25, 1982, 12.

"Microcomputers in Computer-Assisted Instruction." Computers and People, September-October, 1982, 27.

Miller, James G. "Deciding Whether and How to Use Educational Technology in the Light of Cost-Effectiveness Evaluation." To Improve Learning: An Evaluation of Educational Technology. Edited by Sidney G. Tickton. Vol. II. New York: R.R. Bowker Co., 1971.

Miller, Inabeth, "The Micros are Coming." Microcomputers in the Schools, edited by James L. Thomas. Phoenix: The Oryx Press, 1981.

Milner, Stuart. "Teaching Teachers about Computers: A Necessity for Education." Microcomputers in the Schools. Edited by James L. Thomas. Phoenix: The Oryx Press, 1981.

Molnar, Andrew R. "Critical Issues in Computer-Based Learning," The Computer and Education, The Educational Technology Review Series. IX. Englewood Cliffs: Educational Technology Publications, 1973.

_____. "National Policy Toward Technological Innovation and Academic Computing." The Journal; Technological Horizons in Education. IV (1977), 39-43.

_____. The Next Crisis in American Education: Literacy. Arlington: Computer Microfilm International. Co.: Jan. 1981.

Mondale, Walter F. "The Challenges Facing American Education." Academic. 68 (September-October, 1982), 8-12.

Naiman, Adeline. Microcomputers in Education: An Introduction. Cambridge: Technical Education Research Centers, Inc. 1982.

Needs of Elementary and Secondary Education in the 1980's. A Compendium of Policy Papers. 96th Congress, 2D Session. Congress of U.S., Washington, D.C. House Committee on Education and Labor. Jan. 1980.

Nomeland, Ronald. "Some Considerations in Selecting a Microcomputer for School." Microcomputers in the

<u>Schools</u>. Edited by James L. Thomas. Phoenix: The Oryx Press, 1981. 17-19.

Oettinger, Anthony G. <u>Run, Computer Run; The Mythology of Educational Innovation.</u> Cambridge: Harvard University Press, 1969.

Papert, Seymour. <u>Mindstorms</u>. New York: Basic Books, 1980.

_____. "Personal Computing and Its Impact on Education," <u>Computers in the Schools: Tutor, Tool, Tutee</u>, ed. Robert Taylor, New York: Teachers College Press, 1980. 197-202.

Poirot, James L. <u>Computers and Education.</u> Austin: Sterling Swift, 1980.

Postman, Neil. "Curriculum Change and Technology," <u>To Improve Learning: An Evaluation of Educational Technology</u>, II. ed. Sidney G. Tickton. New York: R.R. Bowker, Co., 1971. 281-288.

Ringle, Martin. "Computer Literacy: New Dimensions and New Aspects," <u>Computers and People.</u> (Nov.-Dec. 1981), 12-15.

Sanger, David E. "The Computer Develops Some Glitches," <u>New York Times Education Winter Survey</u>, January 9, 1983.

Schimming, B.B. "A Case for Information Literacy," <u>Proceedings of NECC/2 National Educational Computing Conference 1980</u>. Iowa: The University of Iowa, 1980, 58-61.

Seidel, Robert J. "Is CAI Cost/Effective? - The Right Question at the Wrong Time," <u>The Computer and Education</u>, Englewood Cliffs: Educational Technological Publications, 1973.

Shane, Harold G. "Social Change and Educational Outcomes 1980-2000," <u>Education: A Time for Decisions</u>, ed. Kathleen Redd and Arthur M. Harkins. Washington, D.C.: World Future Society, 1980.

Shane, Harold G., and M. Bernadine Tabler. <u>Educating for a</u>

New Millenium. Bloomington: Phi Delta Kappan Educational Foundation, 1981.

Sharken, William and John E. Goodman. "Improving the Climate for Educational Technology," Instructional Innovator, 27 (1982), 12-13.

Sheingold, Karen. Issues Related to Implementation of Computer Technology in Schools: A Cross-Sectional Study. New York: Bank Street College of Education, 1981.

Shostak, Arthur. "The Coming Systems Break: Technology and Schools of the Future," Phi Delta Kappan. January 1981, 357.

Silberman, Charles E. "Technology is knocking at the Schoolhouse Door," The World of the Computer, ed. John Diebold. New York: 1980, 203-225.

Starkweather, John A. "Adaptive Machine Aids to Learning," To Improve Learning: An Evaluation of Educational Technology, I. ed. Sidney G. Tickton. New York: R.R. Bowker, Co., 1971.

Suppes, Patrick. "The Teacher and Computer-Assisted Instruction," The Computer in School: Tutor, Tool, Tutee. Edited by Robert Taylor. New York: Teachers College Press, 1980.

Sweetman, George. "Computer Kids, The 21st Century Elite." Science Digest, November, 1982, pp. 84-88.

Toong, Hoomin D. and Amar Gupta. "Personal Computers," Scientific American, 247 (1982), 87-107.

Watt, Daniel H. "Computer Literacy: What Should Schools be Doing About It?" Microcomputers in the Schools. Edited by James L. Thomas. Phoenix: The Oryx Press, 1981.

Willett, Edward J., Swanson, Austin D., and Eugene A. Nelson. Modernizing the Little Red Schoolhouse. Englewood Cliffs: Educational Technology Publications, 1979.

THE RELEVANCE OF A LIBERAL ARTS EDUCATION

IN A POST-INDUSTRIAL SOCIETY

David McCarthy

A liberal arts education is relevant for a post-industrial society because of its practicality.[1] One can obtain the skills and confidence needed to deal with problems as well as the ability to continue learning for the rest of one's life from a liberal arts education. While the aims of liberal arts education may seem a bit idealistic or perhaps unfounded, the current evidence on its effects shows that students are affected in measurable ways. Besides training the mind for analytic and logical reasoning, liberal arts education also affects non-classroom behavior. It helps citizenship, economic productivity, and family life. Evidence shows that it can even help in the stabilization of values. Current evidence suggests that liberal arts education is not only relevant but necessary in a post-industrial society.

The paradox of theoretical knowledge being practical dates back to the first piece of legislation enacted in America dealing with education. Passed in 1642, The Old Deluder Satan Act was the first law in the English-speaking world ordering children to be taught to read. Children were specifically required to read the Bible; this world "delude" the influence of Satan in the new world. They were also required to cipher (learn arithmetic) and learn to survey. Because of the vast tracts of uncharted land in the colonies, surveying was a necessary (practical) skill. The debate between practical and theoretical knowledge has never been resolved and the pendulum continues to swing between the two poles. The emphasis on practicality and relevance that was so demanded by students in the 1960's is now being replaced by the call for a more traditional common educational experience as witnessed by Harvard's reintroduction of a core curriculum and distribution requirements in 1979.

Educators claim that the purpose of a liberal arts education is to "liberate" the individual by honing his capacity for rational thought as well as by rounding out his personality. Liberal arts education should aid in the development of logical and precise thinking and foster a clear use of language - both spoken and written. The student should also acquire an understanding of the vast range and diversity of human customs, pursuits, ideas, values, and longings. A rigorous introduction to the various intellectual disciplines - the sciences, the social sciences, and the humanities - and an immersion in one discipline is elemental to a liberal education. The student of liberal education should come to the realization that truth is not universal; rather it is subject to change over time and according to perspective. This realization will help the student to free himself fron narrowness and provincialism and hopefully lead him to the joys of lifelong intellectual strength and growth. Exposure to the diversity of reality will make the student more open to the new and foreign and make him less likely to engage in rash generalization or narrow-mindedness. A sense of human fallibility and wise skepticism will characterize the student of liberal education. He will be prepared for a life of effective management of new situations and demands. A sense of creativity based on historical perspective will also mark the student. Finally the values of open-mindedness, personal responsibility, mutual respect for others, empathetic and aesthetic understanding, and even playfulness will characterize the student of liberal arts education.[2]

The human mind must be exercised in order to grow.[3] While this involves an immersion in a number of disciplines, the important thing is not what one knows but how one knows. Daniel Bell insists that college is the only place where a student can gain self-consciousness, historical consciousness, and methodological consciousness. One must be concerned with the "grounds of knowledge," and college must teach "modes of conceptualization, explanation, and verification of knowledge."[4] Finally, the student of liberal education should be given a sense of confidence; when he completes his undergraduate education he must have confidence in his capabilities to understand and cope with the world.[5] He should be able to make connections and see the interrelations between various disciplines and have an overall understanding of the world.

While the aims of liberal arts education seem to be quite liberating and enlightening, it remains to be seen

363

whether these aims are in fact realized. To claim that a liberal arts education frees the intellect from narrowness, simplicity, and provincialism is fine, but without concrete proof these claims remain unfounded. John Dewey adequately expressed this problem when he stated, ". . . if such a search is not (to) lead us into the clouds (we must discover) what actually takes place when education really occurs."[6] We need to ascertain what changes, if any, happen to people who study at a four-year liberal arts college.

Perhaps the best data collected on the changes one undergoes while attending a four-year liberal arts college comes from a series of tests conducted by the psychologists Winter, McClelland, and Stewart. This data clearly shows that students are changed by a liberal arts experience and is documented in the book, A New Case for the Liberal Arts. To date this is the most thorough information available on the subject and it is consistent with the data collected by other researchers like Heath, Astin, and Bowen. Winter, McClelland, and Stewart developed two tests to measure the salient effects of liberal education. The researchers hoped that these two tests would cover the wide range of intended effects of a liberal education. It was also hoped that these effects could be measured in later life, i.e. beyond the college years.

The first test is the "Test of Thematic Analysis" and is used to measure the ability for critical thinking.[7] Critical thinking involves differentiation and discrimination, formation of abstract concepts, integration of these concepts and the ability to make relevant judgments. It also involves evaluation of evidence and revision of the concepts and hypotheses, articulation and communication of concepts, differentiation and discrimination of abstractions, identification of concepts, and comprehension of the logics governing the relationships among these abstract concepts.[8] Taken as a whole these abilities constitute the skill of "advanced concept formation."[9] Coinciding with the capacity to discriminate is the ability to communicate in clear language. A liberally educated person is one who should be able to assimilate a large influx of data, organize it into categories, and then communicate the findings or patterns.

Winter, McClelland, and Stewart are very careful in

drawing a distinction between concept attainment and concept formation. Attainment is simply remembering a concept that someone has already defined; formation is developing and defining a concept. The latter process is much more complex and should be nurtured by a liberal education.[10]

The Test of Thematic Analysis is like a comparison and contrast essay test. It involves integration of a wide range of material as well as selection, organization, and interpretation. To keep the test as fair as possible, the researchers chose two groups of stories that were rather bland. The stories were designed not to favor any particular discipline - either history (by being about a certain period) or biology (by being about organic structures or processes). At the same time the stories had to be complex enough to allow the researchers to "construct an adequate measure of the effects of liberal education."[11] In the Test of Thematic Analysis, subjects were given two different groups of short, imaginative Thematic Apperception Test (TAT) stories with four stories in each group. Subjects were asked to read and then formulate and describe the differences between the two groups. About thirty minutes were given to complete the test. Examples similar to those used in the test are listed in Appendix 1 on page 23.

The test was given to freshman and seniors in the fall of 1974 at Ivy College (Ivy College is described as a small, highly selective, co-educational liberal arts college located in a large city on the east coast). Nine categories were developed to differentiate the freshmen and senior responses. Six categories were more often found in the senior responses and were scored +1 while three categories were more often found in the freshmen responses and were scored -1. The categories are listed in Appendix 2 on pages 23-24.

The nine categories were used to measure a student's critical thinking ability. Possible scores ranged from a -3 to a +6. To keep the scoring as objective as possible two trained scorers worked independently on the same material and reached over 85 percent category agreement.[12] The six senior categories are usually present in "good" college essays and would distinguish them from "bad" essays. The assumption is that over the course of four years, these qualities would be inculcated in the student's rational

365

capacities.

The second test is that of "Intellectual Flexibility in Analysis of Argument." Disciplined flexibility in analytic thinking involves the ability to see all points of view while being committed to or arguing for one point of view. It is the ability to "keep cool under fire" and to realize that truth often varies from one person or group to another. To measure this capability, Winter, McClelland, and Stewart developed the Analysis of Argument Test. Subjects were given a 300-400 word quotation expressing a "strong, extreme position on a controversial and emotional issue."[13] In this particular case, subjects were given excerpts of a 1968 sermon by Norman Vincent Peale, attacking the "permissive" child rearing advice of Dr. Benjamin Spock and linking his advice to moral laxity, race riots, crime, and opposition to the Vietnam War.[14] The researchers assumed that this statement would arouse a strong negative response from the students. After reading the statement the subjects were given five minutes to write a response: "Assume that you are to argue against it, your response can be of whatever form and nature that you think are most satisfactory and appropriate."[15] After completing this assignment the subjects were asked to "Write a defense of the article - including, if you wish, argument against your criticism that you have just written. Again, your defending response can be of whatever form and nature that you think are most satisfactory and appropriate."[16] This latter assignment forced the subject to take an alternate point of view and defend it under the pressure of time and consistency.

As with the first test, this one was given to both freshmen and seniors at Ivy College in the fall of 1974. Again certain categories appeared more often in the senior responses than the freshmen responses. Five categories, three for criticism and two for defense, were common to the senior responses and were scored +1 while five categories, three for attack and two for defense, were more common in the freshmen responses. Each was scored with a -1 value. The response categories are listed in Appendix 3 on page 24.

Again two trained scorers worked independently of one another on the same material and this time reached 94 percent category agreement for these categories. The possible range of scores was from -4 to +6.

The two tests were the primary source of information for assessing the effects of a four year liberal education. They have been thoroughly described here because they are the basis for much of the information on the effects of liberal education and because they are interesting in themselves. The conclusions reached by Winter, McClelland, and Stewart give some good concrete evidence that a liberal education does affect its students. To assess whether these effects occur at other academic institutions, they gave the tests at a two-year community college and a four year state-teacher's college. The results indicate that only a four year liberal arts college is capable of producing these changes in its students. Two areas of cognition are covered by the tests. One is that of intellectual abilities like concept attainment, divergent thinking, sensitivity, and the ability to learn new material. The other area covers qualities of mind such as achievement motivation, fear of success, and self-definition and maturity of adaptation.

In the area of critical thinking and analysis it was found that Ivy College seniors showed greater skill than the freshmen. There was a significant increase from freshman to senior year. Data was collected from the two groups of stories given in the Test of Thematic Analysis. The results from the other colleges showed that there was no significant difference between freshmen and seniors at the state-teachers college or between the freshmen and sophomores at the two year community college. A longitudinal study of the Ivy College freshmen showed that their scores increased, when they were given the test again in their senior year. This suggests that the results of the first test were not simply a difference of intelligence between the freshmen and seniors but that critical thinking and analysis are developed over the course of a four year liberal arts experience.[17]

In the area of concept attainment, the researchers found that there was little difference between the seniors and the freshmen. Both groups scored very well and this would suggest that concept attainment is a product of previous experience.[18]

The Ivy College seniors were more likely to be accurate in their responses in that they contained correct elements, issues, and dichotomies.[19] They were more likely to

367

organize their responses in terms of central themes or ideas. One might raise the question as to whether the differences in correct responses were related to natural intelligence as measured by SAT scores (if indeed SAT scores are indicative of natural intelligence). For the freshmen there was no correlation between SAT scores and thematic analysis while for seniors there was a correlation. Those seniors with high SAT scores had high Thematic Analysis Test Scores. This suggests that at Ivy College those students with high SAT scores gain more (in terms of thematic analysis) than those students with lower SAT scores.[20]

Results of the Analysis of Argument test clearly show that a student's capability to analyze an argument increases over the course of a four year liberal arts education. While scores also increased at the state-teachers college and the community college, these increases were not as great as those at Ivy College. There was an increase in both the attack and defense subscores within the overall scores from the freshmen to seniors. These results are particularly significant in that Ivy College freshmen scored the same or even lower than the freshmen at the other two schools. Longitudinal results show that the Ivy College freshmen increased their scores both for the total score and in the defense subscore. The researchers concluded that the Analysis of Argument test "reflects a generalized ability to reason well in a variety of controversial and emotional situations."[21] The results of the test show that this ability is increased from freshmen to senior year.

The ability to think for one's self is measured by the Self-definition score. This information was gathered from the TAT stories. Individual's who score high in Self-definition view the world in terms of cause and effect relationships. They also see people as acting in ways that have effects on the world around them. At all three schools the older students scored higher than the freshmen. Therefore, the scores could be the result of maturation or of any education. However at Ivy College the difference was the greatest of all the schools. Longitudinal studies done at Ivy College in 1978 support the results. It seems that a liberal education increases the belief that people act in ways that have effects on their world.[22]

Divergent thinking is a type of creativity and it was found that there were no significant differences in gains across time between the three colleges. There is some evidence that indicates divergent thinking to be linked with high intelligence. However at Ivy College there were no significant gains from freshmen to senior year. Perhaps this is because Ivy College does not stress the creative and performing arts as much as it does more traditional forms of thought. Therefore it would be rash to conclude that the liberal arts hinder creative thinking.[23]

Achievement motivation also did not increase at Ivy College. The scores for freshmen and seniors were nearly identical. Achievement motivation is characterized by innovative thought, resistance to conformity, and the use of feedback to improve outcomes. The researchers concluded that achievement motivation might be influenced by social climate and the individual's early life experiences.[24]

The results of the Analysis of Argument test showed that Ivy College students were able to argue and defend a point of view which was different from their own. This is a form of empathy or, more specifically, a form of intellectual empathy. When students were tested for emotional empathy it was found that there were minor differences between the senior and freshmen responses. Perhaps this is because Ivy College students have reached a level of empathetic maturity before entering the college or perhaps Ivy College just produces latter day sophists. Studies should be made at other liberal arts colleges to ascertain whether a liberal education enhances the capacity for emotional empathy.[25]

To measure self-assurance the researchers analyzed the fear of success scores obtained from the TAT. Fear of success was lower in senior year than freshmen year at all three colleges. However there was no change from freshmen to senior year for Ivy College Women. The researchers concluded that there was not enough evidence to explain this finding.[26]

Maturity of adaption (one's psychological adaptation to the environment) showed large and significant differences at all three colleges. This suggests that maturity of adaptation develops with age. There are four levels of adaptation - each one more "mature" than the preceding one. First is the receptive stage where one sees authority as benevolent and sees wishes as immediately gratified. Second is the autonomous stage where authority is not seen as benevolent and wishes are not gratified. Third is the assertive stage where one opposes authority and feels alienated from society. Finally in the fourth stage, the integrative stage, one reaches a sense of balance where authority is seen as limited and one feels a part of the world around him.[27] At Ivy College the increase in adaptation was greater than at the other schools and Ivy College seniors were higher on the scale than the state college seniors and the second year community college students. The Ivy College seniors were very near the fourth or integrative stage. Winter, McClelland, and Stewart concluded that college "contributes in a major way to the development of personal maturity."[28]

In their conclusion, the researchers found that education at Ivy College has various cognitive benefits. It increases students' abilities to form and arrange concepts as well as to communicate them and it enables students to assess an argument from a number of perspectives. Since Ivy College has motivational and emotional effects it gives students the ability to cope with society and also helps them mature. Education at Ivy College affects self-knowledge by teaching students how to take stock of their strengths and weaknesses.[29]

The results of the Winter, McClelland, and Stewart research are rather heavily emphasized here (perhaps too heavily emphasized) because they are one of the few primary sources available on the effects of a liberal education. The research is also up to date (1981) and clearly shows that students are changed by a liberal education. Much of the evidence gathered here is cognitive or theoretical in that it measures qualities of the mind. There is evidence to suggest that liberal education has practical effects as well. Howard R. Bowen, an economist specializing in the economics of higher education, believes there are four practical traits of the liberaly educated person. These traits are the needs for achievement, future orientation, adaptability and leadership.

Althouth the research gathered by Bowen deals with higher, i.e. college education, in general, the findings can be used to support the argument that liberal arts education produces effects in its students that are needed for a post-industrial society. Bowen does cite research showing that the effects of college education are greatest at small liberal arts colleges and this serves to strengthen the data used here. In the following paragraphs "college education" can be read "liberal arts education," although one should realize that the effects produced are the results of college education, not just liberal arts education.[30]

The need for achievement involves such qualities as motivation toward accomplishment, persistence, energy, and drive. Bowen cites the studies of Feldman and Newcomb (1969) showing that college seniors value opportunities for leadership, creativeness, and independence more than freshmen. However, the research in this area is not extensive and it would be premature to conclude that a liberal education increases the need for achievement. Still there is some evidence to suggest that it might.[31]

Bowen suggests that college-educated people are more oriented to the future than are those people without a college education. His conclusion is based on a number of test findings. Morgan, Sirageldin, and Baerwaldt constructed an "index of planning and time horizon" based on a family's decisions about vacations, retirement, and children's education. It was found that in families where the head of the family was college-educated, the index was higher. This suggests that college-educated people are willing to make plans about the future.[32] A study by Solmon (1975)found that college-educated people were more likely to save their money. Another sign of future orientation is the willingness to assume reasonable risk. Bowen found that college seniors were less concerned with job security than freshmen and he suggests that seniors are more willing to try a number of jobs before settling down. Finally studies by Morgan, Sirageldin, and Baerwaldt (1966) and Withey (1971) found that college graduates were more prudent in practical affairs, like family planning and investment of money, which implies a concern for long term well-being. All of the studies seem to point out a concern with the future - future orientation.[33]

College education seems to enhance the trait of adaptability. Bowen cites a number of studies (Morgan, Sirageldin, and Baerwaldt, 1966; Lansing and Mueller, 1967; and Spaeth and Breely, 1970) showing that college-educated people are indeed more adaptable than non-college-educated people. Some of the defining characteristics of adaptability are willingness to change jobs and geographic location, openness to new products, attraction to scientific development, and a concern with progress.[34] Lansing and Mueller studied the relationship between education and geographic mobility and found that those people with higher levels of education were more likely to change geographic location. Morgan, Sirageldin, and Baerwaldt (1966) found that college-educated people were more open to new developments in applied science like flouridation of water and space exploration.[35]

To date there is no conclusive evidence on the effects of college on leadership. Educators like to claim that higher education helps to cultivate leadership but no studies have actually verified this claim. There is no conclusive evidence in either the Bowen or in the Winter, McClelland, and Stewart research. Yet most of our leaders today are college-educated and one would assume that their education helped to mold them into good leaders. Some interesting findings on leadership are given by Winter, McClelland, and Stewart. They found that participation in varsity athletics during college significantly increased leadership motives in women but not in men.[36]

A great deal of evidence has been gathered on the effects of education on citizenship in America. Studies going back to 1938 (Murphy and Likert) show that during the college years students tend to shift toward more liberal political views. This would include endorsement of civil liberties, opposition to discrimination, concern for conservation and foreign affairs, acceptance of activism and civil disobedience, and suspicion of big business and big government.[37] Curtin and Cowan (1975) found that there was a positive correlation between education and support for education, mass transit, pollution abatement, welfare, medicare, and housing for low income families.[38] While some studies find that college students tend to become more politically active during college (Powers, 1976; Yankelovich, 1974), the overwhelming evidence shows that college alumni are much more politically active than college students. Studies by Verba and Nie (1972), Nyman, Wright, and Reed (1975) and Pace (1974) show that college alumni are more likely to become actively involved in political parties and civic affairs than either college students or non-college educated people. If not actually involved, many college alumni keep a close watch on political affairs by reading and watching television. Polls by Gallup (1975) and Yankelovich (1974) show that college-educated people are more likely to vote than those people without a college education.[39]

College education seems to increase community participation, particularly volunteer work. Morgan, Sirageldin, and Baerwaldt (1966) found that an important factor determining social work was college education. Studies by Solmon (1973) suggest that increased education raises the intensity of participation. Finally, a survey of college alumni (Powers, 1976) found that they were more likely than other adults to become involved in service clubs, Parent Teachers Associations, and other school organizations. In conclusion, it appears that college education increases political and community involvement and interest.

Bowen finds that higher education is a boon for economic productivity. He lists six ways in which college education increases worker productivity. First, workers are able to produce more goods per unit of time through greater dexterity or greater drive as the result of increased aspirations or better health. Second, the quality of the product may increased or in the realm of services there may be increased sensitivity in human relations. Third, college-educated workers may be able to produce goods and services that are more highly valued by society. Fourth is an increased desire to become a part of the work force, and time lost from unemployment and illness may diminish. Finally, the fifth area is allocative ability. Bowen defines this as being able "to adjust appropriately and promptly to changing conditions underlying demand and supply and thus to bring about more efficient resource allocation."[40] Finally, job satisfaction may be increased.[41]

The benefits to economic productivity or job competence can be direct. While a liberal arts education does not always give one specific job skills, it does enable one to adapt to new demands and have confidence in one's ability to understand the demands. Frank Machlup (1970) assesses the effects education can have on increased productivity:

It is with regard to . . . improvements in the quality of labor, that education can play a really

373

significant role. Positive effects may be expected on five scores: (a) better working habits and discipline, increased labor efforts, and greater reliability; (b) better health through more wholesome and sanitary ways of living; (c) improved skills, better comprehension of working requirements and increased efficiency; (d) prompter adaptability to momentary changes, especially in jobs which require quick evaluation of new information and, in general, fast reactions; and (e) increased capability to move on to more productive occupations when opportunities arise. All levels of education may contribute to improving the quality of labor.

To the use of better machines, education can contribute in at least two ways: (a) by making people more interested in improved equipment, more alert to its availability, and more capable of using it; and (b) by training people in science and technology and expanding their capacity for the research and development work needed to invent, develop, adapt, and install new machines.[42]

In general terms education contributes to economic productivity by providing confident workers who are able to adjust to new markets and technology and who are willing to move around and be flexible in their career aspirations.

Closer to the home front it has been found that higher education has a strong influence on the family. Education helps to narrow traditional differences between the sexes in interests, attitudes, and behavior patterns and as educational levels rise sex roles become less stringent. College education also affects marriage attitudes with college-educated people tending to marry later in life than non-college-educated people and tending to marry people of similar educational levels. Some researchers suggest that marital happiness is greater among college-educated people.[43]

Divorce rates tend to be lower among college-educated people and family sizes are smaller (U.S. Bureau of the Census, 1975).[44] College-educated people seem to be more concerned about their children and how they are reared.

More time, thought, and money are put into childrearing. Certainly this is a form of future orientation. The children of parents with a college education are more likely to go on to college and graduate school than those children whose parents did not have a college education (Trent and Medsker, 1968).[45]

Douglas Heath has studied the effects of liberal education on the development of values. He found that the principal effects of liberal education were the "stabilization and integration of values."[46] His research was done at Haverford College in Philadelphia, a predominantly male liberal arts college, through the 1960's and 1970's. By studying successive classes of "articulate and reflective" Haverford men and then studying them ten years after graduation, he concluded that the distinctive effect of liberal arts education at Haverford was to permanently alter the motives of many of its students. This was done by fostering a hopeful vision of what the students and society could become that was realistically compatible with the expectations and values of the faculty.[47] The men consistently identified their college education as having a lasting effect on their lives. If this happens at other liberal arts colleges, then the claims that a liberal education can enhance value development can be justified.

In spite of this wealth of information on the effects of liberal education, the question still waiting to be answered is "Does a liberal arts education remain relevant for a post-industrial society?" The answer is yes. Current information shows that students are affected by liberal arts education in measurable and positive ways. The skills and confidence needed to cope with an ever-changing and increasingly complex world can be developed with a liberal education. It is effective training for the future or, as Dr. Samuel Mudd, professor of psychology at Gettysburg College, states, "liberal arts education is a Sancho Panza for future society."[48] It provides a broad-based educational background that is highly adaptable and multifarious enough to meet new demands. Finally, it serves to link our present society with the past epochs of Western Civilization and keeps us firmly rooted in this heritage.

In a 1968 publication, Toward the Year 2000, Harold Orlans spoke of the need for a broad-based undergraduate educational experience (liberal education) to keep communication open between the various and ever-increasing and ever more complex specialties.[49] Liberal education can provide a common language to keep discourse open between the specialties. It can also serve to keep political power in balance. In an age of technology the technocrats can become very powerful because they wield the technologies. A broad-based educational experience enables society at least to understand the technology and then keep the political structure as democratic as possible.

The important word here is communication. Without it the threat of provincialism becomes acute. Our world is becoming increasingly interdependent and the ability to communicate is essential for understanding the complexity and contradiction of world events; communication prevents narrowness, rash generalizations, and ignorance. To be able to communicate one must first be able to understand and then convey the information in understandable, common language. The evidence gathered on cognitive ability shows that a liberal education does increase this ability.

In his book, Art As Experience, John Dewey explains that the "good" art critic helps the spectator to see the work of art in the totality of his (the spectator's) experience as some type of unified whole. The art critic does this through analysis and synthesis. First, he explicates the various parts of the work (analysis) and then he relates the parts to one another and then to the whole work (synthesis). But the good critic goes beyond this; he places the art work in an overall historical context, both within the art world and within the world at large. This not only helps the spectator to understand the work and appreciate it but helps him to understand the world as a unified whole.[50] At its best liberal education can produce "good" critics, i.e. people capable of communicating complex and confusing information in understandable language so that the rest of society can appreciate their world and feel a part of it.

Futurists like to tell us that the future will be bright and rosy - a "horn of plenty" thanks to the wonders of science and technology. Yet as any student of history will point out, the "golden" ages of Western Civilization were often based on exploitation and gross inequalities in the political and economic structure. Usually it was a case of a few "haves" and many "have-nots." There is no reason to believe that post-industrial society will be different. Indeed if one accepts the premise that we are now living in a post-industrial society it becomes obvious that the problems of inequality, exploitation, and poverty are going to be part of the new age. While industrial countries like Japan and the United States will in all probability experience continued economic growth, such third-world nations as Zimbabwe will probably remain poor. As stated above, a liberal education can help by increasing communication and hopefully providing answers to these problems. It can also help by stabilizing and strengthening the family structure (pp. 16-17). Perhaps by rebuilding (strengthening) the basic family unit, society as a whole will become more stable. The positive correlation between liberal education and citizenship will also help to stabilize society.

It seems increasingly clear that in a society where specialization in the work force is necessary, there will be a need for people with a broad overview of the society. These people, to use a familiar cliche, will have to see the "forest" rather than the "trees" or even the "branches and leaves" as may be the case. Daniel Bell maintains that the largest class in the new economic structure will be the professional (specialized white collar workers) and they will not have the overall view of the economic structure that will be needed.[51] At the highest levels of the economic and political structure, in the corporate board rooms and in the congress and senate, there will have to be people with a broad scope of vision. These are the people who will see society as a unified collection of seemingly disparate elements. Their role will be to see the links between these elements and give society a sense of unity. The broad scope of knowledge gained from a liberal arts education would be a good foundation for the corporate elite. They must be trained to handle a vast range of information rather than concentrating on one field.

While there will be an increase in specialization, there will also be an increase in political participation on the local levels. Again Daniel Bell points this out and current trends in American politics show an increase in community action groups.[52] Citizens want to participate in the political process and they want to have a sense of control and even power within society. This will help to change the political structure from a vertical hierarchy to

377

a more horizontal one. The current structure of the government will probably not change, but more people will be involved on the local and perhaps state level. Again a liberal arts education will be needed to give the public a sense of perspective and to give them the skills needed to communicate and make reasonable judgments. It has already been shown that there is a positive correlation between liberal education and citizenship and this should be strengthened, if anything.

On both the federal and local levels there is and will be a need for goals - for a sense of purpose. It is here that a liberal education must have an impact. This can be done by giving one a hopeful outlook for the future by reminding one of the past achievements of man. If one can be given a sense of confidence not only in one's self but also in man, then the future can be embraced with the hope that it will be bright.

The move into post-industrial society will bring an increase in leisure time and with this will come an increase in the need for self-actualization. It will be important that people know how to structure their time so as to move toward self-actualization. They will need to feel a sense of personal fulfillment and self-worth. The ability to assess one's needs and capabilities fostered by liberal education is needed for self-actualization. Also a broad educational experience will expose one to a multiplicity of fields and this cannot hurt the drive for personal fulfillment.

Finally, for liberal education to remain relevant for post-industrial society, it must speak the language of this society. This would mean an increased emphasis on science and information processing. It is now generally accepted that students should have an understanding of computers and computer languages. Still, we should never foresake the belief that the promise of a better future for man rests on his humaneness and this hope of a better life is nurtured by a liberal education.

378

Appendix 1.

Group A Story

A business office - head of business sitting at desk surrounded by his executive staff. News has been received that business has failed. Head of business has been cheating and swindling and the letter he is holding has brought it out into the open. The others are horrified, but at the same time they like their "boss" and want to help him, as well as them, so that news doesn't leak out and ruin firm, their boss, and their jobs. How will it be covered up? An impossible situation. No matter how they cover up, boss and firm can't be saved.

Group B Story

Five men - the speaker is attempting to improve his speech and the four men are helping to give him advice. This man is a public official who has been asked to express his views on a controversial topic. He is taking great care to be prepared. The man looks like he's thought for a moment, hoping the other men could possibly help him with advice. The problem will be resolved - by his work and his associates the speech will be given in clear and accurate form.[53]

Appendix 2.

Senior categories:

1. Parallel Comparison: Some element is ascribed to one group of stories, and in the other group, either it is explicitly not ascribed or a contrasting element is ascribed.

2. Exceptions or Qualifications to any ascription.

3. Examples are used to illustrate an ascription.

4. Overarching Issues: Similar to category 1 except that an overarching issue that unifies the two contrasting ascriptions is mentioned.

5. Redefinition: Redefining an element in order to

379

improve coverage.

6. Subsuming Alternatives: Defining an element disjunctively, with nonsynonymous but functionally equivalent options. An example is a "strike" in baseball, which can be a missed swing, a pitch in the strike zone, or a foul hit.

Freshmen categories:

7. "Apples and Oranges" Nonparallel Comparison: Comparison between groups using two elements that are unrelated (John is tall and Jerry is smart.").

8. Affective Reaction: Comparison based on emotions.

9. Subjective Reaction: Comparison based on the writer's subjective reaction, using first-person singular pronoun.[54]

Appendix 3.

Senior categories:

Attack

1. Central Organizing Principle of the criticism.

2. Focus on Logic and logical errors of the statement.

3. Proposing Distinctions and Exposing Contradictions among elements of the statement, some of which may be accepted while others of which are rejected.

Defense

4. Modified Endorsement: Reworking or delimiting the original statement so as to defend it.

5. Acceptance of Particular Arguments: Singling out of some elements (rather than all) of the original statement for defense.

Freshmen categories:

Attack

6. String of Criticisms: A series of criticisms that are not organized around a central principle, insight, or focus.

7. Focus of Attack on Facts: Debate about whether the contents of the statement are true.

8. Proposing Counter-Facts: Opposing arguments are given, but without evidence or other support.

Defense

9. Total Endorsement: Simple reversal from previous criticism to a fairly global endorsement (without regard to any inconsistencies thereby created).

10. Proposing New Arguments: Introduction of new arguments not even given in the original statement.[55]

ENDNOTES

[1]A liberal arts education is a broad-based undergraduate experience that includes the study of language, philosophy, history, literature, physical and social science, and mathematics. It provides general knowledge and develops general capacities of the mind like reason and judgment rather than developing specific professional or vocational skills.

[2]Gettysburg College Catalogue 1982/1983, (Gettysburg: Gettysburg College, 1982), pp. 14-17.

[3]Theodore Hesburgh, "The Future of Liberal Education," Change, April 1981, p. 38.

[4]Daniel Bell, The Reforming of General Education, (New York: Columbia University Press, 1966), p. 8.

[5]Samuel A. Mudd, from an informal discussion with the Senior Scholars' Seminar at Gettysburg College on November 4, 1982.

[6]Douglas Heath, "What the Enduring Effects of Higher Education tell us about Liberal Education," Journal of Higher Education 47 (March/April 1976): 173.

[7]David Winter, David McClelland, and Abigail Stewart, A New Case for the Liberal Arts, (San Francisco: Jossey-Bass, 1982), p. 27.

[8]Ibid., p. 12.

[9]Ibid., p. 27.

[10]Ibid., p. 28.

[11]Ibid., p. 29.

[12]Ibid., p. 32.

[13]Ibid., p. 32.

[14]Ibid., p. 33.

[15]Ibid., p. 33.

[16]Ibid., p. 32.

[17]Ibid., p. 63.

[18]Ibid., p. 63.

[19]Ibid., p. 64.

[20]Ibid., pp. 64-65.

[21]Ibid., p. 66.

[22]Ibid., p. 70.

[23]Ibid., pp. 70-71.

[24]Ibid., p. 71.

[25]Ibid., p. 72.

[26]Ibid., p. 75.

[27]Ibid., p. 48.

[28]Ibid., p. 79.

[29]Ibid., p. 81.

[30]Howard Bowen, _Investment in Learning_, (San Francisco: Jossey-Bass Publishers, 1978), p. 248.

[31]Ibid., pp. 138-139.

[32]Ibid., p. 139.

[33]Ibid., pp. 139-140.

[34]Ibid., pp. 140-141.

[35]Ibid., p. 141.

[36]Winter, McClelland, and Stewart, _A New Case for the Liberal Arts_, p. 131.

[37]Bowen, _Investment in Learning_, pp. 143-144.

[38]Ibid., p. 148.

[39]Ibid., p. 154.

[40]Ibid., pp. 159-60.

[41]Ibid., pp. 159-60.

[42]Ibid., pp. 162-63.

[43]Ibid., pp. 189-191.

[44]Ibid., p. 192.

[45]Ibid., p. 197.

[46]Heath, "Enduring Effects of Higher Education," p. 173.

[47]Ibid., p. 180.

[48]Mudd, Senior Scholars' Seminar lecture.

[49]Daniel Bell, gen. ed., _Toward the Year 2000, Work in Progress_, (Boston: Houghton Mifflin Company, 1967), _Educational and Scientific Institutions_, by Harold Orlans, pp. 191-192.

[50]John Dewey, _Art As Experience_, (New York: Paragon Books, 1979), pp. 298-325.

[51]Daniel Bell, _The Coming of Post-Industrial Society_, (New York: Basic Books, Inc., 1973), pp. 121-65.

[52]Ibid., pp. 339-69.

[53]Winter, McClelland, and Stewart, _A New Case for the Liberal Arts_, pp. 29-30.

[54]Ibid., pp. 30-31.

[55]Ibid., p. 34.

BIBLIOGRAPHY

Astin, Alexander W. _Four Critical Years, Effects of College on Beliefs, Attitudes, and Knowledge._ San Francisco: Jossey-Bass Publishers, 1977.

Bell, Daniel. _The Coming of Post-Industrial Society, A Venture in Social Forecasting._ New York: Basic Books Inc., Publishers, 1973.

_____. _The Reforming of General Education._ New York: Columbia University Press, 1966.

Bevan, John M. "Reflections and Projections: 4-1-4." _Liberal Education,_ Vol LIX, No. 3, (October 1973), pp. 336-348.

Bowen, Howard R. _Investment in Learning, The Individual and Social Value of American Higher Education._ San Francisco: Jossey-Bass Publishers, 1978.

Boyer, Carol M. and Ahlgren, Andrew. "'Visceral Priorities' in Liberal Education." _Journal of Higher Education,_ March/April 1982, pp. 207-15.

Brodbelt, Samuel. "Education For An Interdependent Future." _The Social Studies,_ January/February 1979, pp. 11-15.

Brubacher, John S. _On the Philosophy of Higher Education._ San Francisco: Jossey-Bass Publishers, 1978.

Bullough, Robert V., Jr. "General Education: Bode and the Harvard Report." _The Journal of General Education_ 33 (Summer 1981): pp. 102-112.

Carnegie Commission on Higher Education. _A Digest of Reports of the Carnegie Commission on Higher Education._ New York: McGraw-Hill, 1974.

_____. _Missions of the College Curriculum._ San Francisco: Jossey-Bass Publishers, 1977.

_____. _More Than Survival, Prospects for Higher Education in a Period of Uncertainty._ San Francisco: Jossey-Bass Publishers, 1975.

Chandler, John W. "The Liberal Arts College in Post-Industrial Society." Educational Record 55 (Spring 1974): pp. 126-130.

Change Magazine. The Third Century, Twenty-Six Prominent Americans Speculate on the Educational Future. New York: Change Magazine Press, 1977.

_____. "General Education: A Proper Study." Change, September, 1981, pp. 29-58.

Cobb, Wm. Daniel. "Manifold Reality and Integrative Assumptions: A Perspective on Liberal Education." Liberal Education LVIII (October 1972): pp. 370-380.

Combs, Arthur W. "What the Future Demands of Education." Phi Delta Kappan, January 1981, pp. 369-372.

Commission on the Humanities. The Humanities in American Life. Berkeley: University of California Press, 1980.

Daedalus. Toward the Year 2000, Work in Progress. Edited by Daniel Bell. Boston: Houghton Mifflin Company, 1967.

Dewey, John. Art As Experience. New York: Paragon Books, 1979.

Dewey, John. Democracy and Education. New York: The MacMillan Company, 1917.

Gettysburg College 1982/83, Catalogue Issue 1982/83. Gettysburg: Gettysburg College, 1982.

Hanson, Desna W. "New Directions in General Education." Journal of Higher Education 33 (Winter 1982): pp. 249-262.

Harmon, Christopher C. "Liberal Education Should Do More Than Just Liberate." The Chronicle of Higher Education. October 14, 1981, p. 24.

Harvard Committee. General Education in a Free Society. Cambridge: Harvard University Press, 1945.

Heath, Douglas H. Growing Up in College, Liberal Education and Maturity. San Francisco: Jossey-Bass Publishers, 1968.

_____. "What the Enduring Effects of Higher Education tell us about Liberal Education." Journal of Higher Education 47 (March/April 1976) pp. 173-189.

Hesburgh, Theodore M. "The Future of Liberal Education." Change, April 1981, pp. 36-40.

Miller, William C. The Third Wave and Education's Futures. Bloomington: Phi Delta Kappa Educational Foundation, 1981.

Mood, Alexander M. The Future of Higher Education. New York: McGraw-Hill, 1973.

Morse, H. T. General Education in Transition, A Look Ahead. Minneapolis: The University of Minnesota Press, 1951.

Newman, John Henry. The Uses of Knowledge. Arlington Heights: AHM Publishing, 1948.

Nicoll, G. Douglas. "Liberal Learning, History and the 'Bell Thesis,'" Liberal Education LVIII (October 1972): pp. 317-327.

Pelikan, Joroslav. "The Liberation Arts." Liberal Education LIX (October 1973): pp. 292-297.

Tetlow, Joseph A. "A Principle of Determinacy For Liberal Education." Educational Record 55 (Winter 1974): pp. 23-28.

Wegener, Charles. Liberal Education and the Modern University. Chicago: University of Chicago Press, 1978.

Winter, David G., McClelland, David C., Stewart, Abigail J. A New Case for the Liberal Arts. San Francisco: Jossey-Bass Publishers, 1981.

CHRIST AND POST-INDUSTRIAL SOCIETY:

IS FAITH RELEVANT?

Brian Martin

PREFACE:

The question this study attempts to answer has been largely ignored by students of post-industrial society. For many social scientists, the demise of religion and the triumph of the secular have become foregone conclusions. Christian theologians, on the other hand, have almost always taken Christ's timeless relevance for granted. The purpose of this study is to give an affirmative answer to the question posed in the title, and to persuade students of post-industrial society to consider Christian belief and practice as important aspects of that society.

As a Christian I stand with Christian theologians who contend that belief in Jesus Christ has relevance for all peoples and for all ages. I do not intend this to be simply a statement of faith, but seek to provide a reasonable explanation why I believe that Jesus Christ is relevant to post-industrial man. My approach will be on both a sociological and theological level.

This study will be limited to Christian religion in the United States. The primary reason for focusing on this country is that it is the most generally accepted example of a post-industrial society. A secondary reason is that there are more Christians in the United States than in any other nation, thus allowing for a better sociological examination of the thesis question.

Although several scholarly writings have some bearing on the subject, little synthesis has been done which directly addresses the question. This lack of synthesis has forced me to erect my own conceptual framework. At times I have found this framework awkward, but I have tried to provide as systematic an approach as possible. My use of sociological and theological jargon has been unavoidable, but I have attempted to define the most salient terms. In the course of the study I have pursued several apparent digressions, such as an examination of some of the secular theological thought of the nineteen-fifties and sixties. Yet, I hope these departures from the main line of argument will enhance the understanding of the larger issue. Despite its exploratory nature, this study seeks to convince students and scholars alike that to ignore the Christian faith as a significant part of post-industrial society risks offering an incomplete conceptual rendering of that society.

Does religion in general, and Christianity in particular, have a place in post-industrial America? Such a question cannot be approached in a purely objective manner, simply because religion is viewed subjectively from either a sociological or a theological point of view. Many scholars apparently have taken the triumph of the secular as a given when developing a conception of post-industrial society. Christianity, the most prevalent religion in American post-industrial society, is not merely a residual aspect from a previous society. A case can be made from both the sociological and theological perspectives for the persistence of Christianity, not simply as an ailing social institution, but as a vigorous part of our post-industrial society.

Scholars have found it difficult to reach an agreement on either a sociological or a theological definition of religion. For the purpose of this study, "religion" will refer to that which provides meaning to mankind's existence in relation to that which is both within and beyond man. This description of religion can be seen in a homocentric or theocentric configuration.

The homocentric view essentially states that it was man who invented God, and therefore it was man who provided the impetus for religion. Man invented religion in order to answer those questions of existence that could not be explained in other ways. Sigmund Freud saw religion as an illusion created by man, which fulfilled an important functional role in society. However, he also concluded that man must eventually replace the religious illusion with scientifically defined reality. Ludwig Feuerbach conceived of religion as man's projection of himself into the beyond. Karl Marx saw religion as an opiate used by oppressors to pacify the oppressed in society. Many sociologists of religion follow Emile Durkheim's functional approach, which regards religious structures as being based upon man's need for the legitimizing functions of religion. Religion, as defined from a homocentric perspective, is based upon man's perceived need for something or someone beyond himself to provide meaning to some aspect of his human existence. The corollary to this perspective is that when man's perceived need disappears, so does religion.

The theocentric view of religion begins with the premise that "We did not invent God; God invented us."[1] Christianity, as a religion, is man's response to God's progressive revelation of himself in history, and is therefore theocentric in nature. Christians believe that "Ever since the creation of the world his [God's] invisible nature, namely, his eternal power and deity, has been clearly perceived in the things that have been made," and that this same God ". . . became flesh and dwelt among us."[2] Helmut Thielicke, a noted theologian, has commented that "Concepts of God may very, but for all believers God himself is independent of the consciousness."[3] Thus, a corollary to the theocentric conception of religion is that God is not captured and possessed by man's religious response, but exists beyond the consciousness of man.

Religion, viewed in either a homocentric or theocentric light, provides meaning to mankind's existence. However, Herman Kahn has contended that ". . . sensate, secular, humanist, perhaps self-indulgent criteria become central" in defining meaning in a post-industrial society.[4] It is not simply that secular thought dominates today's society, but more particularly, that the process of secularization is believed to have progressively forced religious belief and practice into a privatized compartment of society. Many regard a compartmentalized religion as powerless to exert any significant influence in a post-industrial society. In providing much of the conceptual schema for the post-industrial society, the triumph of the secular over the sacred has been accepted as a basic underlying assumption.

This assumption of the ultimate triumph of the secular has its roots sunk deep into the history of Western civilization. A common starting point is the thought of the Enlightenment, in which reason came to be deified. The high-priest of rationalism, Voltaire, called for the disengagement of the church from areas of culture and society over which it held hegemony, such as the arts, literature, and the state. Voltaire eventually made his peace with the church upon his death-bed, but his thought, along with that of several other philosophes, paved the way for more refined theories of the demise of religion. Auguste Comte, the father of sociology, developed a positivist philosophy, based upon his law of three stages. This law is a theoretical construct which seeks to demonstrate reason's triumph over religion. In Comte's system, human thought passes from the childish theological stage, where phenomena are explained in terms of the supernatural, to the youthful metaphysical stage, where phenomena are explained in terms of natural forces, and it finally reaches its mature stature in the positive stage, where all phenomena are explained by laws derived through man's reason. For Comte, religious thought was to be eventually superseded by positivistic thought as the sole provider of meaning for man's existence. Karl Marx, like Comte, was a nineteenth century scholar. He excluded religion from his concept of the classless society, the supposed culmination of mankind's history of conflict, because religion would then be an obsolete tool of oppression. In the eighteenth and nineteenth centuries, the theoretical foundations for the assumed triumph of the secular were laid, but the meanings of the terms "secular," "secularization," and "secularism" are more deeply rooted in the history of Western Civilization.

The modern English word secular is derived from the Latin word saeculum, meaning 'this age'. Secular, therefore, originally had a temporal connotation. Prior to the Enlightenment, the term secularization, a derivative of the word secular, came to mean the process by which certain responsibilities and functions of the ecclesiastical authorities were transferred to the sphere of the political authorities. During the disintegration of the feudal system in Western Europe, secularization was a term that was particularly applied to the takeover of church lands by the temporal authorities.

Today secularization has taken on the larger meaning of being ". . . the process by which sectors of society and culture are removed from the domination of religious institutions and symbols."[5] Secularization on the societal level, then, is the disengagement of religion from the political, educational, economic, and other social institutions in a given society. On the cultural level, it is the disengagement of religion from literature, art, philosophy, and other expressions of a particular culture. The process of secularization also manifests itself in the rise of science as the central interpreter of human experience, while religion becomes more and more a compartmentalized aspect of society that deals with only certain private aspects of individual lives.

Secularization is an historical process, while secularism ". . . is the name for an ideology, a new closed world-view which functions very much like a new religion."[6] The key tenet of the secular world-view is ". . . that the world is a closed nexus of experience which does not include God as a perceptible object."[7] As an ideology, secularism presupposes that God himself has been disengaged from human society and culture. In saying this, secular thought goes beyond the sociological assumption of the societal compartmentalization of religion, to the theological assumption that God no longer participates in the realm of human existence.

To the degree that secularization and secularism affect the individual and social consciousness of Western man, they are considered part of post-industrial society. Peter Berger, a sociologist of religion, contends: "As there is a secularization of society and culture, so is there a secularization of consciousness. Put simply this means that the modern West has produced an increasing number of individuals who look upon the world and their own lives without the benefit of religious interpretations."[8] Post-industrial society, in both its institutional infrastructure and its overall consciousness, appears to be based upon empirical, rationalistic, and secular foundations, rather than superempirical, traditionalistic, and religious grounds.

Ironically, Christianity, as the dominant religion of the West, has been seen as the impetus behind the secularization process. In the words of Peter Berger, ". . . Christianity has been its own grave digger."[9] Harvey Cox, another sociologist of religion, has pointed out the biblical roots of the secularization process.[10] The Genesis account of creation frees God from nature, and differentiates nature from man. The Exodus from Egypt was a movement which liberated the Jews from the sacral-political order of the Pharaohs. Never again could rule by divine right go unchallenged. The pronouncement of the Decalogue, with its interdict forbidding the worship of idols, placed man's projections of himself (including his values) in relation to a God that stood beyond man. Seen in this light, secularization is a liberalizing process which frees man to take an objective look at his world and himself. The theologian, Thielicke, would agree that "The gospel, then, makes possible a rational and objective investigation of the world. It opens up the possibility of a scientific and technological relation to it."[11] Yet, he emphatically states that "Secularization is not the child of the church."[12] Thielicke points to the process of secularization as it manifests itself in the non-Christian world (i.e. Japan), and concludes that while Christianity can be viewed as a trigger or carrier of the process, the process itself ". . . seems to be insolubly bound up with empirical science and technical civilization."[13] Thus, while it may be true that Christianity has dug its own grave by aiding the secularization process, it has always correctly regarded secularization as its wayward offspring.

In the decades of the nineteen-fifties and sixties, many believed the secularization thesis was on the verge of fruition, if not already an accomplished fact. The disengagement of religion as a functional social structure and God as a theological concept was thought to be an imminent occurrence. It was also during these decades that the concept of post-industrial society was undergoing its initial scholarly development. Thus, it is not surprising that the basic tenets of the theory of secularization, if not the ideology of secularism, were incorporated in the notion of a post-industrial society. As Herbert Muller points out, ". . . the subject of religion has been ignored in almost all the studies of a technological society," and "The main reason for this neglect is doubtless simply that religion is no longer a vital concern for most contemporary thinkers."[14] Scholars, on the whole, had come to believe that empirical science, the ultimate manifestation of the secular, had replaced religion.

Daniel Bell, in his original conception of post-industrial society, concluded that such a society "cannot provide a transcendent ethic--except for those who devote themselves to the temple of science," and that this "is the cultural contradiction of the society, the deepest challenge to its survival."[15] Apparently Bell had

393

reluctantly submitted to the idea that post-industrial society, which he had described, had found no room for religion. In only a few short years, however, Bell found himself arguing for a return of the sacred. Something had changed with the advent of the mid-nineteen-seventies.

Langdon Gilkey, a theologian at the University of Chicago Divinity School, put his finger on the nature of this change. The essence of his observation is that from the ambiguities in the scientific method has come a resurgence of the sacred, not just a reappearance of a moribund religion, but a powerful, vigorous resurrection of the sacred in modern society.[16] Like the prodigal son, what was thought to be lost has been found. The accepted belief of the fifties and sixties, that "religion would gradually dissipate as an effective force in personal and social life alike," was challenged by the resurgence of religious activity of all varieties during the decade of the seventies.[17] It is not simply that the resurgence of the sacred had caught the scholars unawares (They had anticipated that religion would have residual effects and slight come-backs), but the strength of the resurgence was not foreseen by either social scientists or most theologians. The scholars assumed that the secular had triumphed over the sacred, and that there were only a few pockets of ineffectual religious resistance remaining.

A statistical look at religion in today's post-industrial America reveals some interesting trends in a society that scholars have labeled secular. According to figures compiled by the Princeton Religion Research Center, nine out of every ten Americans express a Christian religious preference.[18] These figures reveal an overwhelming Christian religious preference in American society. This is not to say that American is a 'Christian nation,' or that there is not diversity within American Christianity, but this data simply lends credence to the assertion that ". . . it is the sheerest sort of snobbery to reject the religion of the majority of the population as irrelevant to the analysis of contemporary religion."[19]

Although the religious preference data is astounding for a supposedly secular society, the figures on church membership and church attendance give an even clearer picture of the status of the church in America. For thirty-nine years church membership has been declining, yet in 1981 it appears to have leveled out. Sixty-nine percent of Americans claim membership in a religious institution.[20] Many might assume that religion and reason are antithetical, and thus also assume that those individuals with greater education would be less likely to be involved in a religious organization. Significantly, however, education does not appear to be a factor in church membership. Seventy percent of those with a college or grade school background and sixty-eight percent of those with high school training claim membership.[21] Church attendance has been relatively stable since 1973, with approximately four in every ten Americans attending a house of worship during any particular week.[22] These statistics appear to make the claims of the triumph of the secular sound like a hollow victory.

However, there are some negative statistical findings that are important to this study. Probably the most important negative statement about the predominantly Christian population of the United States is that while "The vast majority of Christians say they believe in the divinity of Jesus Christ, . . . many, it would appear, have not come to terms with Jesus Christ in their lives--they have not examined their <u>understanding</u> of Jesus Christ, their <u>relationship</u> and <u>commitment</u> to Him."[23] Another significant negative figure is that forty-six percent of the population believe that religion is losing influence in their society, while only thirty-five percent see religion's influence increasing in America's post-industrial society.[24] These negative findings deal more with the church's failure to be the church, rather than its failure to exist as a viable and influential institution in a post-industrial society.[25]

Statistics can be misleading, especially when they are based upon self-reported data that deal with a topic as subjective as religion. However, the data on Christian churches in America seem to uphold Gilkey's contention that ". . . it has been precisely those forms of religion believed in one way or another to be antithetical to a secular world, and so vulnerable to the 'acids of modernity,' that have sprouted up everywhere and have grown at an astounding rate."[26] Since 1967, there has been a general decline in the membership of the major Christian denominations in the United States, including Catholic, Lutheran, Episcopalian, Methodist, Presbyterian, and some

395

Baptist groups.[27]   This decline has been coupled with a
rise of more theologically conservative and evangelical
Protestant groups.[28]

The foregoing data help to give a rather accurate
measure of the number of people who have religious
convictions and participate in an organized church, but
statistics fail to distinguish the strength or weakness of
an ˙ individual's religious commitment. Man is not a good
judge of the level of religious commitment, because ". . .
the Lord sees not as a man sees; a man looks on the outward
appearance, but the Lord looks on the heart."[29]   There are
apparently insurmountable difficulties that face anyone who
attempts to measure religious commitment. Theology can only
shed some light on the nature of the religious commitment of
a particular school of theologians.

During the fifties and sixties, Christian theology, in
its most radical forms, sought to come to terms with what
other scholars were calling the triumph of the secular. The
so-called radical theologians expounded a new theology,
which they thought would be relevant to a secular world.
They ". . . seemed to feel that the demand for precision in
scientific expression in general made it necessary to
restate traditional theological concepts. Behind this lay
the conviction that modern man would reject traditional
concepts out of hand as totally outdated."[30] In their
search for relevance in a new theology, they strove to
demythologize the Christian gospel along the lines of Rudolf
Bultmann, a renowned German biblical scholar, or they simply
sought to purge the Christian gospel and theology of its
religious language. W. Warren Wagar, who believes that the
new theologians did not go far enough, concedes that "To
demythologize, secularize, or otherwise purge an ancient
faith of its imagery, its thought forms, and even the moral
and philosophical ideas assimilated from its historical
milieu, is not to purify such a faith, but to kill it."[31]

The most significant construction of the radical
theologians of this era was the death-of-God theology. This
theology will be discussed in greater depth later in this
study, and some general statements about its content will
suffice for the present. The death-of-God theology, along
with almost every other theological statement made during
these decades, tried to define itself in terms of Dietrich
Bonhoeffer's 'religionless Christianity'. In its cleaest

form the death-of-God theology bases itself on a totally immanent view of God in Christ. God as a trascendent being is dead, yet his son, Jesus Christ, is alive today as a fully immanent being. The assertion is that only God-as-fully-man is relevant for today's secular world. In their claim that Jesus Christ has no God-like transcendent characteristics, the death of God theologians are anything but original. A similar view of Jesus was held by a group of Jewish-Christians called the Ebionites as early as the first century A.D., and is also present in Socinianism, a rationalistic theology which appeared in the sixteenth century. The death-of-God theology attempts to make God relevant to secular man, by making God a secular man.

Another important theological development during this period was a shift from personal ethics to societal ethics. As Rodney Stark and Charles Glock stated at the time, "The long Christian quest to save the world through individual salvation has shifted to questions of how to reform society directly."[32] This shift was also influenced by Bonhoeffer's thought, and became a subsidiary aspect of the death-of-God. Such a shift would require churches to make institutional changes in order to accommodate the revolution in theological thought. Stark and Glock regarded the tension created by "institutional inertia" on one hand, and "theological revolution" on the other, as the crisis of the church.[33]

It may be that what Stark and Glock saw as a crisis is now only irony. The radical theological revolution of the fifties and sixties never captured the support of the mass of Christians in the United States. The death-of-God theologians were theologians without churches. The shift in emphasis from issues of personal salvation to those of collective salvation was not as complete as some had hoped it would be, and today the fastest growing churches are generally those that emphasize personal salvation. The crisis which the radical theologians sought to manage with a new theology is not too dissimilar from the crisis Hermann Ridderbos, a conservative and reformed theologian, seeks to resolve with a more orthodox Christian gospel today. Ridderbos describes the crisis by saying, "The way God is involved in what happens in the world is more and more becoming a problem for many people."[34] The difference is that, today, there appears to be a return to religious, rather than secular, theological constructs as a means of

397

resolving this problem of belief.

The church, as an institution in society, has tried to confront the crisis. One of the prevalent ways it has attempted to show how God is involved in the world is by patterning its structure and organization after that of the world. As one commentator on today's church puts it, "In a world of big, impersonal institutions, the church often looks like just another big, impersonal institution."[35] In the early sixties, Peter Berger called for a church structure, which he labeled the "ecumenical parish", to supersede the local congregation as the basic structural unit of the church in the United States.[36] This shift mirrored the greater society's move towards increasing institutional size, as well as the subtle theological shift towards issues of collective salvation. As the church has taken on larger structural manifestations in society, it has come to resemble other large social institutions in its degree of routinization.

Another factor contributing to the growing routinization of the church's charismatic structures has been its emphasis on success. Berger commented in the early sixties that, ". . . there emerges a fairly clear image of what 'success' means in the local congregation. In brief, 'success' means expanding membership, expanding budget, and an expanding program," and ". . . this pattern is well in accord with the secular culture in general."[37] Martin Marty, a respected church historian and commentator on contemporary religion in America, warned of the dangers of such a success orientation when he wrote the following in 1973: "When religious forces forget all else in order to grow and when they place the highest premium on success, they have to overlook emphases that had once been dear to them."[38] The success orientation of the church, like its move toward bigness, has led to its inheriting the same bureaucratic ailments of the other routinized social structures.

The emphasis on bigness and success has caused changes to occur in the role of the clergy, and made that role increasingly stressful. While the church, in some of its institutional forms, is an almost typical Weberian bureaucracy, many of the clergy's functions have escaped, until the present time, the forces of routinization. The emphasis on bigness, with its larger budgets and more complex programming, has forced the clergy to become better administrators and organizers. These bureaucratic skills were not a part of the typical seminary curriculum until the mid-sixties, when it became apparent that the clergy were not prepared for an aspect of their calling which had been of only minor importance prior to that time. The emphasis on success has also provided role strain for the clergy. The criteria for worldly success (i.e. a comfortable standard of living, a responsibility for large groups of people, and recognition for achievement within a particular field of endeavor) are the accepted standards by which many of today's pastors are judged. To a greater degree than ever before, clergymen must meet the success-oriented expectations of their congregations, their peers, and even their families. The demands of increasing size and the increasing emphasis on success have combined with other factors to cause an increasing number of clergy drop-outs.

The foregoing brief examination of the church in America over the last twenty years reveals somethings more than the inconsequential dimension of society that some scholars have assumed it to be, but it also reveals an imperfect social structure. Many would think it almost too simplistic to assume that since the church has continued to exist as a legitimate social structure up to the present day, there should be no reason for it not to continue into the post-industrial future. In order to make such an assertion, both the homocentric and theocentric criteria for Christianity's persistence must be satisfied. Once Christianity is demonstrated to be an abiding presence in post-industrial society from both the homocentric and theocentric perspectives, the church's failure to be the church remains to be scrutinized.

In order to satisfy the homocentric criteria for persistence, Christianity, as a religion, must perform some necessary and legitimate function in society. Emile Durkheim saw religion from a functional perspective. Writing in 1912, he said,

> There is something eternal in religion which is destined to survive all the particular symbols in which religious thought has successively enveloped itself. There can be no society which does not feel the need for upholding and reaffirming at regular intervals the collective sentiments and

the collective ideas which make its unity and its
personality.[39]

Durkheim could not foresee any replacement for religion in
its functional societal niche. However, even though
Durkheim believed religion would not disappear, to him ". .
. it seems destined to transform itself."[40] Durkheim saw
religion from a truly homocentric perspective, and thus
could say, "There are no gospels which are immortal, but
neither is there any reason for believing that humanity is
incapable of inventing new ones."[41] Durkheim believed
that religion was eternal. However, he would never contend,
as this study does, that Christianity would persist as a
form of religion in post-industrial society.

Andrew Greeley, a modern sociologist of religion,
provides a similar functionalist evaluation of the future of
religion. In the introduction to his book, Unsecular Man,
Greeley writes, "The thesis of this book, bluntly, is that
the basic human religious needs and the basic religious
functions have not changed very notably since the late Ice
Age; what changes have occurred make religious questions
more critical rather than less critical in the contemporary
world."[42] Greeley, after looking at the sociological data
on religion in America (particularly, Guy A. Swanson's
comparison of religious behavior with political behavior),
comments that these statistics ". . . require our being
cautious indeed concerning assertions of the present
irrelevance of religion for the personal lives and
institutional commitments of most Americans."[43] In his
examination of what changes he believes have occurred in
contemporary religion, Greeley points out that religious
commitment in the present age is a more individual matter,
based upon free choice, than it has been in the past.[44]
In today's society religious questions more often require an
individual decision, while in the past they were often
decided for the individual on the collective level. By
making a personal decision, the individual will most likely
have a better understanding of his religious preference and
a stronger commitment to it. An individual decision not
only affects a particular person, but also the society in
which that person lives out the consequences of that
decision. Therefore, religion becomes a more critical,
rather than less critical aspect of a post-industrial
society.

The functional approaches of Durkheim and Greeley espouse the persistence of religion from both a theoretical and a statistical basis. The evidence they submit is convincing; however, neither sociologist views religion as an unchanging aspect of society. Neither man goes so far as to say that Christianity will undoubtedly survive into the future in its present form. While they do not deny this as a possibility, Durkheim's thoughts on the transforming of the gospel appear to preclude the survival of an historically-recognizable Christianity.

Martin Marty, an historian with some sociological leanings, employs the functional theses, but takes it beyond Durkheim and Greeley. In 1973, Marty wrote that, "Behind my writing is the assumption that people and culture in the foreseeable future will not be purely secular . . . . People, or at least many of them, will be somehow religious, somehow given to finding meaning and ultimate values."[45] He continues, possibly more as an historian, by saying, ". . . I assume that in western culture and in the United States much of their expression will continue to pick up from Jewish-Christian or biblical themes and traditions."[46] Such an assumption certainly cannot be founded upon a purely empirical theory, but it does take into account the historical resiliency of Judeo-Christian tradition.

Daniel Bell, the sociologist who could not seem to fit religion into his original scheme of the post-industrial society, delivered a lecture in 1977 entitled "The Return of the Sacred? The Argument on the Future of Religion". He hoped to demonstrate that religion does have a future in a post-industrial world. He began by taking religion beyond the societal level and into the realm of culture. In essence he felt that culture is the manner or form of man's response to the existential questions in life (i.e. the meaning of suffering and death, the nature of love, etc. . .), and said that,

> Religion is a set of coherent answers to the core
> questions of life that confront every human group,
> the codification of these answers into a creedal
> form that has significance for its adherents, the
> celebration of rites which provide an emotional
> bond for those who participate, and the
> establishment of an institutional body to bring
> into congregation those who share the creed and

the celebration, and provide for the continuity of
these rites from generation to generation.[47]

This is basically a functionalist approach, in that Bell
regards religion as a legitimate means of providing answers
to the basic questions of existence. However, Bell views
religion as being functionally rooted in culture, rather
than society.

One of Bell's basic assumptions, as his critic Bryan
Wilson points out, is that culture and social structures are
distinct: Social structures are based upon reason, while
culture is grounded in religion.[48] Bell argues that
secularization alone cannot account for the decline of the
sacred since the process of secularization deals with
institutions and not culture. Wilson rightly calls this
thesis into question by asserting that if culture is
autonomous, then it does not influence, nor is it influenced
by the social structures.[49] Such a separation of the
cultural and societal spheres does not even hold true in
Bell's own definition of religion, where the creed and
rites become institutionalized. Wilson's legitimate
contention that Bell artificially separates culture and
society may provide the clue as to why Bell originally
ignored religion in his development of the concept of
post-industrial society. For the purpose of this study, the
removal of the artificial barrier between culture and social
structure does little more than make Bell more of a
traditional functionalist in terms of religion.

Bell, like Marty, moves behind this strictly functional
approach when he takes on a more historical perspective. In
concluding his talk, Bell intimates that religion in the
West will take these "new" forms: "moralizing",
"redemptive", and "mythic and mystical".[50] In actuality,
none of these forms of religion will be truly original.
They will all be, in some respect, a "return to the past,"
and "the resurrection of Memory."[51]

Bell sees moralizing religion as the strongest of the
three forms in the immediate future. He regards it as a
fundamentalist and evangelical reaction to the modernism
which is bent on the secularization of society and the
profanation of culture. It may be assumed that Christianity
already embodies this form of religion.

Redemption religion, Bell's second form, is slightly more complex and less precedented. It is not so much a reaction to modernity, as it is a "retreat" from its excesses. Bell also links this redemptive form with Peter Berger's idea of mediating institutions. Berger and Richard John Neuhaus define these structures ". . . as those institutions standing between the individual in his private life and the large institutions of public life."[52] In this form, the Christian church has functional relevance, in that it provides caring in the private sphere of life, and gives meaning to the large, impersonal public sphere of existence.

Bell's mythic or mystical form of religion is more "diffuse" than those forms already described. It can be understood as a reaction to the demythologizing and rationalization of Western society and culture. Bell sees that Western culture is becoming the "pupil" of Eastern culture with its emphasis on a mysticism that should appear alien to the contemporary post-industrial society.

This turn to the East is an important aspect in the form of religion that Robert Bellah describes as "revolutionary" in his essay, "The New Religious Conscienceness and the Crisis in Modernity".[53] Bellah also speaks of a "biblical" form of religion that has similarities to Bell's historically-based moralizing religion, but it is more clearly associated with the Christian tradition. Bellah's third form is neither a reaction to, nor a retreat from the modernity of Western culture and society; rather, it is a continued embracing of that modernity. The turn to the East, which both Bell and Bellah incorporate in their projections on the future of religion, is a trend which the death-of-God theologians also saw as important to the future of religion in the West. Thomas J. J. Altizer, the foremost death-of-God theologian, comments on the turn to the East when he says that ". . . an originally alien form of mysticism is increasingly becoming real to the Western mind and is casting its spell upon a contemporary and seemingly post-Christian sensibility."[54] Martin Marty dissents from this opinion that the turn East will become a widespread phenomenon in the future, when he intimates that the ". . . ultimate adoption and assimilation of new types of religious experience" will be fairly uncommon in the forseeable future.[55]

The possible growth of Eastern religions in Western societies raises the questions of religious truth and power. In 1975, a group of Christian theologians asserted that the following theme was becoming more prevalent in American theological circles: "All religions are equally valid; the choice among them is not a matter of conviction about truth but only of personal preference or life-style."[56] They then proceeded to

> . . . repudiate this theme because it flattens diversities and ignores contradictions. In doing so, it not only obscures the meaning of the Christian faith, but also fails to respect the integrity of other faiths. Truth matters; therefore differences among religions are deeply significant.[57]

To ignore the truth in a religion is to take away its power. If, in trying to open a religious dialogue, the power of either tradition is limited, then the result is not a dialogue between representatives of diverse faiths. Instead, a discussion ensues about the faith of others. The voluntary turn to the East will not provide the impetus for the massive growth of Eastern religions in America, because such growth can result only from the convincing power of a particular religion. The opening to the East will only grow if the East can demonstrate the power and truth in their mystical beliefs to the West.

It is interesting to note that Marty, Bell, and Bellah all speak of an historically-based, biblical form of religion as being a part of future society. If their projections hold true, and statistics showing this form of Christianity to be the fastest growing religious form in the United States lend weight to their argument, then it appears that Christianity will remain a significant part of post-industrial America. Christianity can also embody the redemptive and mystical forms which Bell spoke of. The church, as the largest network of voluntary associations in the United States, is already an important mediating structure in post-industrial society. As for mysticism, Christianity has always had its mystical elements. The most notable of these may be the Apostle Paul's notion of "in Christ".[58] The Christian faith is already a viable part of our society, and is presently clothed in various institutional forms.

To see religion from the homocentric perspective is to believe that religion will exist as long as man has need of it to answer life's existential questions. The social scientists cited above have contended that religion continues to perform this function, and that legitimate secular, empirical, and technological answers to these questions have not been widely accepted. A few of these scholars have gone a step further to contend that for Americans, Christianity, in some recognizable form, will continue to provide viable answers to these ultimate questions well into the foreseeable future.

Further evidence of Christianity's persistence in a post-industrial America can be brought to light by looking at it from a theocentric perspective. Christianity is theocentric, because its foundation is the living God.[59] The assertion here is that Christianity, as man's response to the living God's revelation of himself in history, will continue to exist, because God continues to reveal himself to mankind. God reveals himself in the fullness of his creation. In man's history, God reveals himself most fully as the incarnate Christ, who was crucified, resurrected, and who has promised to return. Since the ascension of Jesus Christ, God has revealed himself as the Holy Spirit, the comforter, the paraclete. The reality of God, in these three manifestations of himself, is the object of Christian faith and practice.

This study cannot begin to prove the reality or existence of God in a purely empirical manner, because God is both an immanent and transcendent reality. He is immanent, in that he confronts mankind within the limits of human history. However, he is also transcendent, in that he was in the beginning and evermore shall be. Only that which is confined within certain limits of space and time can be explained on a purely rational and empirical level. God transcends human history, yet the Bible is a record of God's progressive revelation of himself in human history. The Bible is a rendering of who God is, yet it does not fully capture the transcendent God. This can be seen in Paul's proclamation to the church at Rome:

O the depth of the riches and wisdom and knowledge of God! How unsearchable are his

> judgements and how inscrutable his ways! 'For who
> has known the mind of the Lord, or who has been
> his counselor?' 'Or who has given a gift to him
> that he might be repaid?' For from him and
> through him and to him are all things. To him be
> glory for ever. Amen.[60]

God has made himself known to man, but he is not fully known
by men.

What sets Christianity apart from its roots in Judaism,
and from the other religions of the world, is its belief
that Jesus Christ, the God-man, is the ". . . way, the
truth, and the life; no one comes to the Father but by
[him.][61] In orthodox Christian theology, Jesus Christ is
both fully man and fully God at the same instant. Jesus
Christ took the form of man in order to reconcile man, the
creature, to God, the Creator. Sin, the manifestation of
man's proclamation of autonomy from the Creator, was the
cause of the estrangement between man and God. Sin is one
of humanity's common traits, yet it cannot be overcome by
human initiative. God, in Christ, took the initiative to
overcome sin in the historical acts of the crucifixion and
the resurrection, thus reconciling the sinful world to
himself. Mankind participates in this reconciliation
through the means of faith. Once the creature is reconciled
to the Creator, he participates in his maker's transcendence
as part of his new life "in Christ".[62] Jesus Christ has
relevance for all men in all times, because he alone offers
freedom from sin and its consequences (i.e. guilt,
suffering, condemnation, and death).

The ideology of secularism has challenged the Christian
conceptions of both God and man in the exposition of a
secular theology. Since the Second World War, secular
theology has appeared under the labels "new theology,"
"radical theology," "post-Christian theology," but its most
striking form has been the death-of-God theology. The
death-of-God theology is secular, because it takes a closed
view of the world by denying the existence or relevance of a
transcendent beyond. The contention that God, in
transcendent form, is dead, challenges the orthodox belief
that God is in the world, but not of it.

Much of the theological thought of the post-World War
II period, both secular and orthodox Christian, makes some

connection to the theology of Dietrich Bonhoeffer. Bonhoeffer, a German-Lutheran theologian and pastor, implicated in the plot to assassinate Adolf Hitler, was executed in a Nazi death camp after a long imprisonment. A collection of his letters and other writings composed during his internment have become the most influential theological corpus in the post-war period.

Two of Bonhoeffer's biographers, including the same Eberhard Bethge who edited <u>Letters and Papers from Prison</u>, issue a <u>caveat</u> in light of his untimely death, warning that these writings must not be understood as Bonhoeffer's mature theological thought.[63] However, many theologians during the fifties and sixties failed to heed this warning. In fact, many of the more radical theologians read their own theological perspective into the more radical statements of Bonhoeffer. In an outline for a book that was never written, Bonhoeffer put forward the following idea:

> Our relation to God is not a 'religious' relationship to the highest, most powerful, and best Being imaginable--that is not authentic transcendence--but our relation to God is a new life in 'existence for others', through participation in the being of Jesus. The transcendental is not infinite and unattainable tasks, but the neighbor who is within reach in my given situation.[64]

Altizer, the death-of-God theologian, goes well beyond Bonhoeffer's critique of the 'religious' conception of a transcendent God, when he proclaims that, "With the death of the transcendent God, every transcendent ground is removed from all consciousness and experience, and humanity is hurled into a new and absolute immanence."[65] There is a clear relation between these two theological statements, but the martyr would almost certainly not have agreed with his successor.

Altizer's theological thought rests upon the claim that the transcendent God of Christian orthodoxy is dead, while the fully immanent Jesus Christ remains. The transcendent God must be dead, because there is an ". . . infinite distance separating the creature and the Creator."[66] This transcendent God is no longer relevant to the secular man who has 'come of age', to use Bonhoeffer's expression. The concern of the death-of-God theologians is no longer their estrangement from the transcendent God, but is now their fear of a transcendent God. Altizer contends that the death of God is a liberating experience, when he writes, "If we can truly know that God is dead, and can fully actualize the death of God in our own experience, then we can be liberated from the threat of condemnation and freed from every terror of a transcendent beyond."[67] In essence this is an attempt to gain freedom from guilt and condemnation by killing the judge, and stands in marked contrast to the orthodox contention that freedom from these consequences of sin is given, in the voluntary death and resurrection of the judge, to those who confess their sin before him.

God is dead, because man has decided that he is better off without a transcendent God. Thielicke, an evangelical theologian, sees a contradiction in this idea:

Either the God who is now dead never really was God, so that this death is in fact only the death of an earlier illusion, or the death of God means simply that he is dead for us, that a certain experience of God has gone, that a prior certainty has been extinguished, that a recognized concept of God has been weakened or revised, so that God himself is not really dead, but only a form of our faith or our view of God. If God is dead, he cannot die, for there has never been a God and Feuerbach is right. Only belief in God can die, and it can do so only if there is no God. For if God is, he will constantly find recognition and kindle new faith.[68]

What has died is the experience of a transcendent God by the death of God theologians. They fail to experience the Holy Spirit at work in the lives of men. They have chosen to work out their own salvation, but no longer believe that God is at work within men, ". . . both to will and to work for his good pleasure."[69]

The Jesus Christ that remains after God is dead is the fully immanent Christ. He is a man, with no transcendent characteristics. He is not the Jesus Christ who claimed ". . . that before Abraham was, I am."[70] He is not the Jesus Christ who is the same yesterday, today, and forever", and who "can never be overtaken by time . . . is never at a loss whatever changing times may bring."[71] He is no longer the

God who forgives sin, because man is supposedly no longer concerned about sin. The immanent Christ is supposedly relevant to a world 'come of age', but the immanent Christ lacks the transforming power which makes Jesus Christ, the God-man, truly relevant to man. The Christian gospel asserts that ". . . only that which transcends the world [God as revealed in Christ] can make us worldly."[72] Man is truly human only in relation to a transcendent God.

Dietrich Bonhoeffer expounds on the relevance of Christianity from the theocentric perspective. He points out that it is wrong ". . . to use God as a stop-gap for the incompleteness of our knowledge."[73] In essence, he is critiquing the homocentric view of Christianity, in which God exists solely as an answer to the existential questions of life. Bonhoeffer continues:

> We are to find God in what we know, not in what we don't know: God wants us to realize his presence, not in unsolved problems but in those that are solved . . . God is no stop-gap; he must be recognized at the centre of life . . . The ground for this lies in the revelation of God in Jesus Christ. He is the centre of life, and he certainly didn't 'come' to answer our unsolved problems.[74]

Jesus Christ is relevant to a post-industrial society, not simply because he gives meaning to the lives of men. Faith in Christ is germane, because he is the giver of life, and he gives it abundantly.[75]

Christ continues to be relevant in a post-industrial America on both the homocentric and theocentric levels. Christian theology is a functional part of American society, as evidenced by the sixty-five percent of Americans who believe that religion is able to answer all or most of today's problems.[76] More important, it is relevant, because God in Christ continues to give life to those who believe. It cannot be assumed that the secular has triumphed over the sacred in American society. The questions of the future are not how post-industrial man will discover meaning apart from religion, but "The more interesting questions for the future have to do with how some people will be religious and how they will be Christian."[77]

The theology of the Christian church has withstood the onslaught of secularism for at least one hundred years, but the institutional structures of the church appear to have fallen victim to the process of secularization. The church, as a social institution, has become somewhat compartmentalized. It has tried to conform itself to the world of which it is a part, rather than living out its calling to be in the world, but not of it. The crisis of the Christian church in post-industrial society is not on a theological level, but on an institutional level. In conforming itself to the other structures in society, the church is failing to be the church.

Daniel Bell speaks of the post-industrial society as being "a game between persons", or a "communal society".[78] If this is true, then the church has the potential to be the most functional aspect of this society. The church is a community of faith. It is the body of Christ, and the people of God.[79] What is important about the church, as part of the greater society, is its people, not its structures.[80] The church should be people who relate to one another, and to those outside the church, on the basis of their relationship to God as revealed in Jesus Christ. To many, the church has become simply another social institution, a privatized compartment in society and, therefore, of no real relevance to those who exist outside its compartmentalized sphere of influence. The people of God have bound themselves to a secularized institution.

George Gallup, commenting on the future of religion in America, says that a decline in the church's relevance will not be the result of a triumph of the secular, but more likely will occur because of the church's failure to fulfill its role as the prophet, teacher, and comforter of society.[81] The future relevance of the church depends upon its willingness to become the dynamic people of God, rather than a static social institution. A prophetic voice has proclaimed that "It is time for judgement to begin in the house of God. Let the church break with the customs that have stifled its living service of the Lord."[82]

Howard Snyder, in his book The Problem of Wineskins: Church Structure in a Technological Age, sets forth a thesis which might be the cure for the church's structural ailments. Snyder takes the following verse from Luke's

Gospel as his point of departure: "And no one puts new wine into old wineskins; it he does, the new wine will burst the skins and it will be spilled and the skins will be destroyed. But new wine must be put into fresh wineskins."[83] Snyder sees the gospel of Jesus Christ as the new wine, and the church structures as the wineskins.[84] The gospel wine is always new and essential, but the structural vessels that hold the gospel wine are temporal and subsidiary. In his book, Snyder suggests the types of church structure which might by the most effective vessels for carrying the gospel wine to a thirsty post-industrial world.

Industrial society has been characterized by large bureaucratic institutions, in which hierarchical modes of organization have been almost universally employed. Today's church in America is characterized by large bureaucratic denominational structures, in which each individual fellowship is hierarchically related to other fellowships. Such a characterization is nothing new for the church, yet it has already been suggested that such denominational structures are facing a decline in membership. Growth is occurring in churches that are not governed by a denominational hierarchy, but rather in those churches that are related to other small individual church fellowships on a horizontal level. This shift from hierarchical to horizontal modes of communication and social organization in the churches can be seen as part of a larger shift from vertical to lateral communication in the post-industrial society.[85] This shift in social organization makes small groups, which emphasize community, the most viable forms of church structure.

Small groups of Christians who form a community of faith and fellowship may be the wineskins of the post-industrial society. Martin Marty contends that, "If mainstream religious forces are to reassume their responsibility, they will have to work with less complex organizations than they employed in the recent past," and small groups will be an organizational structure that will surely be a part of the church in the future.[86] These small groups will perform important social and religious functions as manifestations of Berger's mediating institutions. They will offer individuals an experience of close community, and will also serve to relate the smaller community to the larger society and to the church universal.

411

The small group is not new to the church; for instance it was used by John Wesley during the period of England's transition to the industrial society. It is a structure that has proved itself a viable vessel for the gospel of Jesus Christ.

The small group appears to be the best wineskin for the gospel wine in a post-industrial society, but it should not become the identifying mark of the Christian church. The gospel of Jesus Christ is the timeless message of the people of God, and it is <u>him</u> with whom the church must identify. The structural vessels of the gospel are limited expressions of a particular time and culture, and are therefore secondary considerations for the people of God. Bonhoeffer has said that,

> Love that is really lived does not withdraw from reality to dwell in noble souls secluded from the world. It suffers the reality of the world in all its harshness. The world exhausts its fury against the body of Christ, and the church must be willing to risk its existence for the sake of the world.[87]

The church that is free to be the church, is a community of faith that is willing to sacrifice its institutional form of existence in order to bring the gospel message of reconciliation to the world.

Christ, the Son of the living God and the head of the church, is relevant to man in post-industrial society. The transcendent God, who sent his Son to take away the sins of men, is still an important part of the lives of many in post-industrial America. To say that this fact has little or no influence on the rest of society, on both the personal and institutional levels, is to make a very dubious assumption indeed. If the people of God who comprise the church can identify themselves more closely with the gospel of Jesus Christ, rather than with the institutional trends of the larger society, then they will pursue an even more influential position in post-industrial America. The secular has not triumphed in post-industrial America, but as the church seeks to escape their secularized compartment, they cannot avoid continued conflict with secularism. Post-Industrial man will be confronted with Jesus Christ. He may choose to believe that Christ has personal and social relevance as the risen Lord and Savior. At the very least, he must consider the fact that faith is relevant to other members of post-industrial society.

ENDNOTES

[1]Peter L. Berger and Richard John Neuhaus, eds., Against the World for the World, The Hartford Appeal and the Future of American Religion (New York: Seabury Press, 1976), p. 2.

[2]Romans 1:2, John 1:14 (Revised Standard Version).

[3]Helmut Thielicke, The Evangelical Faith, vol. 1: Prolegomena: The Relation of Theology to Modern Thought Forms, trans. and ed. Geoffrey W. Bromiley (Grand Rapids, MI: William B. Eerdmans Publishing Company, 1974), p. 224.

[4]Herman Kahn and Anthony J. Wiener, The Year 2000: A Framework for Speculation on the Next Thirty-Three Years (New York: The MacMillan Company, 1967), p. 186.

[5]Peter L. Berger, "Secularization and the Problem of Plausibility," in Sociological Perspectives: Selected Readings, eds. Kenneth Thompson and Jeremy Tunstall (New York: Penguin Books, 1971), p. 447.

[6]Harvey Cox, The Secular City (New York: The MacMillan Company, 1965), p. 21.

[7]Thielicke, p. 221.

[8]Berger, "Secularization and the Problem of Plausibility," p. 447.

[9]Ibid., p. 450.

[10]Cox, pp. 17-37. Cox discusses the biblical sources of secularization under the three headings, "disenchantment of nature," "desacralization of politics," and "deconsecration of values." I have attempted to give a summary of his discussion of these topics in the text.

[11]Thielicke, p. 324.

[12]Ibid., p. 323.

[13]Ibid.

[14]Herbert J. Muller, The Children of Frankenstein: A Primer in Modern Technology and Human Values (Bloomington: Indiana University Press, 1970), p. 317.

[15]Daniel Bell, The Coming of Post-Industrial Society (New York: Basic Books, 1973), p. 480.

[16]Langdon Gilkey, "Theology for a Time of Trouble," The Christian Century, April 29, 1981, p. 475.

[17]Ibid., p. 476.

[18]Princeton Religion Research Center, Religion In America, 1981 (Princeton, NJ: Princeton Religion Research Center, 1981), p. 4.; Bradley R. Hertel and Hart M. Nelsen, "Are We Entering A Post-Christian Era? Religious Belief and Attendance in America, 1957-1968," Journal for the Scientific Study of Religion 13 (December 1974): 417-418.

[19]Andrew M. Greeley, Unsecular Man: The Persistence of Religion (New York: Schocken Books, 1972), p. 8.

[20]Princeton Religion Research Center, p. 25.

[21]Ibid.

[22]Ibid., p. 5.

[23]Ibid., pp. 3-4.

[24]Ibid., p. 5.

[25]The church fails to be the church when it fails to relate Christ to the world. In the Sermon on the Mount (Matthew 5-7) Christ calls his followers to be "the salt of the earth" and "the light of the world." Christians are called to flavor their environment with the gospel message, and to bring the light of understanding to those who do not have a clear vision of their relationship to Jesus Christ.

[26]Gilkey, p. 475.

[27]Princeton Religion Research Center, p. 14.; The membership figures for the American Lutheran Church provide a good example of this phenomenon. In 1970, they recorded 2,543,293 members, and by 1980 this figure had declined to

2,353,229. Constant H. Jacquet, Jr., ed., <u>Yearbook of American and Canadian Churches</u>, <u>1982</u> (Nashville: Abingdon, 1982), pp. 244-45.

[28]The Assemblies of God exemplify this growth pattern. In 1970, their members numbered 625,027, and by 1980 their numbers had increased to 1,064,490. Jacquet, pp. 244-45.

[29]1 Samual 16:7b (Revised Standard Version).

[30]Harold B. Kuhn, "Hunger Pains After a Decade of Secularism: But Christians are Still Working Up a Menu," <u>Christianity Today</u>, November 20, 1981, p. 82.

[31]W. Warren Wagar, <u>Building The City of Man</u>: <u>Outlines of a World Civilization</u> (New York: Grossman Publishers, 1971), p. 55.

[32]Rodney Stark and Charles Y. Glock, <u>American Piety</u>: <u>The Nature of Religious Commitment</u> (Berkley: University of California Press, 1968), pp. 26-27.

[33]Ibid., p. 212.

[34]Herman Ridderbos, <u>Studies In Scripture And Its Authority</u> (Grand Rapids, MI: Wm. B. Eerdmans Publishing Company, 1978), p. 1.

[35]Howard A. Snyder, <u>The Problem of Wineskins</u>: <u>Church Structure in a Technological Age</u> (Downers Grove, IL: Inter-Varsity Press, 1975), p. 89.

[36]Peter L. Berger, <u>The Noise of Solemn Assemblies</u>: <u>Christian Commitment and the Religious Establishment in America</u> (Garden City, NJ: Doubleday & Company, 1961), p. 168. The "ecumenical parish" refers to the entire Christian community within any single locality, as opposed to the individual denominational congregations within that locality. Berger hoped that the ecumenical community could come together as one body to serve as the church in a particular area or parish.

[37]Ibid., p. 163.

[38]Martin E. Marty, <u>The Fire We Can Light</u>: <u>The Role</u>

of Religion in a Suddenly Different World (Garden City, NJ: Doubleday & Company, 1973), p. 96.

[39]Emile Durkheim, "On the Future of Religion," in Sociological Perspectives: Selected Readings, eds. Kenneth Thompson and Jeremy Tunstall (New York: Penquin Books, 1971), p. 441.

[40]Ibid., pp. 444-45.

[41]Ibid., p. 442.

[42]Greeley, p. 1.

[43]Ibid., pp. 8-10. Here Greeley ties into the study made by Swanson, which showed Americans more active on various levels in religion, than in politics.

[44]Ibid., pp. 14-15.

[45]Marty, pp. 193-94.

[46]Ibid.

[47]Daniel Bell, "The Return of the Sacred? The Argument on the Future of Religion," British Journal of Sociology 28 (December 1977): 428-29.

[48]Bryan Wilson, "The Return of the Sacred," Journal for the Scientific Study of Religion 18 (September 1979): 271-75.

[49]Ibid., p. 277.

[50]Bell, "The Return of the Sacred? The Argument. . .", pp. 444-47.

[51]Ibid., p. 444.

[52]Peter L. Berger and Richard John Neuhaus, To Empower People: The Role of Mediating Structures in Public Policy (Washington, D.C.: American Enterprise Institute for Public Policy Research, 1977), p. 2.

[53]Robert N. Bellah, "The New Religious Consciousness and the Crisis in Modernity," in The New Religious

<u>Consciousness</u>, eds. Charles Y. Glock and Robert N. Bellah (Berkley: University of California Press, 1976), pp. 349-352.

[54]Thomas J. J. Altizer, "A Wager," in <u>Toward</u> <u>a</u> <u>New</u> <u>Christianity</u>: <u>Readings</u> <u>in</u> <u>the</u> <u>Death</u> <u>of</u> <u>God</u> <u>Theology</u>, ed. Thomas J. J. Altizer (New York: Harcourt, Brace & World, Inc., 1967), p. 314.

[55]Marty, p. 206.

[56]Berger and Neuhaus, <u>Against</u> <u>the</u> <u>World</u> <u>for</u> <u>the</u> World. . ., p. 3.

[57]Ibid.

[58]Galatians 2:20, 2 Corinthians 5:17 (Revised Standard Version); Albert Schweitzer comments in his study <u>The</u> <u>Mysticism</u> <u>of</u> <u>Paul</u> <u>the</u> <u>Apostle</u>, trans. William Montgomery (London: A. & C. Black, Ltd., 1931), p. 3, that "The fundamental thought of Pauline mysticism runs thus: I am in Christ; in Him I know myself as a being who is raised above this sensuous, sinful, and transient world and already belongs to the transcendent; in Him I am assured of ressurection; in Him I am a child of God."

[59]Exodus 3:13-15, John 8:58 (Revised Standard Version).

[60]Romans 11:33-36 (Revised Standard Version).

[61]John 14:6 (Revised Standard Version).

[62]2 Corinthians 5:17 (Revised Standard Version).

[63]Harold B. Kuhn, "But Which Bonhoeffer?" <u>Christianity</u> <u>Today</u>, April 14, 1972, p. 50.

[64]Dietrich Bonhoeffer, <u>Letters</u> <u>and</u> <u>Papers</u> <u>from</u> <u>Prison</u>, enlarged edition, ed. Eberhard Bethge (New York: The MacMillan Company, 1971), p. 381.

[65]Altizer, p. 315.

[66]Ibid., p. 304.

[67]Ibid., pp. 311-12.

[68]Thielicke, p. 224.

[69]Philippians 2:12-13 (Revised Standard Version).

[70]John 8:58 (Revised Standard Version).

[71]Hebrews 13:8 (Revised Standard Version); Ridderbos, Studies In Scripture And Its Authority, pp. 8-10.

[72]Thielicke, p. 385.

[73]Bonhoeffer, p. 311.

[74]Ibid., pp. 311-12.

[75]John 10:10 (Revised Standard Version).

[76]Princeton Religion Research Center, p. 5.

[77]Marty, p. 194.

[78]Daniel Bell, The Cultural Contradictions of Capitalism (New York: Basic Books, 1976), pp. 147-48.

[79]Colossians 1:18, 1 Peter 2:9-10 (Revised Standard Version).

[80]Snyder, p. 161.

[81]Princeton Religion Research Center, p. 6.

[82]Edmund P. Clowney, "The Politics of the Kingdom," The Westminster Theological Journal 41 (Spring 1979): 310.

[83]Luke 5:37-38 (Revised Standard Version).

[84]Snyder, p. 13.

[85]As cited by Snyder, pp. 187-88.

[86]Marty, pp. 151, 204.

[87]As cited by Heinrich Fraenkel and Rodger Manvell, The Men Who Tried to Kill Hitler (New York: Coward-McCann,

Inc., 1964), p. 72.

BIBLIOGRAPHY

Altizer, Thomas J. J. "A Wager." In <u>Toward</u> <u>a</u> <u>New</u> <u>Christianity</u>: <u>Readings</u> <u>in</u> <u>the</u> <u>Death</u> <u>of</u> <u>God</u> <u>Theology</u>, pp. 303-20. Edited by Thomas J. J. Altizer. New York: Harcourt, Brace & World, Inc., 1967.

Bell, Daniel. <u>The</u> <u>Coming</u> <u>of</u> <u>Post-Industrial</u> <u>Society</u>. New York: Basic Books, 1973.

_____. <u>The</u> <u>Cultural</u> <u>Constradictions</u> <u>of</u> <u>Capitalism</u>. New York: Basic Books, 1976.

_____. "The Return of the Sacred? The Argument on the Future of Religion." <u>British</u> <u>Journal</u> <u>of</u> <u>Sociology</u> 13 (December 1977): 419-449.

Bellah, Robert N. "The New Religious Consciousness and the Crisis in Modernity." In <u>The</u> <u>New</u> <u>Religious</u> <u>Consciousness</u>. Edited by Charles Y. Glock and Robert N. Bellah. Berkley: University of California Press, 1976.

Berger, Peter L. <u>The</u> <u>Noise</u> <u>of</u> <u>Solemn</u> <u>Assemblies</u>: <u>Christian</u> <u>Commitment</u> <u>and</u> <u>the</u> <u>Religious</u> <u>Establishment</u> <u>in</u> <u>America</u>. Garden City, NY: Doubleday & Company, 1961.

_____. "Secularization and the Problem of Plausibility." In <u>Sociological</u> <u>Perspectives</u>: <u>Selected</u> <u>Readings</u>. Edited by Kenneth Thompson and Jeremy Tunstall. New York: The MacMillan Company, 1974.

Berger, Peter L. and Neuhaus, Richard John, eds. <u>Against</u> <u>the</u> <u>World</u> <u>for</u> <u>the</u> <u>World</u>, <u>The</u> <u>Hartford</u> <u>Appeal</u> <u>and</u> <u>the</u> <u>Future</u> <u>of</u> <u>American</u> <u>Religion</u>. New York: Seabury Press, 1976.

Berger, Peter L. and Neuhaus, Richard John. <u>To</u> <u>Empower</u> <u>People</u>: <u>The</u> <u>Role</u> <u>of</u> <u>Mediating</u> <u>Structures</u> <u>in</u> <u>Public</u> <u>Policy</u>. Washington, D.C.: American Enterprise Institute for Public Policy Research, 1977.

Bonhoeffer, Dietrich. <u>Letters</u> <u>and</u> <u>Papers</u> <u>from</u> <u>Prison</u>. Enlarged edition. Edited by Eberhard Bethge. New York: The MacMillan Company, 1971.

Clowney, Edmund P. "The Politics of the Kingdom." The Westminster Theological Journal 41 (Spring 1979): 291-310.

Durkheim, Emile. "On the Future of Religion." In Sociological Perspectives: Selected Readings. Edited by Kenneth Thompson and Jeremy Tunstall. New York: Penguin Books, 1971.

Fraenkel, Heinrich and Manvell, Rodger. The Men Who Tried to Kill Hitler. New York: Coward-McCann, Inc., 1964.

Gilkey, Langdon. "Theology for a Time of Trouble." The Christian Century, April 29, 1971, pp. 474-80.

Greeley, Andrew M. Unsecular Man: The Persistence of Religion. New York: Schoken Books, 1972.

Hertel, Bradley R. and Nelson, Hart M. "Are We Entering a Post-Christian Era? Religious Belief and Attendance in America, 1957-1968." Journal for the Scientific Study of Religion 13 (December 1974): 409-419.

Jacquet, Constant H. Jr., ed. Yearbook of American and Canadian Churches, 1982. Nashville: Abingdon, 1982.

Kahn, Herman and Wiener, Anthony J. The Year 2000: A Framework for Speculation on the Next Thirty-Three Years. New York: The MacMillan Company, 1967.

Kuhn, Harold B. "But Which Bonhoeffer?" Christianity Today, April 14, 1972, pp. 49-50.

_____. "Hunger Pains After a Decade of Secularism: But Christians Are Still Working Up a Menu." Christianity Today, November 20, 1981, p. 82.

Marty, Martin E. The Fire We Can Light: The Role of Religion in a Suddenly Different World. Garden City, NJ: Doubleday & Company, 1973.

Muller, Herbert J. The Children of Frankenstein: A Primer in Modern Technology and Human Values. Bloomington: Indiana University Press, 1970.

Princeton Religion Research Center. Religion In America,

1981. Princeton, NJ: Princeton Religion Research Center, 1981.

Ridderbos, Herman. Studies In Scripture And Its Authority. Grand Rapids, MI: Wm. B. Eerdmans Publishing Company, 1978.

Snyder, Howard A. The Problem of Wineskins: Church Structure in a Technological Age. Downers Grove, IL: Inter-Varsity Press, 1975.

Stark, Rodney and Glock, Charles Y. American Piety: The Nature of Religious Commitment. Berkley: University of California Press, 1968.

Schweitzer, Albert. The Mysticism of Paul the Apostle. Translated by William Montgomery. London: A. & C. Black, Ltd., 1931.

The Bible. Revised Standard Version.

Thielicke, Helmut. The Evangelical Faith, vol. 1: Prolegomena: The Relation of Theology to Modern Thought Forms. Translated and edited by Geoffrey W. Bromiley. Grand Rapids, MI: William B. Eerdmans Publishing Company, 1974.

Wagar, W. Warren. Building The City of Man: Outlines of a World Civilization. New York: Grossmann Publishing, 1971.

Wilson, Bryan. "The Return of the Sacred." Journal for the Scientific Study of Religion 18 (September 1979): 268-80.

INTERRELIGIOUS DIALOGUE IN AN ERA OF COMMUNICATION:

A PHILOSOPHICAL CHALLENGE

Douglas Brouder

We live in an era of communication. Every day, information from all over the world bombards our senses through the media of television, radio, and the press. Events in places which were formerly heard of only in geography classes now affect our daily lives.

Of course, our economic interdependence with other countries is nothing very new; but in the last decade it has become an established fact. The demands of "modern existence" have turned the world into a "global community," each part influencing and being influenced, and we Americans can no longer (and in fact do not) pursue a course of economic or political isolationism.

What does the communication revolution have to do with the study of religion? I would like to propose that we can no longer pursue a course of religious isolationism, either. Other cultures, to which we are now being exposed more than ever before, bring with them other religions. These religions, which sometimes seem seductively familiar, sometimes exotic, and sometimes primitive, are usually dismissed quickly by Western minds, or misunderstood. In our era of world communication and intercultural exchange the field of interreligious studies is receiving more attention as an essential link in promoting true understanding between East and West. My purpose in this paper is not to dictate the attitudes we should hold toward other faiths, but to outline the process by which one comes to question his or her own attitudes, and to stimulate questions which should arise from the exposure to other religions. These questions must be dealt with if any religion is to maintain its claims of universality and salvation, however these may be interpreted. And it should become obvious that if we are to live what can be called religious lives in the modern world, we will have to deal seriously with these questions on a personal level. We must challenge our own faith, and perhaps lose it in the process. Out of this may come a greater understanding of our most deeply held beliefs as we compare our religion with that of others.

The process of communications' influencing religious patterns has been described by Philip Ashby, a professor of religion at Princeton and a participant in several exchange programs with India, as follows:

> ...the one primarily new condition of human existence is the pre-eminent reality of intra-world relationships. These relationships constitute the present pattern of the fabric of human society, a composite of not two or a few of the traditional societies of the past, but of all groups of mankind. Even those groups which have been classified as pre-literate or primitive are now a part of this unity, a unity which will not allow them to stay isolated nor to retreat to the conditions of their immediate past. Even many fringe societies, obviously destined to be eradicated by the present process, are partners in shaping the process as they combine to participate in it and bring to bear their own interests and peculiar insights.[1]

Here in the United States, we have become accustomed to hearing of the debate (or conflict, or dialogue, or what have you) between "religion" and "science," or of the raging battle between "religion" and "secularism," whatever these terms may be interpreted to mean. However, in the United States the term "religion" is usually meant to signify the Christian faith, however loosely defined or organized, or whichever denomination, or sect, to which the speaker happens to adhere.[2] The ongoing conflict between creationists and evolutionists is, in effect, confined to one particular group of people: Christians in the United States (and not even all people calling themselves Christians see this as a major issue).

In an era of communication, a new conflict or challenge is arising. It is one which American Christian churches will have to confront in the years to come, and will have profound and far-reaching effects on believers and non-believers alike. Every 'true believer' of any denomination will personally have to confront new religious questions, and attempt some kind of answers for himself or

425

herself, if religion is to survive as a meaningful experience.

I believe the greatest challenge within the Christian Church in America in a post-industrial society will not be the challenge of science, but that of religions. I say "religions" because it is becoming clear that Christianity is not the only system claiming to be a 'world religion.' Other religions, just as valid in their own time and place, and with similar claims of universality, are bringing more people to critical examination of their own beliefs.

> People have woken up to the importance of the problem.... Vaguely or not so vaguely, they sense that Christianity and the validity of its claim are in a tight spot, with the appearance on the scene of the great non-Western religions. Not a few conscientious Christian believers feel themselves plunged by all this into an atmosphere which fills them with an indefinable sense of anxiety and arouses all sorts of misgivings; and they do not see how these doubts are to be ... laid to rest. To a greater or lesser degree, they are jolted out of the confident assurance in which they had been accustomed to live.[3]

It is at once the universal claims of these alternate systems, as well as their religious character, that poses a challenge for believers of the Christian faith.[4] It seems that a religious person could easily defend himself against the charges of a religious skeptic, or one who followed no religious tradition.[5] A different situation arises when a religious person confronts another who is truly religious, but whose religion differs radically from that with which he is familiar. This situation, and the criticism of one's own beliefs in a religiously pluralistic society, will be an important and inevitable step in the development of religion in a post-industrial society.

The fact that our world has produced and nurtured many different religious traditions has never been a very carefully guarded secret.[6] Even in ancient times, writers and prophets, monks and wise men were aware that there existed people outside of their own traditions, otherwise there might have been fewer restrictions or pressures within most traditions to renounce all others and cling to one doctrine alone. In the Jewish faith, there is an absolute insistence on monotheism, and an abhorrence of idols or other gods, especially in the writings of the prophets.[7] In Islam, the greatest sin a man can commit, and the only unforgivable one, is that of shirk--association-- which is "the failure to recognize that the final truth and power of the universe is one."[8] This Islamic monotheism, with its insistence on the rejection of other gods and doctrines, is just as strict as that of the Old Testament prophets.[9] Even some of the Eastern religions exhibit the trait of an exclusive claim to the truth.[10]

Some people in these earlier simpler times were aware of other traditions due to trade with foreigners, and other means of contact. In these times it was possible, and logical, for the adherent of a particular faith to regard his or her faith as being the one universal religion. In the absence of any mass communication of cultures or ideas, a religion of these times could call itself 'religion' without leading anyone to ask which religion was meant.

> Communication between the different groups of humanity was then so limited that for all practical purposes men inhabited a series of different worlds. For the most part, people in South America, in Europe, in India, in Arabia, in Africa, in China, were unaware of the others' existence. There was thus, inevitably, a multiplicity of local religions that were also cultures. Accordingly, the great creative moments of revelation and illumination occurred within different cultures and influenced their development, giving them the coherence and impetus to expand into larger units, thus producing the vast religious-cultural entities that we now call the world religions.... the broader picture is one of religions developing separately within different historical and cultural settings.[11]

For example, at a time when the Roman Empire considered itself the whole of the civilized world, there was no question of the youth's turning to Buddhism as an alternative, or even of their acquaintance with it. Perhaps this is an exaggerated example. The Catholic Church in the Middle Ages, however, clearly had no room for the followers of any other tradition. In the society of that time and place, it was possible to believe (perhaps it would be more appropriate to say that it was necessary to believe) that the Church was the only fount of wisdom and the only means to salvation. Pope Boniface VIII (1294-1303) stated the issue as follows:

> We are obliged by the faith to believe and hold--and we do firmly believe and sincerely confess--that there is one Holy Catholic and Apostolic Church, and that outside this Church there is neither salvation nor remission of sins.... In which Church there is one Lord, one faith, one baptism....

> For this authority, although given to a man and exercised by a man, is not human, but rather divine, given at God's mouth to Peter and established on a rock for him and his successors in Him whom he confessed, the Lord saying to Peter himself, "Whatsoever thou shalt bind," etc.... Furthermore we declare, state, define and pronounce that it is altogether necessary to salvation for every human creature to be subject to the Roman pontiff.[12]

In the Catholic Church of the Middle Ages, the believer could well believe that the whole world worshiped with him. Places like Jerusalem or Mecca (the 'land of the infidel') were distant enough so that a religious person could feel secure in his beliefs, and could also be assured of the infidels' eternal damnation.[13]

Those days of religious isolationism are, for the most part, gone. Today's world does not allow for such provincial views, in which one's own religion is thought of as 'the only one.' Our world's religious diversity was brought to the forefront of this country's consciousness recently when Islamic students seized the American embassy in Iran. That event was not a very positive way to introduce American minds to Islamic ideas, but it called attention to the fact that the world outside of the United States is not a religious vacuum. Most of all, it exposed the resentment which those devoted to Islamic ideals felt toward the forced Westernization of their society, and pointed out the need to understand their religious ideals.

428

Another way in which we are exposed to other religions involves the "turn toward the East," which has become an important factor in American religion in recent years. Harvey Cox, in Turning East, has investigated the phenomenon of neo-Oriental religious groups in the United States. He proposes that many people today are drawn to the Eastern religions because these religions fulfill a need in their lives which Western religions were not able to satisfy.[14] The popularity of Zen (in what is usually an Americanized version) has been an important trend in the "turn to the East."[15] Many blacks in the United States have adopted their own particular version of Islam, changing their names and their ways of life in the process as a way of renouncing the "racist" Christian religion which their ancestors were forced to accept.[16] The economic growth of some of the non-Western countries, such as Korea and Japan, has brought our respective business communities together, and the resulting cultural exchanges have presumably influenced both participants in the communication process.

These examples show a few of the ways we are exposed, in this era of international interdependence and rapid world-wide communication, to religious values or systems different from the Christian system which has prevailed in the United States for so long.[17] There are many other ways in which other religions enter our lives, but it is not my intention to list them all here. I hope that it has been made clear that the educated person today can no longer operate with an attitude that his is the only religion, since contact with other cultures has become unavoidable in this age of communication.

The understanding of other cultures on a philosophical-religious level is a necessary task which should be pursued as fervently as we pursue our financial and political interests in other countries.

We have yet to learn our new task of living together as partners in a world of religious and cultural plurality. The technological and economic aspects of 'one world', of a humanity in process of global integration, are proceeding apace, and at the least are receiving the attention of many of our best minds and most influential groups. The political aspects also

429

are under active and constant consideration, even
though success here is not so evident, except in
the supremely important staving off of disaster.
The ideological and cultural question of human
cohesion, on the other hand, has received little
attention, and relatively little progress can be
reported; even though in the long run it may prove
utterly crucial, and is already basic to much
else. Unless we can learn to understand and to
be loyal to each other across religious frontiers,
unless we can build a world in which people
profoundly of different faiths can live together
and work together, then the prospects for our
planet's future are not bright.[18]

The importance of a basis of understanding between East and
West cannot be overestimated, and I think that this
understanding can be sought in a common phenomenon:
religion.

The extent of the understanding that arises from any
interreligious communication is dependent upon the capacity
of a person to see another's religion as that other person
sees it. Therefore, the attitudes which one holds regarding
other religions may need examination and revision. The
individual nature of religion should be understood, because
religion is a phenomenon which obviously has a different
effect on everyone. Many of us attend church services, but
we do not often stop to put our thoughts regarding the
nature of God or salvation, or the actual nature of "eternal
life" into words. Religious truth is seen differently by
different people and different cultures. For example,
consider the various ways of interpreting the Bible. One's
conception of truth may be exclusive or inclusive, as can be
seen in the world's different religions.[19]

I think the individualistic nature of belief will
become more apparent if a person is confronted by another
system of belief. In this confrontation of faiths, the
believer in a religious system, in order to defend his
position, must first define his position.[20] Trying to
define one's own position brings usually unexamined
thoughts under close scrutiny. Defining one's own position
is a necessary step, however, in any attempt to understand
others.

Many different levels of understanding are possible in the study of religions.[21] Crucial for the purpose of international understanding and dialogue, however, is the attitude with which we view the religious beliefs of other people, and our consequent views of those other people.

There are many ways of distinguishing between oneself and "others", and many attitudes one may feel toward those others. In the field of interreligious studies, three broad categories have been proposed by Raimundo Pannikar, Professor of Religious Studies at the University of California at Santa Barbara, to describe religious attitudes: exclusivism, inclusivism, and parallelism.[22]

What are these attitudes regarding other religions? I have already mentioned one possible attitude, exclusivism, which I believe is no longer a viable option. Let us examine the attitude of religious exclusivism as it might appear in the United States, and see to what conclusions this attitude would lead us, regarding other faiths. From there we will examine the other attitudes regarding various faiths, and consider each one as a basis for interreligious understanding.

In a way, religious exclusivism is a position which is very easy to defend, if one truly believes in it. One simply dismisses all other claims regarding religious truth as false, if they do not agree with one's own. For example, in the United States of the seventeenth century, any unorthodox views were condemned as heresy. The people holding such views were condemned as witches or heretics, and religious uniformity was protected. This group behavior may have strengthened the sense of community, especially in its dealings with outsiders, who could be seen as religionless (or worse, as heathen or pagans!).[23] One's brushes with outsiders at this time could be kept to a minimum (as far as religious understanding was concerned, anyway) and thus an attitude of exclusivism was not challenged.

This attitude seems to persist to some degree today, and should therefore be investigated further. Religious exclusivism as we can observe it in the United States tends to reduce the world to two opposing camps: Christians and non-Christians. It is usually claimed by these Christians that all of the non-Christians can be converted, thus uniting the world under one God.[24] However, another trend can be seen underlying this religious exclusivism: the tendency to equate good with what is familiar (and American) and evil with what if foreign or unfamiliar. This trend was at a high point during the recent Iranian crisis and manifested itself in frequent attacks on the Islamic faith and peoples, or caricatures of the religion and people.

Hence, a problem is apparent in Christianity, which involves a contradiction implicit in its teachings. The contradiction is between the moral and intellectual 'sides' of the Christian teachings, as they are seen today. On the moral side, there is

> ... an imperative towards reconciliation, unity, harmony, and brotherhood. At this level, all men are included: we strive to break down barriers, to close up gulfs; we recognize all men as neighbors, as fellows, as sons of the universal father, seeking Him and being found by Him. At this level, we do not become truly Christian until we have reached out towards a community that turns all mankind into one total 'we'.[25]

However, the idea of brotherhood is in opposition to the intellectual side of Christian doctrine, which divides the world into 'the saved' and 'the damned':

> At this level, the doctrines that Christians have traditionally derived have tended to affirm a Christian exclusivism, a separation between those who believe and those who do not, a division of mankind into a 'we' and a 'they', a gulf between Christendom and the rest of the world: a gulf profound, ultimate, cosmic.

> ... the traditional doctrinal position of the Church has in fact militated against its traditional moral position, and has in fact encouraged Christians to approach other men immorally. Christ has taught us humility, but we have approached them with arrogance.[26]

The contradiction between exclusive claims to truth, and the goal of universal brotherhood, appears in other religions besides Christianity, but Christianity, since it is most

432

familiar to us, will serve as an example. Every religion will, of course, claim to possess the truth. This claim sets the followers of any tradition apart from the rest of humanity. When coupled with a position of exclusivism, however, in which one's religion is seen as the only possible truth, the claim to possess the truth becomes an alienating factor, in that it becomes easy to see other traditions (especially foreign traditions, and in the West, Eastern religions) as less than religion. The religions in question can be thought of as primitive or savage, and the use of such terms as pagan, heathen, or infidel can be used to dismiss these rival systems, and their followers, as inferior. Sometimes the religions of the East, such as Buddhism or Hinduism, are seen as idolatrous by Americans, while the Islamic religion is characterized by its primitive qualities and apparently bloodthirsty insistence on Islamic law ('an eye for an eye', which also characterizes Old Testament law). At any rate, the Christian religion is seen as the only way to salvation, and there are no other systems worthy of the title 'religion," since none of them bring about the salvation of man through Christ, which should, according to Christian doctrine, be man's ultimate goal.

Seeing other systems as less than religion is not an acceptable position in this day and age, where international cooperation is more necessary than ever before. But there are other dangers in exclusivism, besides the misunderstanding of other religions. Under certain conditions, this exclusivist position can lead a person to either become alienated from those of his own tradition, or to reject the moral position of Christianity as it was outlined above. This is explained clearly by Wilfred Cantwell Smith, in The Faith of Other Men:

> From the notion that if Christianity is true, then other religions must be false ... it is possible to go on to the converse proposition: that if anyone else's faith turns out to be valid or adequate, then it would follow that Christianity must be false--a form of logic that has, in fact, driven many from their own faith, and indeed from any faith at all. If one's chances of getting to Heaven ... or of coming into God's presence, are dependent upon other people's not getting there, then one becomes walled up within the quite intolerable position that the

Christian has a vested interest in other men's damnation....

When an observer comes back from Asia, or from a study of Asian religious traditions, and reports that, contrary to accepted theory, some Hindus and Buddhists and Muslims lead a pious and moral life and seem very near to God by any possible standard, so that, so far as one can see, in these particular cases at least faith is as 'adequate' as Christian faith, then presumably a Christian should be overjoyed, enthusiastically hopeful that this be true.... Instead, I have sometimes witnessed just the opposite: an emotional resistance to the news, men hoping firmly that it is not so, though perhaps with a covert fear that it might be.... It will not do, to have a faith that can be undermined by God's saving one's neighbors; or to be afraid lest other men turn out to be closer to God than one had been led to suppose.[27]

It is obvious that an exclusivist attitude which condemns other religious systems is not conductive to interreligious understanding. In order to deal with people of other faiths, we will have to refrain from judging their religions as superior or inferior as compared to our own. Exclusivism sometimes appears in obvious forms; however, sometimes attitudes may appear to be liberal and open-minded while actually remaining narrow-minded and self-centered.

One such outlook, which appears at first examination to answer the questions arising from the conflict of differing religions, is the attitude of inclusivism. Rather than excluding foreign faiths from consideration, the inclusivist simply affirms that there are many different levels on which the world's religions can be interpreted, and that they all can be true simultaneously. Sometimes, inclusivism involves redefining other religions so that they agree with one's own. In this way, the believer can hold fast to his beliefs without questioning them, at the same time affirming every other doctrine with which he is confronted. If one adheres to this policy regarding other religions, he may remain secure in his original beliefs while appearing to maintain an attitude of religious tolerance and understanding.[28] It is even possible to truly maintain a sincere acceptance of other faiths as different levels of the same truth. For example, a person with some knowledge of Buddhism may read into it his own Christian viewpoint, and see the Buddha as a kind of Savior. In actual practice, such an 'inclusivist' may reinterpret Buddhism in light of his Christian experience, and the final goal, Nirvana, may be equated with heaven. In a similar way, the Hindu gods Vishnu, Brahma and Shiva may be seen as the Christian Trinity in order to bring unfamiliar gods into familiar terms. The adjustment of religious doctrines to make them more easily understood can be appealing, but anyone studying other religions must be careful not to distort the tradition beyond the point of recognition (that is, the recognition of someone from within that tradition).

This type of distortion and assimilation, however, is not always present in inclusivism. As mentioned earlier, a person can see truth at different levels. When encountering a new religion, it may be possible to see various doctrines of that religion as occurring at different levels, while maintaining the truth of all of them. The truth of these doctrines, at closer inspection, is not as concrete as the truth associated with the inclusivist's own religion. In an inclusivist viewpoint, the truth of other religions is usually seen as metaphysical or mystical while the details of the religion are passed over. In this way an inclusivist outlook becomes inclusive of form, but not of content.[29] Contradictions in doctrines will be overlooked on the basis that "underneath, the meanings are the same." This idea leaves the inclusivist with the problem of justifying his belief in the specific content of any particular religion, and in effect, makes the possibility of belief in any faith in all its depth and detail an act of hypocrisy. Superficially, it is easy to be an inclusivist whenever one is directly confronted by another religion. Then when the threat of the opposing faith is gone, one merely retreats into his own religion again, and forgets about the other religions. The tendency to reinterpret other religions is a strong one, and is difficult to avoid. So it is this type of inclusivism which is most often encountered.

In a way, inclusivism is again based on an idea of one's own religious beliefs. But in this case, the contact with other religions brings one to assimilate them into the faith which is already believed, or to claim that it contains the same truth which one already believes, at a different level. We say "I already know that"--in effect, the inclusivist affirms only that the foreign or new religion has nothing further to offer him which he does not know already (in his own terms and concepts).[30] In this way, the new religion is reduced to a replica of the familiar religion, while the inclusivist's own religion is given top priority due to its familiarity and its being seen as the most valid interpretation of the fundamental 'truth'. Inclusivism also reduces truth itself to a relative, subjective phenomenon, in the name of universality, as well as putting the inclusivist in a position of being the only person to know the true nature of religions, and the place which each faith must take, in his or her view of religion.

An inclusive type of attitude regarding other religions can be a tempting one. However, if we engage in dialogue with those of different faiths, our misconceptions of the other religion (which we could formerly assimilate into our own doctrines) will inevitably be removed or corrected by the follower of that other faith. (This can be in the form of an actual dialogue, or may take the form of any type of communication regarding the nature of another religion, as interpreted by a follower of that religion.) For that reason, the challenge to thinking Christians will come, not from those who reject religion for various reasons, but from those genuinely religious people of other faiths. These people will be sympathetic to religious claims or experience but will interpret their experiences differently in light of their different backgrounds. Thus, any attitude which relies on ignoring other doctrines, or reinterpreting and then claiming them as one's own, cannot survive in a climate of interreligious communication, or one where ideas are easily exchanged.

A climate of interreligious contacts, which I believe is being realized in our age of rapid communications and increasing cultural exchange, has forced a new religious attitude on us: one which Raimundo Pannikar calls "parallelism."[31] The attitude of parrallelism may be a genuine improvement over the attitudes previously outlined, since it involves the recognition of another's religion as exactly that: another's religion.

> If your religion appears far from being
> perfect and yet it represents for you a symbol of
> the right path and a similar conviction seems to

be the case for others, if you cannot dismiss the religious claims of others nor assimilate it completely into your tradition, a plausible alternative is to assume that all are different creeds which, in spite of meanderings and crossings, actually run parallel to meet only in the ultimate, in the eschaton, at the end of the human pilgrimage. Religions would then be parallel paths and our most urgent duty would be not to interfere with others, not to convert them or even to borrow from them, but to deepen our respective traditions....[32]

Parallelism appears as a logical and simple alternative to the others presented, but in practice it involves a difficult commitment: the commitment to accept another's religion as a valid expression of his or her beliefs, and as a valid and true religion. This commitment is an extremely difficult one for the religious man or woman to accept, because it involves acknowledging that another religion may be valid, in the same sense that one's own is valid. It raises the question of exactly what is the purpose or validity of any religion, and it may start the believer on a path of questioning his or her beliefs. I see this as a positive step.

This understanding and acceptance of other religions in a parallelistic way involves a few problems. It assumes that each religion is a self-sufficient whole, and disregards the influences which have been exercised in the past, as a result of the development of one religion as an offshoot or reaction to another (i.e., Christianity developing out of Judaism, Buddhism as a reaction to Hinduism, Islam's debt to both Judaism and Christianity, etc.). It denies the possibility of growth due to interaction between faiths, as if all of human experience was contained (and every possible insight explained) in every religious tradition. In effect, it "splits humanity into watertight compartments."[33] Nonetheless, as a basis for interreligious understanding, this viewpoint appears to be superior to the previous ones as a 'stepping-off point' for interreligious dialogue. It is the first attitude which recognizes the value of other religious as religions, and with all their differences, it still recognizes that other faiths have value and meaning.

437

I believe that it is the attitude of parallelism to which the present 'communications revolution' will first lead us. In a world where religions abound, and where we in the United States are exposed to more and more people of different religious backgrounds, it is not feasible to maintain the attitude that people of different backgrounds are pagans, savages, or 'heathen'. Nor will it be possible, in this era of communication, for the educated and informed person to cling to misconceptions about other religions, and to assimilate these other religions into his or her own views. Since we are still faced with interaction with other religions, we must come to accept them on their own terms, without ideas of superiority, without the submerged motive of converting their adherents to our own faith, but with a recognition of their status as unique religions with distinctive beliefs and histories. Most important, we must recognize them as means by which their followers pursue their religious goals, whether these involve a moral life, a pleasant afterlife, escape from suffering or any other goal.

Understanding and acceptance of other religions may sound like a worthwhile goal, and one which a person can just decide to attempt, but in practice it may be difficult to achieve. I say that it is a difficult goal, because to accept others in their own religions, a Christian or Muslim (or whatever) must abandon his or her mission of evangelism, and realize that a Hindu or a follower of the Buddha is already a religious person. He or she does not need conversion, and furthermore probably resents our attempts at evangelism as an insult to his or her religious sensibilities. (I have already mentioned the widespread anti-Christian feeling in Africa and the Near East. Perhaps the insistence on 'evangelism' was an important cause of it.) Acceptance of other religions as valid religions, in the same sense and with as much force for their believers as ours for us, imposes radically new demands, but it is a necessary step if we are to approach people from other cultures and other faiths without the 'ugly American' tendency of condescension toward an 'inferior' group of people. It is a necessary step in understanding other faiths, a goal which is becoming more and more crucial in today's 'world community.'

What are the goals of religion? Are they the same for all religions, and is the Christian God the same as the

Hindu Brahman, or the Islamic Allah? These are some of the types of questions which arise when we begin to study other religions, or when we are confronted by other faiths in any way. It is the business of those engaged in interreligious dialogue to delineate areas of agreement and disagreement, in hopes of arriving at a common core of understanding, and also in hopes of understanding, those differences which are inevitable.

Once we have progressed past the barrier of acceptance of other religions as living traditions with something to contribute to our own understanding, an unbelievably vast wealth of insight and experience is opened to us. From the doctrine of dukkha in the Buddhist tradition (all in life is suffering[34]) to the central affirmation of Hindu thought ("that thou art" or "you are that," which affirms the ontological and underlying unity of all things)[35] many different doctrines can be found and explored. Where does one start, when confronted by such a large field? Usually interreligious dialogue starts with a fundamental question or a particular topic. Questions regarding the place of man in the world, the existence or nonexistence of a Supreme Being, and the nature of this Being, and the interrelation between these two topics are usually at the center of religion. Therefore, it is in these areas that the dialogue must begin.

It is not the purpose of this paper actually to present examples of his dialogue, because any examples would necessarily reflect bias and would be instead, a monologue. Therefore, I will conclude with a broad overview of one possible set of the questions which arise when we are confronted with other religions, and suggest further directions which an actual dialogue would attempt to follow.

In general terms, every religion may be said to have a goal toward which one is striving, and an outline of the method which the follower of that religion must use to obtain that goal. For a Christian, this might be a state of salvation, or a certain relationship with his God, through belief in Christ, or moral actions. For the Hindu, the goal may be a certain relationship with his God (although a different name would be used) through the recognition of the essential oneness of everything. Other religions pursue different goals, through different methods. The question

439

which arises is, are the goals the same? In what ways are they different? If the goals are similar in some ways, or if they are essentially identical, should we insist on our own means, or method, in achieving that end?

The type of dialogue which is desperately needed at the present time is one which builds on the similarities between the major world systems, rather than emphasizing the differences in form or content. Only by beginning with the similarities, and building on them, will it be possible to arrive at an understanding which will encourage good will between nations of different faiths, instead of those nations remaining in a state of ignorance or misunderstanding.

In our age of growing international interdependence and rapid world-wide communications, people all over the world are awaking to the fact that theirs is not the only religion in the world. The attitudes arising from this realization range from isolationism and rejection of the other faiths to the acceptance of these faiths as valid expressions of religious truth. In a post-industrial society, with its communications networks encompassing the world, it will be difficult for anyone to maintain an attitude of religious isolationism, and the believer in any of the world's various traditions will have to answer many questions regarding the world's other religions. If true international understanding is desired, dialogue between members of different faiths is necessary, and the result may be a deeper sense of religion than any achievable today.

It happened after some days, perhaps as the fruit of an ardent and sustained meditation, that a vision appeared to this ardently devoted Man. In this vision it was manifested that by means of a few sages versed in the variety of religions that exist throughout the world it could be possible to reach a certain peaceful concord. And it is through this concord that a lasting peace in religion may be attained and established by convenient and truthful means.

Nicholas of Cusa (1400-1464)[36]

ENDNOTES

[1]Philip Ashby, <u>History</u> <u>and</u> <u>Future</u> <u>of</u> <u>Religious</u> <u>Thought</u> (Englewood Cliffs, N.J.: Prentice-Hall, Inc., 1963), p. 151.

[2]For our purposes, a religion will be roughly defined as any system incorporating a world view which attempts to bring its adherents to an improved state of existence, through beliefs, actions, or ritual. Most important, a system is a religion if its followers perceive it as such.

[3]Hendrik Kraemer, <u>Why</u> <u>Christianity</u> <u>of</u> <u>All</u> <u>Religions</u>? (Philadelphia: The Westminster Press, 1962), p. 22. In referring to the various world systems of religion, it will be necessary to make use of the terms usually used--Christianity, Buddhism, etc. Although these terms tend to reify schools of thought or to refer to monolithic institutions which do not exist, their use is necessary. Also, because it is obvious that most of the major world religions have undergone substantial transformations throughout their various histories, I will try to limit any references to their specific beliefs to those which are usually seen as the most essential, or basic, to that faith.

[4]Examples of this are included in Mircea Eliade, <u>From</u> <u>Primitives</u> <u>to</u> <u>Zen</u> (New York: Harper & Row, 1967), pp. 41, 75, 623.

[5]At least to his own satisfaction--the point is that the believer can see a dichotomy between himself and non-believers, but when it becomes obvious to a Christian that a Muslim is also a religious person, the situation becomes more complicated.

[6]Seventeen <u>major</u> world religions are listed in David E. Sopher's <u>Geography</u> <u>of</u> <u>Religions</u> (Englewood Cliffs, NJ: Prentice-Hall, Inc., 1967), p. 13.

[7]Especially in the books of Ezekiel, Jeremiah, Isaiah, and elsewhere in the Old Testament.

[8]Wilfred Cantwell Smith, <u>The</u> <u>Faith</u> <u>of</u> <u>Other</u> <u>Men</u> (New York: Harper & Row, 1972), p. 55.

[9]See Eliade, p. 75.

[10]Ibid., p. 41, 478.

[11]John Hick, "The Outcome: Dialogue into Truth," in John Hick, editor, Truth and Dialogue in World Religions: Conflicting Truth-Claims (Philadelphia: The Westminster Press, 1974), p. 150.

[12]Quoted in John A. Hutchison, Paths of Faith (New York: McGraw-Hill, 1975), pp. 508-509.

[13]Implicit in one's belief in salvation was the corresponding fact that those outside the faith were damned forever.

[14]Harvey Cox, Turning East (New York: Simon & Schuster, 1977), pp. 12FF. Also see Part 1 of The New Religious Consciousness, ed. by Charles Glock and Robert Bellah (Berkeley: University of California Press, 1976).

[15]Abraham Kaplan, The New World of Philosophy (New York: Vintage Books, 1961), p. 335.

[16]Islam commands submission to Allah alone and is therefore seen as non-conducive to slavery or one person's domination over another; this would be shirk. See Smith p. 61, and Sopher p. 10.

[17]We should not forget that Christianity was not the first religion in America. Some examples of American Indian religion are given in Eliade, especially pp. 11-13.

[18]Smith, p. 127.

[19]The Western intellectual tradition tends to see truth in an exclusive way; and important example of this is the exclusively Western concept that no statement can be both true and not true. See Hick, p. 1ff. An Eastern philosopher would probably not agree with this Western idea.

[20]One point which should be mentioned is the nature of religious belief as central to a person's point of view. It is not so much that the believer constantly affirms the

truth of religious doctrine; as that he or she takes them for granted, and proceeds from that viewpoint. See Smith, p. 58.

[21]It is possible that in order to truly understand a religious tradition, we must actually believe its doctrine. Augustine once wrote, "I believe in order that I may understand." Perhaps this should be our practice.

[22]Raimundo Panikkar, The Intrareligious Dialogue (New York: Paulist Press, 1978), p. xvi.

[23]The American Indians were such a group, with a religious tradition which developed separately from Christianity. They were looked upon as religion-less savages and therefore inferior, and alternating attempts were made between trying to convert them, and trying to exterminate them. Neither attempt succeeded.

[24]Or so they say. The emphasis in many Christian churches is on conversion, evangelism, and spreading the "Good News."

[25]Smith, p. 129.

[26]Ibid., pp. 129-130.

[27]Ibid., pp. 131-132.

[28]One may even deceive himself into believing in his own religious liberalism.

[29]Panikkar, p. xvi.

[30]For example, there is the view held by some people that "we all worship the same God, we just call him by different names."

[31]Panikkar, p. xvi.

[32]Ibid., p. xviii.

[33]Ibid., p. xix.

[34]This concept is explained in the Dhammapada, one of the central Buddhist scriptures.

[35]See Smith, Chapter two.

[36]Nicholas of Cusa, _De Pace seu Concordantia Fidei_, I, 1, quoted in Panikkar, p. xi.

Bibliography

Argyle, Michael and Beit-Hallahmi, Benjamin. The Social Psychology of Religion. Boston: Routledge and Kegan Paul, 1975.

Ashby, Philip. History and Future of Religious Thought. Englewood Cliffs, NJ: Prentice-Hall, 1963.

Bellah, Robert and Glock, Charles. ed. The New Religious Consciousness. Berkeley: University of California Press, 1976.

Bradley, David G. Circles of Faith. Nashville: Abingdon Press, 1966.

Cogley, John. Religion in a Secular Age. New York: Frederick Praeger, Publishers, 1968.

Cox, Harvey. Turning East. New York: Simon & Schuster, 1977.

Deloria, Vine. The Metaphysics of Modern Existence. New York: Harper & Row, Publishers, 1979.

Dhammapada. P. Lal, trans. New York: Farrar, Straus & Giroux, 1967.

Eliade, Mircea. From Primitives to Zen: A Thematic Source-book of the History of Religions. New York: Harper & Row, Publishers, 1967.

Francoeur, Robert. Evolving World, Converging Man. New York: Holt, Rinehart & Winston, 1970.

Graham, Dom Aelred. The End of Religion. New York: Harcourt Brace Jovanovich, Inc., 1971.

Hick, John, ed. Truth and Dialogue in World Religions: Conflicting Truth-Claims. Philadelphia: The Westminster Press, 1974.

Hutchison, John A. Paths of Faith. New York: McGraw-Hill Book Company, 1975.

Johnson, Roger A., ed. Critical Issues in Modern Religion.

Englewood Cliffs, NJ: Prentice-Hall, 1973.

Kaplan, Abraham. The New World of Philosophy. New York: Vintage Books, 1961.

Kraemer, Hendrik. Why Christianity of All Religions? Philadelphia: The Westminster Press, 1962.

Panikkar, Raimundo. The Intrareligious Dialogue. New York: Paulist Press, 1978.

Slater, Robert Lawson. World Religions and World Community. New York: Columbia University Press, 1963.

Smith, Wilfred Cantwell. The Faith of Other Men. New York: Harper & Row, Publishers, 1972.

Sopher, David E. Geography of Religions. Englewood Cliffs, NJ: Prentice-Hall, Inc., 1967.

Wels-Schon, Greta. Portrait of Yahweh as a Young God. New York: Holt, Rinehart and Winston, 1968.

Wilson, Bryan. Contemporary Transformations of Religion. Oxford: Clarendon Press, 1976.

Yinger, J. Milton. The Scientific Study of Religion. London: The Macmillan Company, 1970.

THE DEHUMANIZATION OF MANKIND

Flora Darpino

Preface

This paper is a theoretical essay on distinct social trends throughout history. It follows the logic of socio-economic patterns of organization in any given period of time, and deliberately raises troubling questions about what it means to be "human" in the present age. To avoid diluting the impact of the argument, footnoting will be employed for the benefit of the reader who desires to view the evidence of these trends.

Counterevidence not considered in this paper provides hope that man will be saved from the plight I will describe. The pessimistic view presented is intended to provoke an emotional response from society so that man will recognize the possible danger of dehumanization in the future.

Introduction

Hominidae .. Homo .. Homosapiens .. Human Being .. Mankind. There are many words to describe what we define as our fellow man. We say with irreproachable conviction that we are human. We look around us and comment with the same justification that the man or woman next to us is undeniably human. But, what if we were asked what we base our decision on? Would we look the inquisitor in the eye and say, "Of course, he is human. After all, he walks erect. He has a large brain and is most definitely an athropoid."? Certainly not.

So how does man define a fellow human being? How can one say with conviction that one is human? What makes one person human and another inhuman? First, one would say that man is a rational, logical, thinking animal. This fact is what makes him human and not like other creatures in the world. However, can we honestly make that distinction mans' personal asset? Advanced computer technology has yielded an artificial intelligence beyond any single man's capacity. Yet, no one man would venture so far as to comment that a future Apple III with "artificial intelligence" is human.

447

Man does not look like a computer. He does not have an outside force that controls his life as the programmer controls the computer. Man is human because of his "god made", natural, biological makeup. Artificial hearts, artificial limbs, kidney machines, respirators are needed by some humans to survive. But are those persons who rely on such an "outside force" less human? No. Modern technology has simply redefined the physiological makeup, not the man.

Although all three factors are part of making us human, what does it mean to be human? In a recent motion picture, "The Blade Runner", a young man is faced with this same question. He must hunt "Android" robots who look and act human. When he and a robot fall in love, he asks a new question of himself. If they can love, do they not have a soul? Is it not the soul that makes us truly human?

The soul is not some biblical fantasy but a word given to the "human" ability to love. Love, the caring and need of one man for another, is what makes us essentially human. Therefore, for the purpose of this paper, what it means to be human is to love, need and care about another man. When these characteristics vanish, we will still be genetically human but dehumanized.

As we look around our world today, we become aware of events that are not "human". For example, the rising crime rate, Tylenol poisonings, civilian bombings. However, these same situations are the result of human action. As history progresses, humans are becoming less "human". Each individual man is evolving into an autonomous being. The autonomous man becomes alienated from the fellow "humans" surrounding him. His compassion dissipates along with his 'human'ity.

The cause of this autonomy is the changing economic structure. The Post-Industrial Society holds grave consequences for "humankind." Total dehumanization is more than possible--IT IS PROBABLE.

The Pre-Industrial Age

In the beginning, there was man. This man was part of a family that formed, with other families, a human community. The human community was created by intermarriage or combinations of similar communities. Together these "humans" became a tribe. However, these people had no control over their surrounding world. They had to migrate, as herds, to the areas best suited for their survival. They were unsettled by nature. As they roamed, they depended upon each other for companionship, protection, and survival with the family being the most important institution within the system. "Hence, the tribal community, the natural commmon body, appears not as the consequence, but as the precondition for the joint appropriation and use of the soil."[1]

Eventually, man realized that he could settle down in one location and work the land to his benefit. These communities still had the family as the central organization, but no single family owned or possessed the land on which he lived. The "humans" were tied together as communal proprietors.[2] There was no state, no organized government to control the people. They were united for survival.

The economic system was centered around a division of labor based on sex and age. The main unit, the family, was a cooperative. Men, women, and children pooled their resources. The men did most of the hunting while the women gathered food. Marriage was necessary to insure the individual's survival.[3]

As technology progressed, survival became increasingly assured. Irrigation for crop production allowed communities to enjoy permanence. Man came to view the land as "his workshop, his means of labours, the object of his labour and the means of subsistence."[4] The hunter and gatherer evolved into an agricultural society.

Even with the increasing importance of the land for survival, man still was dependent upon the community for health and protection. He did not have the attitude that the land was his to own. The land still belonged to the community, to the natural order.[5]

These communities were still kin settlements. Limited mobility caused the extended family to become prevalent. Man saw a need to work at the production of the population. THe increasing numbers within his family unit gave him the economic labour force he needed to survive. These large

449

families yielded a more stable economic situation. The division of labour within the kin was dictated by the authoritarian head of the household. The children were needed to help work the field and perform simple chores. The men were needed to harvest the fields, till the soil and protect the family unit and community. Women were needed to raise the children, feed the laborers and mind the household. The elderly's place was to instruct and advise. This structure yielded kinship ties that were stronger than the marital ties.[6] The economic structure propagated an interdependent need based family system for survival. Individual autonomy would destroy the intricate structure.

Craftmanship was an outgrowth of this agricultural system. Humans found a satisfaction in the work that they did together. The need of one man for another made each individual take pride in his work. People knew that their work supported the community system. Work was a "necessity of livelihood and an act of art to bring inner calm."[7] Craftsmanship yields satisfaction through the following six facets of his work.

There is no ulterior motive in work other than the product being made and the processes of its creation. The details of daily work are meaningful because they are not detached in the worker's mind from the product of the work. The worker is free to control his own working action. The craftsman is thus able to learn from his work, and to use and develop from his capacities and skills in its prosecution. There is no split of work and play, or work and culture. The craftsman's way of livelihood determines and infuses his entire mode of living.[8]

Man was one with his work. Yet, because he did not work alone, but with his family, he became one with his kin. Tied together by the same economic bond, the family extended outside of its kinship to the rest of the community. All the bands within the system were dependent upon each other. In the end, the result was an intricately balanced structure. The structure was self-supporting and complementary to the members.

Then something disrupted the balance of the system. This event toppled the basic foundations upon which the community was based. Industrialization was a sleeping satan. Man was once again asked to redefine his existence. History had yielded a man who was what he did. The new system asked him to sell his creativity as work itself. He was asked to perform a task of someone else's creation. For selling his productive capacity, he was given a wage.

The Industrial Age

In the beginning, there was satisfaction. Man abandoned his community in the countryside to sell his labor to the factory. He worked wholeheartedly using new skill to create a product from base to completion. Even though he was robbed of the satisfaction of owning and selling his creation, his wage was compensation.

His factory became his social center as part of a new community. It was not egalitarian for there were rules and regulations that had immediate effects on the old community. There was no longer a need for the extended family. Since the new community fulfilled the needs of the members, the roles of the elders were replaced by company policies and bosses. Numerous children would no longer be needed to work the fields. The human did not depend on the community to sell and/or barter the products he produced. The marketing of products was the company salesman's job. The effect: the marital ties became stronger than the blood ties. The nuclear family was born.

The initial satisfaction of the factory job dissipated as the structure of the work changed. The assembly line employed by Henry Ford soon spread throughout all means of production. This assembly line segmented the job so that the laborer came to engage in repetitive tasks. These tasks seemed useless, since the work was only a portion of the product's production. Insecurity replaced the initial feeling of satisfaction.[9] Therefore, the worker started to turn away from the new community due to the rising alienation from the work.

Mobilization compounded this alienation and insecurity. Because the worker was no longer tied to the land, he was often transferred from one section to another within the same factory or to completely different plants. People stopped forming the strong emotional ties with their fellow workers. Since relationships lost their permanence, humans found they had to learn to forget friends easily.[10] The worker, the "human", lost his security of the new community.

The Present Situation

The final dissatisfaction and alienation of the early industrial worker has worsened with time. "With the growth in the amount of knowledge available ... production has subdivided their ranks ..."[11] This division of labor has alienated man further. In a recent poll by Weiss and Kahn, over three-fourths of the respondents "defined work as activity which was necessary though not enjoyed as an activity which was scheduled or paid for."[12] Many men are no longer satisfied by work activity.

In order to make work more tolerable, leisure time is offered as a reward. The time apart from the job is the time in which man is to self-actualize since the workplace does not offer that opportunity. Self-actualization is when man takes the time to realize his own potential.

Where does man self-actualize? Where is this activity based? Studies by Robert Dublin and Arthur Kornhauser demonstrate that this activity is not centered around the workplace. Dublin found that only one in every three persons felt that the workplace was the area of self-actualization or social experience.[13] Kornhauser delved further into these facts and found that no worker along the production line viewed the workplace as the arena of self-actualization or social experience.[14] The result is that man has become alienated from his work and workplace.

To complicate matters further, increased mobilization has alienated humans from each other. The modern man finds that permanent residence is not a reality in the economic industrial system of today. This factor has become so prevalent that business magazines publish articles to help the man-on-the-move cope. They strictly advise to devote a little time to avoiding the usual get-togethers of people who work alongside each other. He is warned to be firm about limiting entertainment so that he does not become too attached to any one ... set of friends.[15] Man has lost his harmony with his work and his community.

Another breakdown has been a further reduction in the family importance. This can be seen by the rising number of divorces each year. Between 1870 and 1920 the number of divorces increased fifteen times.[16] From that time on, the United States has had the highest divorce rates of any industrialized country. The divorce rate is now over fifty percent for married couples.[17]

A possible reason for this rise is that as people become alienated from their workplace and fellow laborers, they develop an increased degree of autonomy and independence. The autonomy is carried into the family unit. The man of the house, with the aid of modern labor saving devices, finds he does not need a woman to perform household chores. He can function on the labor force, come home and function in the household. He no longer needs the classical division of labor in the home.

The women, on the other hand, can now enter the work force created by the new community. A popular television commercial portrays a lady singing, "I can bring home the bacon. Fry it up in the pan ..." Modern women also no longer need a man to help run a household efficiently. Modern women and men do not need the economic security of the institution of marriage.

Increased economic security compounded by rising divorce rates often instills a fear of marriage in mankind. Even fewer humans may enter the "sacrament" of marriage as they find it less of a need and more of a failure.

The rising divorce rates have caused a marked rise in single parents.[18] In a single-parent home, the one person raising the child is the autonomous role model. The single parent must assume responsibility for both the classic male/female stereotypes. As the child is raised within the single-parent home, he can subconsciously be conditioned to believe that both parents need not live within the same household. The child develops a view of parental autonomy.

Therefore, the children, as the adults, are slowly coming to think that they do not need the opposite sex for children, homes or companionship. The may lead people to turn to "man made sex."[19] Examples of this form of sex are: artificial insemination, frozen sperm and egg banks, embryo transplants, surrogate mothers .... One woman offered her services as a surrogate mother on a recent news program. Her fee was a small $12,000. The implication of these modern technological advances are obvious. Humans are autonomous to the degree that they do not need other humans for reproduction.

Do we still even need reproduction? The industrialized segment of the globe is increasingly deciding that we do not. Not only has the number of children per household decreased, but the number of households with children has declined. Humans no longer need the child to work the field. Instead, "child accounting" has changed from an asset to a liability. People now commonly say "I can't afford a child." Having a child is now an economic decision that many people do not have the time or money to undertake.

Humans who do not want children have many easily available options. Birth control is socially acceptable. Devices are available at every drugstore. Some can be found in public bathrooms for fifty cents. "Child accounting" is a prevalent, personal decision in our world. One author calls birth control a "parapersonal" sexual technology.[20] He explains that sexual technology is commonplace in our life, so common we often do not even notice. "They are out there, so to speak, and do not really impinge directly on each person.... They are tools we use, extensions of our senses and muscles...."[21] These extensions reflect something about the modern human. He has become numb to the fact that by these sexual technologies, he is making an unconscious decision to remain autonomous.

Humans have decided that the right to remain autonomous can justify any means. Therefore, he must make impersonal decisions divorced from his humanity. One area which clearly demonstrates this fact is the abortion issue. Abortion is often the autonomous decision of the woman with child. Legally, she may have the ultimate decision as to whether a certain human will be born. The law has set down guidelines as to the maximum age a fetus can be to be aborted. Besides these guidelines, often any woman of legal age can make the autonomous decision to have an abortion. This may be thought as logical since she is the one who must carry the baby.

There are some underlying reasoning patterns to be

examined in that decision process. The fetus is genetically a human at conception. It can undeniably be nothing but a human being. Therefore, the abortion question asks: when is the human fetus alive? This reveals an interesting aspect of the decision. When one autonomous person makes the decision to have an abortion, she is often making the conscious decision not to allow a human to be born. Yet, is that a decision that reflects love, need and caring for a fellow human being? Is that a "human" decision?

A similar issue which raises the same questions is euthanasia. Once again, humans are deciding if a fellow human life is worth living. This issue raises another interesting aspect of the de"human"ization of man. When discussing the area of euthanasia, the comments usually are: the machine is keeping him/her alive. It is a drain on the family, emotionally and economically.... These are guidelines in a cost benefit approach to end a human life. We are actually deciding which human life is more important: the family or the ailing. We are divorcing ourselves from the emotional aspect and making a clear, rational decision. We are obviously commenting that we have the capability to keep a human alive, but it is an inconvenience we would rather avoid. Is that a "human" decision?

Man's autonomy does not only affect his view of humanity. It has attacked long-standing establishments besides the family. One institution which has been greatly affected is the political structure in the United States. Humans are decreasing their identification with the two major parties in both frequency and intensity.[22] "The significant decline became evident in the 1970's. The fall-off was pronounced among strong party identifiers, down from thirty-five percent to twenty-six percent over two decades."[23] Voters have decided that the political choice should be autonomous from any ties. The decision is an independent choice, alienated from other prerequisites.

American politics has reflected the autonomy of humans in other aspects. Man plays his politics to promote his own personal interests. Groups are formed to support candidates who will further the individual's desires. These legal groups, called Political Action Committees (PACs) are increasing each year. PACs contributions accounted for over 50 percent of the political campaign receipts in the recent federal election.[24] Incumbents are the most heavily supported. It is these candidates who are key committee representatives that can promote the PACs interests.

However, the PACs that are best suited for the non "human" human are single issue groups. These PACs only support candidates who represent the individual;s interests. The Gun Lobby, the Anti-abortion Lobby, and Consumer Groups are examples of single issue groups. Their increasing support demonstrates man's autonomous need to fulfill his own desires.

Man seeks his own desires in other aspects of his life. Without a family, and with fewer friends, he had no established arena to enjoy his increased leisure time. Therefore, he searches increasingly to fulfill his many needs. Some humans advertise in newspapers for companionship, while others use computer dating systems. Many persons, however, gravitate to the "single bars" to mix with other humans.

The search for other humans is also affected by man's autonomy. The goal is personal, selfish gratification. Therefore, the game becomes intercourse-obsessed. One author classifies this as "hot sex" which is for fun, for ego satisfaction, for ego building.[25] Intercourse obsessed humans do not truly care for their partner. They merely look to satisfy their own orgasmic pleasure at the expense of the mate. "In the dark, one hole is the same as any other ... merely move the fig leaf to the face."[26] Hot sex is a physical sex segregated from life and emotions. It views the fellow human as a conquest, a person gained for personal satisfaction. This genital hedonism is in no way correlated to the definition of "human".

Summary of the De"human"ization

Man has changed throughout time. At one early period, he relied upon his fellow man for survival. Together, they worked, socialized, and produced. The family was the core of the structure as each member was interdependent for survival.

When man settled on farm land, the society structure changed little. Each man still depended upon his neighbor to help him daily. The family was based on sex and age. Children were important components of the labor force so the groups had large families. The husband and wife were important also as they divided the work of the field and that of the house.

Industrialization changed most of the established traditions. Man sold himself for wages instead of producing on his own. The work was segmented and did not yield the same satisfaction as the previous craftsmanship. Therefore, man became alienated from his work and the community within. The alienation led to an increased feeling of autonomy which affected every aspect of man's life. There has been a breakdown of the husband-wife relationship, parent-child relationship, male-female relationship, man-man relationship. In all these combinations, the individual makes autonomous decisions which disregard the other person. Man has learned to care less about his fellow man, and more about himself. He has learned to need less from his fellow man, and take more for himself. He has learned to love less of his fellow man, and love himself.

The Future

The future holds grave prospects for humanity. The workplace may have a devastating effect on man's humanism. As the past and present have shown, man has become increasingly autonomous as his economic situation changed. Predictions of man's future labor situation yield a dismal outlook.

Computer technology will be the main vehicle in all industries. The division of labor within the system will be highly specific. The terminal and keyboards will be equipped with all the systems needed for man to function without a common employee office. Due to these modifications, man's workplace will be moved to the home. Consequently, the contact with fellow men will decrease even further. There will be less human interaction and increased autonomy.

These computer systems will also enable man to shop for everything needed by a visual catalog. Entertainment, telephone and music programs will also be included in computer capacities. Therefore, man will no longer need to leave the home to fulfill these needs. Humans will be isolated within the household, autonomous and alienated from other humans.

457

Man will no longer need a mate to cohabitate the home. Labor saving devices and computer shopping will enable one person to sufficiently manage the home. In the beginning, the divorce rate will escalate and remarriages decline further. The single person will become the major status for humans. Finally, the institution of marriage will no longer exist.

One of the main functions of marriage is to propagate the earth. However, with the elimination of families, another institution will be formed to replace the family function. The "child production" industry will fulfill this gap. Test tube babies and other artificial reproduction techniques will produce the optimal number of babies as calculated by the industry. Children will then be raised in "life care" centers by trained professionals and computer systems. The computers will regulate when the child needs food or water for survival. The professional will educate and socialize the children for the world outside the center. Parents will not be needed or missed by the new industry children.

When the child has completed his training, the youth will be placed in an occupation suited to his or her intellectual capacity. The adolescent will move from the center to a computerized home, where job functions will be performed.

If people feel the drive for human sexual stimulation, they may make use of "man-made sex" techniques, which will yield the same satisfaction as intercourse with another person. The act will be reduced to the pleasure principle concept. However, in gathering places, persons may and will request human intercourse with others. If they wish to comply, it means nothing more than that they, themselves, desire pleasure.

Man will be autonomous, alienated from other men except for his own hedonistic pleasure. Humanity, as we now conceive it, will not exist. In fact, it will be an absurdity.

ENDNOTES

[1]Karl Marx, <u>Pre-Capitalist Economic Formations</u>, p. 68.

[2]Ibid, p. 69.

[3]Anna K. & Robert T. Francoeur, <u>The Future of Social Relations</u>, p. 30-32.

[4]Marx, p. 71.

[5]Marx, p. 82.

[6]Gerald R. Leslie, <u>The Family in Social Context</u>, p. 58.

[7]Stanley Parker, <u>The Future of Work and Leisure</u>, p. 8.

[8]Ibid., p. 10.

[9]Ibid., p. 47-49.

[10]Roger Williams, <u>Tomorrow at Work</u>, p. 15.

[11]Ibid., p. 86.

[12]Parker, p. 50.

[13]Ibid., p. 69.

[14]Ibid., p. 69.

[15]Williams, p. 15.

[17]Arlene & Jerome H. Skolnick, <u>Family in Transition</u>, p. 807.

[18]Francoeur, p. 86.

[19]Ibid., p. 5.

[20]Ibid., p. 4.

[21]Ibid., p. 4.

[22]William J. Crotty, <u>American</u> <u>Parties</u> <u>in</u> <u>Decline</u>, p. 28.

[23]Crotty, p. 29.

[24]Ibid., p. 110.

[25]Francoeur, p. 33-34.

[26]Ibid., p. 32.

REFERENCES

Aptheker, Herbert. <u>Marxism</u> and <u>Alienation</u>. (New York: Humanities Press, 1965).

Borland, Marie, ed. <u>Violence</u> <u>in</u> <u>the</u> <u>Family</u>. (Atlantic Highlands, N.J.: Humanities Press, 1976).

Brinton, Crane. <u>The</u> <u>Shaping</u> <u>of</u> <u>Modern</u> <u>Thought</u>. (Englewood Cliffs, N.Y.: Prentice-Hall Inc., 1950).

Crotty, William J. <u>American</u> <u>Parties</u> <u>in</u> <u>Decline</u>. (Boston: Little, Brown and Company, 1980).

De Beus, J. G. <u>The</u> <u>Future</u> <u>of</u> <u>the</u> <u>West</u>. (New York: Harper & Brothers Publishers, 1953).

Francoeur, Robert T. and Anna K., ed. <u>The</u> <u>Future</u> <u>of</u> <u>Sexual</u> <u>Relations</u>. (Englewood Cliffs, N.Y.: Prentice-Hall Inc., 1974).

Gouldner, Alvin W. <u>The</u> <u>Future</u> <u>of</u> <u>Intellectuals</u> <u>and</u> <u>the</u> <u>Rise</u> <u>of</u> <u>the</u> <u>New</u> <u>Class</u>. (New York: Seabury Press, 1979).

Hilton, Rodney, ed. <u>The</u> <u>Transition</u> <u>from</u> <u>Feudalism</u> <u>to</u> <u>Capitalism</u>. (London: Unwin Brothers Ltd., 1976).

Jersid, Paul, ed. <u>Moral</u> <u>Issues</u> <u>and</u> <u>Christian</u> <u>Response</u>. (New York: Holt, Rinehart & Winston, 1976).

Kant, Immanuel. <u>Foundation</u> <u>of</u> <u>the</u> <u>Metaphysics</u> <u>of</u> <u>Morals</u>. (Indianapolis: Bobbs-Merrill Educ. Publ., 1959).

Lesle, Gerald R. <u>The</u> <u>Family</u> <u>in</u> <u>Social</u> <u>Context</u>. (New York: Oxford Univ. Press, 1982).

Marx, Karl. <u>Capital</u>. (New York: The Modern Library, 1906).

Marx, Karl. <u>A</u> <u>Contribution</u> <u>to</u> <u>the</u> <u>Critique</u> <u>of</u> <u>Political</u> <u>Economy</u>. (New York: The International Library Publ. Co., 1904).

Marx, Karl. <u>Pre-Capitalist</u> <u>Economic</u> <u>Formations</u>. (London: Lawrence & Wishart, 1964).

McLuhan, Marshall. Understanding Media: The Extensions of Man. (New York: McGraw-Hill Book Co., 1964).

Owen, John D. The Price of Leisure. (Netherlands: Rotterdam Univ. Press, 1969).

Parker, Stanley. The Future of Work and Leisure. (New York: Praeger Publishers, 1971).

Rothblatt, Ben, ed. Changing Perspective on Man. (Chicago: The University of Chicago Press, 1968).

Skolnick, Arlene & Jerome H., Family in Transition. (Boston: Little, Brown & Co., 1980).

Sorel, George. The Illusions of Progress. (Los Angeles: Univ. of California Press, 1969).

Toffler, Alvin. Future Shock. (New York: Random House, 1970).

Toffler, Alvin. The Third Wave. (New York: Bantam Books, 1980).

Ward, Benjamin. The Radical Economic World View. (New York: Basis Books, Inc., Publishers, 1970).

Williams, Roger, ed. Tomorrow at Work. (London: British Broadcasting Corporation, 1973).

A SENSE OF LIMIT

Nancy Moore

Introduction

Throughout the history of civilization, reactionaries have spoken out against the changes they considered harmful to their society. Many times these concerned men and women have even altered the course of history. One thinks automatically of Jesus, Martin Luther, Gandhi, or the abolitionists and feminists. One genre of literature particularly suitable for performing the role of reactionary in this rapidly changing world is science fiction (hereafter sf). In fact, many sf writers have not hesitated to explore the consequences of technological and scientific advances in what we call the post-industrial society. Their attitudes vary, but one theme seems to emerge in almost all sf concerned with the fate of the human race, the idea that our society is about to reach, or has already reached, a limit where progress is no longer beneficial but instead may destroy our civilization and perhaps humanity itself.

The idea of limit emerges in a variety of different ways in science fiction. Therefore, I feel that the most effective method of exploring this theme is simply to give a broad survey of how sf writers use the idea, incorporating examples from their fiction but also examining how the idea of limit comes through in their essays, interviews, and speeches. Since this discussion is particularly about science fiction writers' attitudes toward the post-industrial society, I will examine works only from 1945 or later, with a few important exceptions. However, the New Wave, which began around 1962, changed the nature of science fiction by aiming for a higher literary quality and at the same time apparently turning sf writers' attention away from social issues. Therefore, the most helpful examples come from the earlier period of the 1950's or early 1960's. Furthermore, in order to concentrate on the post-industrial society as it is emerging in America, I have limited my survey to American writers, although with some exceptions. By obeying these restrictions, the following discussion will show that sf performs an important reactionary function by raising the question of limit to the attention of its readers.

463

There is a limit to how quickly people can assimilate change. In the past, change occurred slowly enough that men could adapt to new experiences and phenomena, but in the present change occurs so rapidly that "our capacity to adapt is being outpaced."[1] The consequences of this failure to assimilate rapid change are the feelings of alienation and insecurity that a person experiences. Alan E. Nourse, author of "Man's Adaptation to Change," says that

> more and more people regard change as threatening and catastrophic. Without sufficient time to assimilate change in an orderly fashion, even the more adaptive individuals have drawn back into a shell of conservatism, consciously or unconsciously fighting change tooth and nail, seeking desperately to maintain the status quo, however precarious it might be, and responding to the changing world with a vast conservative inertia.[2]

One way people react to a world characterized by change is by seizing on the idea of limit. Sf writers use this idea in a variety of different ways in science fiction, but their reason for using it always stems from the notion that there is a limit to how much knowledge humans should acquire, and thus how far they should go with progress. By going beyond this limit, sf writers feel, the human race is courting its own destruction. This belief is as old as the Tree of Knowledge in Genesis, but it also seems to be a dominant attitude among sf writers. For example, one sf writer, Frederik Pohl, implied in a speech at an annual meeting of the Modern Language Association that "the Almighty's patience is exhausted, and . . . the only thing left for us to do is prepare for the worst."[3] Frank Herbert commented once that "the seamless web of our world has come apart at the seams we didn't know it had."[4] Most sf writers have the sense that we are approaching a crisis point in our history, but whether they believe it will be the final catastrophic moment in the history of mankind or whether it will be simply one more crises in a history full of crisis depends on the individual writer.

However, not every sf writer believes that the human race is on the verge of reaching some sort of limit. As a consequence of technological progress, some writers have gradually rejected the idea of an orderly, understandable universe based on physical laws and accepted the idea that the universe is ultimately incomprehensible. In fact, Frank Herbert, somewhat contradictorily, has stated that "the best science fiction and pure science assume an infinite universe where we can look up at the blue sky . . . That sense of infinity (anything can happen) gives us the proper elbow room."[5]

Before proceeding with a discussion of how sf writers illustrate the idea of limit in their writings, we need to describe science fiction and the role it plays in the lives of its readers. Sf is a form of literature more concerned with idea than with style or character, although its quality as a form of literature has been improving in the past few decades. One way sf writers go about the process of writing is by starting with the world as they know it, introducing a certain number of changes, and trying to foresee the consequences of them. In other words, they ask themselves "What would happen if . . ." This method is called extrapolation and does not effectively predict the future because it does not take into account unforeseen circumstances. Nevertheless, extrapolation is still a useful method of exploring the consequences of present trends, as the following discussion will demonstrate. Sf usually does not attempt to predict future occurrences and it should not be judged by its predictive accuracy. As we shall see, science fiction writers describe possible futures but almost never claim that these futures are inevitable.

Even though sf writers include some aspect of science or technology in their stories, the main idea may have nothing to do with either. In other words, some writers use scientific material for reasons other than to describe possible futures. Often they use technology as a device that enables them to make comments about human nature.

Although sf writers may have extensive scientific backgrounds, not all do. Furthermore, their writings are almost always limited in inventiveness by the fact that they are ordinary people living in a certain place and time in history. In fact, Alan E. Nourse states that "too often science fiction's insights into human psychology are embarrassingly naive, and its grasp of sociological implications are distressingly provincial."[6] Sf writers' often narrow views and present-day attitudes are particularly reflected in the way they depict blacks, Jews, and women. Consequently, some critics would argue that science fiction describes the present, not the future. Thus, in any discussion of science fiction, one must keep in mind that in their fiction sf writers simply express their attitudes toward technological and scientific advances and do not necessarily know more than any other intelligent and interested person.

Nevertheless, what makes sf writers different from the average person is that they communicate their views to more people through their literature. Many sf writers take this fact very seriously. They believe that science fiction has an important role to perform because it must help readers adapt to change and yet warn against the dangers of technological and scientific advances. For example, in "The Apocalyptic Imagination, Science Fiction, and American Literature," David Ketterer states that

> the proliferation of science fiction is a response to abruptly changing conditions. During times of stability, when change neither happens nor is expected, or happens so gradually as to be barely noticeable, writers are unlikely to spend time describing the future condition of society, because there is no reason to expect significant difference. With the nineteenth century, things speeded up, and now change is a constant and unnerving factor in our daily lives. If we are to live rationally, and not just for the moment, some attempt must be made to anticipate future situations. Hence writers are drawn to science fiction: it is an outgrowth and an expression of crisis.[7]

Most sf writers feel that science fiction's purpose is to prepare its readers for accepting the inevitability of change, for adjusting to it when possible, and for reducing its negative consequences. For the most part, however, sf writers are concerned with how we should change, not whether we should change. Nevertheless, sf writers are aware of the possibility of a limit which, when reached and passed, may have dangerous consequences for the human race. Therefore, they see sf as being, in the words of Amis, "an instrument of social diagnosis and warning."[8] Many sf writers believe that such an instrument is vitally important in today's world. Furthermore, they believe that sf is the best means of performing this function. Theodore Sturgeon even said that "science fiction is the only possible pill against future shock."[9]

The Man-Machine Relationship

One typical story idea in science fiction is the relationship between man and machine. Since the computer, robots, and similar types of artificial intelligence are increasingly becoming important for the maintenance and survival of our post-industrial society, many sf writers have explored the possible consequences of this development. Their portrayal of the relationship between man and machine is usually negative, since sf writers create images of this relationship by considering the implications of a new computer application and then extrapolating this possibility to the point where it becomes undesirable and threatening. Obviously anything carried to its logical extreme will be depressing.[10] Nevertheless, by extrapolating the present trend in computer technology, sf writers can warn their readers that there is a limit to how much influence and power we should allow the computer to have over us.

Patricia Warrick, author of "Images of the Man-Machine Intelligence Relationship," gives three additional reasons for sf writers' negative attitude toward computers.[11] The first reason seems almost too obvious. Since all stories revolve around a conflict, the sf writer often finds it easiest to cast the computer in the role of the villain in order to create a conflict for his story. Second, computers are expensive and therefore owned by bureaucracies, not individuals. The protagonist in a story is usually an individual who is opposing the bureaucracy, and thus the computer. Third, computer technology is relatively new and not well-understood by most people. Consequently, sf writers often do not know enough about the computer to create imaginative, beneficial uses for it. Furthermore, ignorance of the true nature of computer technology may make sf writers, and others, fearful of it. Therefore, these writers most often depict the computer negatively and imply that we must smash it before it enslaves us.

Stories portraying the computer in the hands of a tyrannical government are common in science fiction. In fact, as early as the 1950's writers like Poul Anderson,

Shepherd Mead, Isaac Asimov, and Bernard Wolfe

foresaw the potential abuses of computerized
record-keeping and decision-making, particularly
in government. They recognized that the
opinion-tapping and opinion-shaping skills
developed by advertisers and politians would be
very dangerous if government controlled, and they
dramatized the power of computers as adjuncts to
the surveillance which so well abets
thought-control.[12]

Most sf stories dealing with the use of the computer in
government are very similar. In some future world, a large
computer is given the task of governing the country, thereby
making the actual leaders of the country mere puppets
carrying out the tasks assigned to them. Since the computer
possesses no emotions or sense of morality, its methods of
keeping order are often insensitive and cruel according to
human standards. Furthermore, it usually maintains peace
and prosperity by forcing citizens to conform to one
pattern. The earliest story of tyranny by computer is Kurt
Vonnegut's Player Piano (1952).[13] The title of this novel
is significant because it emphasizes the story's main idea
that, although a computer may run the country so well it
keeps its citizens from starving, it may also deprive their
lives of meaning.

The computer now performs functions that only the human
brain used to perform. Sf writers have reacted strongly to
the implications of this phenomenon by asking themselves
what makes a person human and by speculating on how the
machine may change human nature. Sf writers cannot help but
believe that man's ability to reason and his high level of
intelligence are uniquely human. For example, Roger
Zelazny's "For a Breath I Tarry" (1966) reflects the opinion
that man has complex and subtle qualities that defy
imitation. In this short story, a unique machine named
Frost is the "builder and maintainer" of the northern
hemisphere. The human race is long since dead and only the
machines are left carrying out the orders given to them
while the human race still existed. Frost becomes curious
about the nature of man and asks another machine to describe
what made the human being different from the machine. The
other tells him that "a machine is a Man turned inside-out,
because it can describe all the details of a process, which

a Man cannot, but it cannot experience that process itself, as a Man can."[14] Frost rejects this idea and attempts to experience reality (beauty, cold, etc.) the way human beings had done but finally admits that he cannot do so without actually becoming a human being.

Mixed with sf writers' belief in humanity's uniqueness is the guilt that by attempting to recreate human qualities in the machine, we are aspiring to be like gods. Thus, many sf stories describe some type of retribution which befalls humans who commit this hubristic action. For example, in "I Have No Mouth and I Must Scream" (1967), the machine develops a consciousness and then a hatred for humanity. After killing everyone on Earth except five people, it keeps these five immortal in order to torture them endlessly.

Many sf writers feel that "in the future man will be radically altered as a result of the machines he has invented."[15] For example, Samuel Delany's Nova (1968) describes how, as a consequence of the alienation that human beings experienced in the twentieth century when automation was increasing, the human beings of the future developed a way of connecting the computer to their bodies by means of neural plugs.

Many sf writers claim that humans take on mechanical qualities as they come to depend more on machines. Mordecai Roshwald's Level 7 (1959) is an excellent story using this idea. The protagonist develops a terrifying indifference to human emotions. When a co-worker commits suicide after pushing a button and thus helping to annihilate everyone on Earth's surface, the protagonist blames his comrade's suicide on the other's inability to detach himself from his own actions:

> I have no liking for the sight of life disappearing, bodies hanging. Like his. But to push a button, to operate a "typewriter"--that is a very different thing. It is smooth, clean, mechanical. That is where X-117 went wrong. For him it was the same thing. Maybe this inability to distinguish between killing with the bare hands and pushing a button was the source of his mental trouble.[16]

The protagonist even comes to wish he were a machine: "If I

469

were a real machine I should be much happier. A happy
gadget!"[17]

Stanislaw Lem approaches the question of man's
relationship to the machine by wondering, not what happens
when humans become more like machines, but what happens when
the machine becomes so much like humans that the two are no
longer distinguishable?

Should it become possible to build a synthetic
being, which nevertheless has all the physical
qualities of a human, it will no longer be
possible to use such a being as a machine . . . A
being so similar physically to a human being is,
considered ethically, a human being.[18]

In science fiction, machines so similar to human beings
are called robots. However, sf writers usually do not use
robots in their stories in order to warn readers of their
potential danger but rather to contrast them to human beings
and thus reveal certain human characteristics which are
either good or bad. For example, Philip K. Dick's short
story "The Defenders" portrays a world where the human race
lives underground while robots fight the war on the
surface. Periodically the robots send down accounts of
terrible battles and urge the humans to stay below for their
own protection. When a party of humans eventually surfaces,
it discovers that the robots called of hostilities as soon
as the last human had disappeared underground. Agreeing
that humans could not be trusted to stay peaceful, the
robots had spent all their time making their reports as
interesting as possible.

Isaac Asimov is probably the most famous writer of
robot stories. In fact, his "Three Laws of Robotics" have
been adopted by many other sf writers. They are

1) A robot may not injure a human being, or,
through inaction, allow a human being to come to
harm,
2) A robot must obey the orders given it by human
beings except where such orders would conflict
with the First Law, and
3) A robot must protect its own existence as long
as such protection does not conflict with the
First or Second Law.[19]

470

Asimov also uses the robot to comment on human nature. However, his faith in the superiority of the machine over the human is nauseatingly optimistic. For example, in Asimov's short story, "Reason" (1941), the robot explains why he cannot believe he was created by a human being:

"Look at you . . . I say this in no spirit of contempt, but look at you! The material you are made of is soft and flabby, lacking endurance and strength, depending for energy upon the inefficient oxidation of organic material--like that." He pointed a disapproving finger at what remained of Donovan's sandwich. "Periodically you pass into a coma and the least variation in temperature, air pressure, humidity, or radiation intensity impairs your efficiency. You are makeshift.

"I, on the other hand, am a finished product. I absorb electrical energy directly and utilize it with an almost one hundred percent efficiency. I am composed of strong metal, am continuously conscious, and can stand extremes of environment easily. These are facts which, with the self-evident proposition that no being can create another being superior to itself, smashes your silly hypothesis to nothing."[20]

Furthermore, in "The Evitable Conflict", the main character explains that in order to carry out the commands implied in the Laws of Robotics, the machines have taken over complete control of the fate of the human race. They are able to direct the course of history and prevent humans from interfering. The main character quite seriously describes this situation as "wonderful."[21]

However, Stanislaw Lem insists that robots, or attempted imitations of human beings, will never actually be used in the real world because

it would be nonsense to people the world with such "supermen" . . . In an automated factory there are no two-legged robots, and they aren't likely to be there in the future. Not the moral, but the technological derivatives point so. The picture

471

of a machine guarded by a machine who, perhaps, after work, will exchange a few words with his electronic colleagues and then go home to his electronic wife is pure nonsense . . . . It isn't worth the effort and never will be, economically, to build volitional and intelligent automatons as part of the productive process.[22]

Furthermore, he claims that Asimov's Laws of Robotics give "a wholly false picture of the real possibilities"[23] because such laws could never by programmed into robots and yet work.

Many sf writers fear the computer because they realize that in the wrong hands it could be used to enslave human beings. They also fear that human beings may become more mechanical as they come to use machines extensively. Finally, sf writers believe that there is something wrong about trying to recreate human qualities in the machine and they fear the retribution that may result from trying to do so. However, not all sf writers view the machine so negatively. In fact, Amis says that the majority of sf writers are not afraid of the machine's potential. They reject the Luddite solution of machine smashing, instead celebrating the progress of technology, despite the problems it may create. Some writers, like Ray Bradbury, reject certain characteristics or uses of the machine without rejecting it entirely. For example, in "The City" (1950), Bradbury shows a still-functioning computerized city on a planet where all the inhabitants have died. The city's sole purpose is to wait until a ship from Earth discovers the planet and then seek revenge against the Earthmen, whose race accidentally killed off all the inhabitants a few thousand years before. In this story, Bradbury objects to how people use the computer, but he does not object to the computer itself. Another story which shows that not all sf writers fear the machine is Frank Herbert's Destination: Void (1966). Here, the author "assumes that in the changed environment of man's future, the computer will be necessary for optimum survival, and so he refrains from simple-minded computer smashing in his treatment of the subject."[24]

Visions of the Future: Science Fiction as Warning

The man-machine relationship is a popular idea in science fiction but it is not the only aspect of the emerging post-industrial society that science fiction writers oppose. Paul A. Carter states in his book, The Creation of Tommorrow, that in the early 1970's many sf writers would have agreed that "man has come to have an impact on the biosphere as profound as that exerted by natural forces, so that it behooves him to act as responsibly toward his environment as if he were a process of Nature, comparable to mountain-making revolutions and ice ages."[25] Only recently have people become aware that the world has a delicate balance which they have disrupted. Consequently, many sf writers have reacted by saying that humanity's ability to influence the future should not be underestimated and that we must give some attention to where we are headed. Carter claims that

> we have eaten from the forbidden tree, and we are stuck with the technology that is one of the fruits thereof. We may pine for a lost Eden where no machines intrude, or dream of another beyond the skies, a theme often picked up and transformed by the authors of interplanetary and interstellar tales . . . but here below we have the exacting task of trying to fashion paradise out of iron.[26]

Most writers accept the necessity of change. They admit there is no going back to a rural, idyllic past, and very few of them would want to. Alan E. Nourse sums up their outlook when he says

> change is going to continue no matter how man reacts to it, and it is going to change lives ever more swiftly and massively, whether men want their lives changed or not. It is no longer even appropriate for us to ask whether change can be prevented, minimized, or controlled; the appropriate question is how change can be dealt with in such a way that individuals and society can survive and prosper in the midst of it.[27]

Nevertheless, sf writers are concerned about how the world is changing. They believe that some present trends will soon reach a limit where they will cause enormous problems which may not be solvable. In order to warn readers about these potential problems, sf writers extrapolate present trends into the future. Some of the issues which concern them are population control, production and consumption, advertising, male/female relations, and religion.

Population control stories became common in the 1960's. These stories reiterate the idea that the Earth can only sustain a certain number of people and that we are rapidly approaching that limit. At a session of the Modern Language Association, Frederik Pohl stated that

> out of every 100 pounds of living matter on the face of the planet--whether it is whales, or porpoises, or redwoods, or bacteria--two pounds are human flesh. . . . This, in turn, has its inevitable consequences, in terms of the degradation of the environment, at least as we perceive it. I mean it's goodbye to beef steaks and all that sort of thing. It is quite certain that we will have inevitable mass famine affecting hundreds of millions of people in our life time, and by inevitable I mean to say it's inevitable . . . we can no longer avoid it.[28]

Stories about overpopulation usually emphasize that the quality of life decreases as a consequence of overcrowding. For example, Frederik Pohl and C. M. Kornbluth's novel, The Space Merchants (1952), shows a world so overcrowded that many inhabitants live on the stairs of Manhattan office buildings. A "spacious suburban roomette" is a home where a partition must be set up to separate the children's nook, and the bed must be folded into the wall each morning. Food is scarce and often is based on some man-made, unappetizing ingredient. Water and fuel are other scarcities.

Since human beings have virtually eliminated such natural methods of controlling population as disease and famine, many sf writers show how future human beings will be forced to take over that role themselves. For example, in Frederik Pohl's "The Census Takers" (1956), the narrator is an area boss who oversees the activities of a group of enumerators taking a census of a certain region. Every three hundredth person they count is an "Over", which means that he or she is excess population and must be killed. Anyone who slows down the census-taking process, for whatever reason, is "Overed". Enumerators who do not do their jobs quickly enough or who begin to crack under the strain are also Overed. The narrator is efficient simply because he is so brutally callous. Furthermore, in Harry Harrison's "A Criminal Act" (1967), the author describes a society where the birth of a surplus child makes it legal for any citizen to kill one of the parents. If the citizen succeeds, or the parent is able to kill him first in self-defense, the population balance is considered restored.

Another issue which many sf writers explore is advanced capitalism. These writers fear that our society is already beginning to force citizens to consume what they neither want or need, simply to keep the system operating. They believe that there is a limit to how far our economic system can develop before it no longer serves us, but we serve it. Consequently, many sf stories portray a future America in which the capitalistic system has come to control the people. At the same time, these writers question the system's ability to satisfy real human needs.

One of these stories is Frederik Pohl's "The Midas Plague" (1954). In this story, the Ration Board regulates the amount of goods each citizen is required to consume. No one works more than a few days because he needs the remainder of his time for consuming. If a citizen fails to consume his allotment per month, or is discovered to have wasted it needlessly, the Ration Board adds to the amount he must consume for the next month.

The author explains that this situation was caused by the use of robots in production. Although robots produced more quickly than the human worker, they could not consume what they produced. Consequently, pressure was placed on humans to keep up with the fast pace of production. The author explains that the solution to this problem is neither to get rid of the robots, since by this time everything is completely mechanized, nor to slow down production, since slowing down is a prelude to stopping completely. The protagonist of the story solves the problem very neatly by programming his robots to do the consuming for him. By adding a satisfaction unit to them, he eliminates the problem of unnecessary waste.

This story is an example of a comic inferno, a literary form which rests

upon conceivable developments in technology. . . .

> Its moral value . . . is that it ridicules notions
> which various heavy pressures would have us take
> seriously: pride in a mounting material standard
> of living, the belief that such progress can be
> continued indefinitely and needs only horizontal
> extension to make the world perfect, the feeling
> that the accumulation of possessions is at once
> the prerogative and the evidence of merit.[29]

Many comic infernos by Frederik Pohl address the issue of production and consumption.

However, Pohl is not the only writer concerned about the future of the present economic system. For example, in "The Subliminal Man" , by J. G. Ballard, the author explores the basic contradictions of advanced capitalism. He shows that the present economic system is completely irrational because it fails to satisfy real human needs and at the same time enslaves the subconscious to it own needs. The title illustrates the author's opinion that advanced capitalism makes humans subliminal, less than human.[30]

One aspect of the production/consumption problem is the fear of how far advertisers will go to sell their products. Many sf writers distrust advertising because they believe that in order to sell people products which they do not need, advertisers somehow get around our conscious, rational minds by invading our unconscious processes. This fear is common in society today. However, sf writers also fear the methods advertisers will invent in the future, when there will be even less need to buy many types of products. Therefore, they try to warn readers that there is a limit to how much we should let advertisers control our needs and invade our mental privacy.

The Space Merchants (1952), by Frederik Pohl and Cyril M. Kornbluth, is a good example of how sf writers portray the dangers of advertising. The protagonist of this novel holds a high-level position in one of the biggest advertising companies in the world. The company, Fowler Schocken Associates, is virtually omnipotent. Furthermore, it will try any means to sell its products. For example, when compulsive subsonics in aural advertising was outlawed, advertisers responded by creating semantic cues. When this method was stopped, they turned to projecting ads on the windows of aircars, and then onto the retina of the eye.

This particular company also put a harmless but habit-forming drug in one of its products, thereby forcing the consumer to keep buying the product because the cure was so expensive. Even though the legal limit for a commercial is three minutes, and the standard is eight, Fowler Schocken Associates intend to create a nine-minute commercial.

Up until recently, most science fiction was sexist. The majority of sf writers were male. The few female writers kept their sex a secret from readers in order to have their works read, since sf readers were mostly male, too. Consequently, sf stories were filled with sexism and male chauvinism. Both writers and readers seemed to think that males and females were two entirely different species, rather than two indispensable sexes of the same animal.[31] The absence of stories trying to represent a more realistic and creative future for male/female relations revealed how limited sf writers were by their present-day attitudes. For example, many writers attempted reversing the roles of the sexes, but when they did, the women were always shown as strong, warrior-like Amazons. Sam Maskowitz, author of Strange Horizons, explains that

> the fictional creation of an Amazon world, as has been documented, is one of the oldest themes in science fiction. . . . Men, in every case, seem uneasy about this "equality," claiming that it will end in domination . . . that women will also insist upon retaining their special privilege which will give them superiority. The intensity with which these feelings are held is underscored by the very high percentage of stories in which women literally exterminate the males upon achieving domination.[32]

Nevertheless, sf is an excellent medium for exploring the possibilities of male/female relations. Consequently, some writers seized the opportunity as early as the 1950's to ask "What if only females existed? What if different, even repulsive, sexual characteristics were discovered? What if the sexes were combined to create a new, androgynous individual?"[33] Many sf writers only play with the idea of male/female relations, although there are some writers who explore the issue seriously. When these more serious writers describe possible futures for the relationship between the sexes, they often depict a movement toward

sexual equality. More creative writers even suggest that
the limit for male/female relations may be androgyny.

Two very good sf stories which explore male/female
relations are John Wyndham's "Consider Her Ways" and Philip
Wylie's The Disappearance (1951, 1966). In "Consider Her
Ways," a mutated virus accidentally kills off all the males.
Through the use of hallucinogenic drugs, the mind of a girl
from the present becomes imprisoned within the body of one
of these surviving women, and then has an extended
conversation with another woman, an elderly historian named
Laura.   Laura is an intelligent, thoughtful person who,
throughout the course of their conversation, wins the
argument that women would be better off without men. She
also makes some damaging criticisms of the contemporary
female role.   In The Disappearance, all the men disappear
from the women's world and all the women disappear from the
men's world.   Both sexes must then rearrange their worlds
to accommodate the loss of the other sex. After four years
they are suddenly reunited, but not until both have learned
from the experience. The point of Wylie's story is that,
though the sexes possess obvious differences, they possess
far more similarities.

Ursula K. Le Guin is a fantasy and science fiction
writer who often examines male/female relations in her
stories. In the introduction to her book, The Left Hand of
Darkness (1969), Le Guin comments that she is not

> predicting that in a millennium or so we will all
> be androgynous, or announcing that I think we
> damned well ought to be androgynous. I'm merely
> observing, in the peculiar, devious, and
> thought-experimental manner proper to science
> fiction, that if you look at us at certain odd
> times of day in certain weathers, we already are.
> I am not predicting or prescribing.   I am
> describing.[34]

The book is about the Gethenians, inhabitants of another
planet, who are neither female or male. Most of the time
they are neutral and have no sexual desire. However, they
do have a monthly fertile cycle called kemmer and during
this time the sexual drive dominates, compelling the
Gethenian to select a sexual partner. Depending on hormonal
secretion, s/he takes either the male or female role.   The

478

advantage of this form of reproduction is that

anyone can turn his hand to anything. This sounds
very simple, but its psychological effects are
incalculable. The fact that everyone between
seventeen and thirty-five or so is liable to be .
. . "tied to childbearing" implies that no one is
quite so thoroughly "tied down" here as women,
elsewhere, are likely to be--psychologically or
physically . . . therefore nobody here is quite so
free as a male anywhere else is.[35]

The narrator of the story is a foreigner who resembles the
American male. At first, he is horrified and repulsed by
the Gethenians. Yet when he returns to his own world, he
sees human beings as incomplete and distorted creatures
constantly concerned about their sexuality. Another book in
which Le Guin addresses the subject of male/female relations
is The Dispossessed (1974). Here, the planet Anarres is so
egalitarian that the inhabitants have invented an artificial
language free of masculine and feminine inflections.

The subject of religion is another issue which sf
writers explore, although their depictions of religion are
usually unfavorable. Surprisingly, sf writers do not
describe futures in which humanity has discarded religion,
but they do depict futures where religion is merely a
remnant of a previous society and has lost its value. Sf
writers seem to think that as we move into the
post-industrial society, religion will reach its limit of
usefulness.

Many sf writers believe that as we venture into space,
religion will no longer be needed to explain the unknown
and, thus, will increasingly become an empty institution.
Consequently, many sf writers describe a future where human
beings have developed religions out of ignorance or
misunderstanding. Other sf writers equate religion with
humanity's return to a savage state.

One reason why sf writers view religion unfavorably is
that science and religion always seem to be struggling
against each other. Even though sf writers often portray
the evils brought about by scientific progress, they still
have more faith in science than in religion. They identify
reason or rational thinking with science, and emotionalism

479

or irrationality with religion. For example, Robert Heinlein states that "we could lose our freedom by succumbing to a wave of religious hysteria. I am sorry to say that I consider it possible. I hope that it is not probable. But there is a latent deep strain of religious fanaticism in this our culture; it is rooted in our history and it has broken out many times in the past."[36]

One short story which explores the conflict between science and religion is "The Quest for St. Aquin" (1951). This story is set in a postatomic world where few Christian believers still exist. In order to attract people to the church, the pope sends out one of his holy men, Thomas, to search for the remains of St. Aquin, who led many people to the church before he died. When Thomas finds him, he discovers that the saint had been a robot. Nevertheless, in order to increase the number of adherents to the faith, the church is forced to canonize it.

Another reason why sf writers treat religion unfavorably is that science and technology are beginning to take over the role of creator. J. Norman King explains that

the technological and behavioral powers born of the new sciences convey the sense of control over the creational process. The idea emerges of man's purposes; of his actively shaping himself, his society, and his history. Instead of inquiring into God's creation of nature and man in the past, attention turns to man's self-creation in the future.[37]

Thus, as man learns to create, an act which before he thought only god could do, he no longer needs the concept of a divine creator to explain how life comes about. Man himself is the creator.

Sf writers also view religion unfavorably because they believe it is limited to Earth. Kingsley Amis says that the common attitude of science fiction writers toward religion is "casual disrespect," adding "it is as if religion were tacitly agreed to have an earthly, or Terrene, limitation when the scale of human activity has become galactic."[38] Thus, sf writers feel that as the human race begins to explore an infinite universe and to see Earth as a tiny, isolated part of that universe, people will lose faith in religion because they will consider it irrelevant in a vast universe with other worlds.

Many sf writers use space travel to different worlds as a means of criticizing religion and religious practices. At the same time, these stories reveal the writer's belief that religion is limited to Earth. For example, in "The Fire Balloons" (1951), by Ray Bradbury, a group of priests is sent to Mars to seek out sin, believing that it will possess different characteristics from earthly sin. The priests arrive, build a makeshift church complete with organ, and wait for the Martians to come to Sunday service. However, when the Martians do arrive, they explain that in the distant past they left their physical bodies and became pure soul. Consequently, they do not sin and thus have no need to be saved. Chastened, the priests humbly ask to learn from them. Another story along these lines is "Apostle to Alpha" (1966), by Bette T. Balke. Here, the church sends an apostle to another planet in order to convert the inhabitants. The apostle discovers, however, that they are a race of highly intelligent birds who never fell from Grace. In order to protect the birds from human beings, the apostle keeps his findings secret.

However, other writers use the space travel motif to give different perspectives on religion, particularly Christianity. For example, Ray Bradbury's "The Man" (1949) describes how the first Earth fleet to land on another planet is ignored because something more important occurred the day before, the coming of Christ. Arthur C. Clarke's "The Star" (1955) is about a group of astronomers, light years from Earth, who are exploring the remnants of a supernova which exploded 6000 years before. On one of the planets, too far away to be dissolved in the explosion, they recover a record of the people who had lived there and who had not been able to escape the blast. By reading the record, the protagonist discovers to his horror that this exploding sun had been Earth's star of Bethlehem.

Visions of the Future: The Fate of Humanity

Many sf writers go beyond extrapolating present trends into the future by speculating on the ultimate fate of humanity and suggesting possible outcomes. Their purpose is still social diagnosis and warning, but they are more concerned about humanity's ultimate survival than in stopping specific social ills from getting out of control. Sf writers who speculate on the fate of humanity obviously reveal concern about the question of limit because humanity's ultimate fate _is_ a type of limit.

One possible fate is the establishment of a utopian or perfect society. David Ketterer claims that "the American imagination is obsessed with dreams of a utopia."[39] Certainly most people would prefer a world where there are no problems, no hungry people, no oppression, no unfulfilling jobs to do, and no reasons to be afraid. Since science fiction is a medium where anything can come true if the author wishes it to, many sf writers have tried to create a utopia. As a result, sf writers, perhaps more than any others, know how difficult a task that is. Paul A. Carter states the problem succinctly when he says "one trouble with a utopia, which by definition is a society that has solved all its problems, is that living in it can be very dull."[40]

Utopias are static societies requiring total conformity in order to succeed. Individuality would upset the balance. Since sf writers value the individual above the group, they assert that a society that does not have a variety of individuals existing within it is a dying society. Furthermore, when people are not allowed to be individualistic, their lives become empty and pointless.

One of the major motivating forces in human beings is the search for meaning. This search is often the reason why people create great works of art, music, and literature. J. Norman King explains in "Theology, Science Fiction, and Man's Future Orientation" that

> man, as science fiction recognizes, is a being who relentlessly searches for meaning and purpose in his life, yet is also threatened by anxiety, death, and meaninglessness. He craves a higher life transcending his present ambiguity. He has at least an inkling that this somehow lies beyond his grasp. The theme of the alien reflects, at its deepest level, man's profound craving that the universe be more than just man and nature. Man is forever in quest of "something more," a prescence and depth of hope, of meaning, of value, which

transcends him. Man is not enough for man.[41]

Utopia prevents the kind of growth necessary for human beings to satisfy their need for meaning. Few people ever end their search for meaning, thus they need an environment where they can continuously explore. The search for meaning must be carried out in a world of dynamic change, a world where people are stimulated and, therefore, a world where pain, fear, and uncertainty exist. King insists that

> it is not just a matter of simple choice between security and courage. . . . Sterile sameness and stable mediocrity are overcome only by unleashing, along with human freedom and creative genius, the forces of anxiety, chaos, and destruction.[42]

Another reason why sf writers oppose the idea of utopia it that one of its characteristics is usually the endless pursuit of pleasure. The thought of how unlimited opportunities to experience pleasure might corrupt a person makes many sf writers reject any situations where that could occur. Also, sf writers oppose situations where people are not required to work. This objection may be a residue from the work ethic. Nevertheless, many writers feel strongly that people should not have too much leisure.

In order to clarify the problems which are inherent in the concept of a utopian society, sf writers depict dystopias, or anti-utopias, which are societies characterized by oppressive conformity and loss of free will. One of these dystopias, which seems to be a model for many of the dystopias created after it, is Yevgeny Zamiatin's We (1924). This excellent novel, written soon after the Bolshevik Revolution, describes a future totalitarian society based completely on mathematics. The main character, a mathematician and engineer named D-503, is thoroughly indoctrinated into his society and thus can describe it in all seriousness as "a victory of all over one, of the sum over the individual."[43]

The citizens of this society believe their totally constricted "unfreedom" is the ideal condition. Those who fail to conform are literally "liquidated" by the Benefactor, the watchful eye of the One State. In fact, in this society passion, creativity, and imagination are nearly forgotten human qualities. Therefore, when the main character develops an uncontrollable passion for another "number," he worriedly goes to a doctor to have his sickness diagnosed and is told that he has an incurable condition known as the "soul." And as more and more citizens begin to feel unrest, the One State Gazette posts a reassuring bulletin:

This is not your fault--you are sick. The name of this sickness is

IMAGINATION

It is a worm that gnaws out black lines on the forehead. It is a fever that drives you to escape ever farther, even if this "farther" begins where happiness ends. . .

The latest discovery of State Science is the location of the center of imagination--a miserable little nodule in the brain in the area of the pons Varolii. Triple X-ray cautery of this nodule--and you are cured of imagination--

FOREVER

You are perfect. You are machine-like. The road to one hundred percent happiness is free. Hurry, then, everyone--old and young--hurry to submit to the Great Operation.[44]

The main thrust of Zamiatin's novel is to warn readers of the dangers of a totalitarian state which takes away the right to individuality and seduces people into believing that the ideal condition is slavery. We is a haunting, frightening illustration of what can happen when the interests of the group are considered over the needs of the individual.

Kurt Vonnegut's Player Piano (1952), on the other hand, is similar to many dystopias in that it portrays a society run by computers. Vonnegut expresses his reason for writing the book through the voice of a minor character: "My husband says somebody's just got to be maladjusted; that somebody's got to be uncomfortable enough to wonder where people are, where they're going, and why they're going there."[45]

484

Player Piano describes a totally automated future America. A computer spanning the Carlsbad Caverns regulates every aspect of the citizens' lives, deciding how many refrigerators, lamps, turbine-generators, hub caps, dinner plates, rubber wheels, televisions, etc. America will produce, who will get them, and how much they will cost. Furthermore, the computer gives education tests to every citizen in order to determine who it will allow to become the engineers and scientists. No other jobs are needed.

In fact, there are only two job choices open to the majority of citizens, either the Reconstruction and Reclamation Corps (Reeks and Wrecks) or the Army. Both are unnecessary occupations because they are merely ways to keep the citizens busy. Some who refuse to do either of these jobs become professional gamblers, trying to guess the song played by an orchestra on television while the volume is turned down, or attempting to be the best match-game player in town. When one character bitterly criticizes the system for allowing computers to compete with people for work, a co-worker reminds him that machines are only slaves which, fortunately, do a better job than people. The former snaps, "anybody that competes with a slave becomes a slave."[46]

Even though no one in this society lacks materially, no one is really happy either. Yet when the citizens of Hometown rebel against their situation by smashing all of the machines, the protagonist discovers them putting everything back together again two days later. Apparently, making machines is the only thing human beings know how to do anymore.

A somewhat different type of dystopia, yet reflecting many of the same ideas, is Ray Bradbury's Fahrenheit 451 (1953). This novel depicts a future America where books are illegal because they make people think. The title of the novel indicates the temperature at which book bindings burn, for in this society firemen set fires instead of put them out.

The author relates that some time after the twentieth century, the population grew so large that of necessity everything "leveled down to a sort of paste-pudding norm."[47] In order to keep everyone happy, conformity became the rule. As one character explains,

we must all be alike. Not everyone born free and
equal, as the Constitution says, but everyone made
equal. Each man the image of every other; then
all are happy, for there are no mountains to make
them cower, to judge themselves against.[48]

The pace of life has also speeded up. The pursuit of
pleasure is the primary aspect of a citizen's life. Each
lives in a fantasy world created by 4-wall television sets
and radios which can be plugged into the ear. In other
words, "life is immediate, the job counts, pleasure lies all
about after work. Why learn anything save pressing
buttons, pulling switches, fitting nuts and bolts?"[49]

Fahrenheit 451 is an obvious attack against certain
aspects of today's society. Bradbury's message is reflected
in what one of his characters says:

Everyone nowadays knows, absolutely is certain,
that nothing will ever happen to me. Others die,
but I go on. There are no consequences and no
responsibilities. Except that there are. But
let's not talk about them, uh? By the time the
consequences catch up with you, it's too late,
isn't it, Montag?[50]

It is definitely too late by the end of the novel, since the
entire country lies glowing in a radioactive haze. Only a
few exiles, trying to preserve the knowledge of past
generations, survive.

More idealistic sf writers believe there is no limit to
how far human beings can progress. They reject the idea of
a dystopian state because they view the future as a
continual movement forward. Stanislaw Lem, for example,
makes fun of the utopian-dystopian concept because he sees
science as "an unending process that throws up new questions
for any problem solved."[51] Frank Herbert expresses the
same idea by pointing out that "now that the Vietnam war has
been brought near a close, we awaken to the realization that
we still live in a world threatened by imminent, totally
destructive, mass conflict. We cure the disease and find we
still suffer from it."[52] And the protagonist of Mack
Reynolds' "Border, Breed, nor Birth" (1962) remarks to
another character, "You know . . . in history there is no

happy ending at all. It goes from one crisis to another, but there is no ending."[53]

Writers who believe that humanity will continue to move forward, experiencing new crises along the way, do not express a pessimistic attitude toward human history. For instance, if someone offered any one of these writers the choice of a utopian world, she or he would most likely reject it in favor of a world where human beings must continually experience new conflicts. In fact, the editor of Astounding magazine once said in an interview that "all science fictioneers are fundamentally optimistic." When the interviewer asked him if he still believed that the future was basically bright, he replied yes, his optimism came from the fact that mankind has had "three billion years of experience."[54] This reply echoes many writers' naive opinion that we can expect the human race to survive any crisis simply because it always has in the past.

Writers who believe that the human race will continue to exist indefinitely have enormous faith in humanity. Their fiction reflects this attitude, constantly portraying human beings overcoming new crises--crises which, for the most part, humans have created--and rising stronger from the challenge. Indeed, if sf stories show someone failing to overcome his problems, usually the problem is too big, but the person is rarely too small. Science fiction often reflects pride in human achievement, a sense of self-congratulation, a belief that both people and society are perfectable, and faith that human beings can master the universe. In fact, Kingsley Amis remarks that this "confidence in human character and abilities, though often of a kind a headmaster says he has in you in the course of a denunciation and threat-session, can at times be almost excessive."[55]

At the same time, sf writers are well aware of humanity's faults. For example, Samuel R. Delany reminds us that "the human animal is potentially capable of any behavior. . . . Not only can the human animal behave in any way, the human psyche can approve or disapprove of any behavior."[56] In order to emphasize this fact, many sf writers contrast aliens to humans in their fiction, showing the aliens to be better both morally and technologically.

Sf writers who have faith in humanity's ability to

survive new crises and continue to progress naturally have great faith in science. They believe that as long as we nurture its growth, the progress of science and technology is open-ended and exponential. Some even believe that humans will eventually overcome many physical laws. For example, Theodore Sturgeon claims that "we'll whip the laws of gravity yet; see if we don't. We'll beat aging and weather and Einstein's "c", I'm convinced of it. It is in our nature to abrogate such laws."[57] However, one writer expresses concern about this attitude by saying that science fiction's

> least reassuring feature is an excessive reverence
> for science. No matter how often we are shown or
> told that disaster may result from unchecked
> scientific discovery, the general feeling seems to
> be that the scientific attitude will and should
> take precedence over any other kind.[58]

Thus, those writers who believe that there is no limit to how far human beings can progress view history as a continual progression forward. They still believe humans will encounter crises along the way, but they have faith that we will overcome them. This view of history is an inherent idea in nearly all science fiction, including most of the works already mentioned in this discussion.

Another view of history held by sf writers, which is less optimistic than the one previously discussed but which still claims there is no limit or end point to human history, is the idea of history as cyclical. Writers who hold this view believe that human history goes through cycles; consequently, prosperity and depression are two equally recurring phenomena. Paul A. Carter explains why sf writers are drawn to the idea of historical cyclism:

> science fiction may have found historical cyclism
> congenial by analogy with the logic of science
> itself; if planetary orbits, star formation, and
> organic life all follow rules of periodicity, why
> not human history as well? But science fiction
> also may have picked up a moral judgment from some
> of the philosophers of history . . . namely, that
> periodic social collapse and barbarian reversion
> are necessary for the rejuvenation--perhaps indeed
> for the survival--of mankind.[59]

Campbell stated in his short story, "Nightfall" (1935), that "evolution is the rise under pressure. Devolution is the gradual sinking that comes when there is no pressure--and there is no end to it." Carter responded by remarking that

> by that reasoning . . . all human progress must be self-negating; man's very success in conquering nature spells his doom, since the conquest would remove the environmental conditions--the "pressure"--which had made that success possible.[60]

Carter's point is a good one. The idea of historical cyclism is different from the belief that the future will be a continuous procession of crises because the former allows for no movement forward. Man is locked into a continuous cycle and there is no way out. With the latter view, however, the future is seen as a spiral staircase in which each crisis provides man with the opportunity to move upward.

Asimov's Foundation (1961) trilogy is one example of how sf writers use the idea of historical cyclism.[61] According to Seldon's Plan, a mathematical method of predicting the future based on the concept of historical inevitability, history is predetermined. The characters in the trilogy, therefore, are faced with a dilemma--given a predetermined outcome, should they act or not? Those few who know and understand Seldon's Plan are able to intervene in any circumstance which may upset the plan and thus prevent the predicted events from occurring. However, those who do not understand the complexity of the plan see human destiny as fixed and inevitable. At the end of the story, despite all that happens, human history is still on its course and Seldon's Plan has won out. Asimov's thinking clearly reflects the belief that human nature never changes and that, given certain circumstances, human beings will always react in similar ways.

Other science fiction writers depict the annihilation of the human race after some world disaster, often nuclear war. Usually these writers describe how the catastrophe affects the world and those who manage to survive. These post-catastrophe worlds reflect a belief that there is a limit to human progress, and they show what may happen if we go beyond that limit. In particular, these writers try to show that the human race is not indestructible, as most people seem to believe.

According to sf writers, there are several reasons why civilization might fall. In "The Iron Years" (1974), Gordon K. Dickson argues that it might happen "neither from external causes such as plague nor by the self-inflicted apocalypse of nuclear war, but simply because 'all the prosaic, predictable things'--crime, inflation, and so on--'had come to a head at once,' beyond any possibility of rational human control."[62] If this statement is true, then those writers who warn about present trends might be doing the world a service.

Another possible cause of civilization's collapse is simply a lack of "challenge-and-response."[63] When human beings are unable to involve themselves in causes greater than their individual lives, they wither, grow bored, and die. This process has nothing to do with science or technology. The breakdown of a civilization may simply be the result of a failure of the will. One might assume, then, that utopia is a prelude to total collapse, since a utopia takes away the challenge which men need in order to grow. The most obvious reason why a civilization may collapse is simply that human beings go too far with their scientific experimentation, their production of weapons, their hatred for each other, their greed, their pleasure, or how quickly they change their world.

Even when sf writers use the post-catastrophe motif they usually allow for some survivors, either because of a technical need to have someone witness the post-catastrophe world, or because they refuse to believe the entire human race will ever cease to exist. Regardless of the reason, most sf writers include at least two survivors, a man and woman as the second Adam and Eve, who escape from the world and go off in another direction to start a new race. However, sometimes the survivors are a small group of people who must try to cope with the world left to them after the disaster. In the Dispossessed (1974), a native of India describes how her society managed:

> My world, my Earth, is a ruin. There are no
> forests left on my Earth. The air is grey, the
> sky is grey, it is always hot. . . . We made a

kind of life in the ruins, on Terra, in the only way it could be done: by total centralization. . . . Total rationing, birth control, euthanasia, universal conscription into the labor force. The absolute regimentation of each life toward the goal of racial survival.[64]

In J. G. Ballard's novel, The Drowned World (1962), the few survivors live in a post-catastrophe world where nothing ever happens. They have no hope or initiative, and simply wait for the death they fear.

Some writers are even more pessimistic, describing the total extinction of the human race. Mordecai Roshwald's Level 7 (1959) is an example. In this novel, the protagonist is a "push-button officer" on Level 7, America's deepest bomb shelter. His job entails always being prepared for the command to push one or all of twelve buttons, thus releasing nuclear weapons powerful enough to destroy the entire world. Since Level 7 is 4000 feet underground and has been prepared to be self-sufficient for 500 years, it should survive a nuclear war even if the rest of the world does not. When the war occurs, everyone but those living on Level 7 die. Ironically, an accident occurs with the nuclear power plant supplying their energy, and they die too. The world becomes a radioactive graveyard.

"There Will Come Soft Rains", by Ray Bradbury, uses the same idea. In this case, however, the story is told from an omniscient point of view because no one is alive to tell the story. The author describes in detail how a computerized house goes through its daily routine, cheerily reciting its wake-up call, cooking breakfast, doing the dishes, letting the dog out, and dusting the furniture, even though no one is there. The story is an eery way of showing that although our machines may be creative inventions and useful, if we destroy ourselves what good will they be?

Finally, John Wyndham's novel, Rebirth (1955), represents another variation on the post-catastrophe motif. This story takes place 1000 years after Tribulation, the atomic disaster of the twentieth century. In a district of Labrador, inhabitants are frantically trying to obey God's command to preserve the purity of all living things by eliminating any mutations. The people believe they are in the process of climbing back into Grace and that when they do, God will restore to them the Golden Age of the Old People. Wyndham makes it clear that when this Golden Age is achieved, human beings will only destroy themselves once again.

Another theme in Rebirth is the idea that humanity must evolve into something higher in order to escape the cyclical rise and fall of history. Some of the mutants in Wyndham's novel are telepathic and are rescued by a group of telepaths from another part of the world. Thus, the novel suggests another possibility for the future of humanity.

The idea of future evolution is shared by more writers than just Wyndham. Often these writers depict man's evolution occurring as a result of a nuclear holocaust which mutates its survivors, as in the case of Rebirth. However, writers may depict man's evolution occurring in the natural course of things. Usually, this development is psychical, involving mental telepathy or other forms of extrasensory perception. The idea of further evolution reflects the belief that humanity has reached its limit in its present form.

More interesting than these writers' conjectures about how we will evolve are their reasons for suggesting our further evolution. One critic gives the reason that

> in the twentieth century . . . writers began to see that the machine technology might make slaves of everyone. Most twentieth-century writers have seen no way to get beyond the enslavement to technology, and we thus find a series of distinguished dystopias . . . that predict a dismal future for humanity. Some writers, however, have tried to get beyond this doom by postulating psychic growth or an evolutionary breakthrough to a race of superpeople. These tactics, of course, presume the possibility of a basic change in human nature; they do not so much see a way beyond technology as around it.[65]

Thus, the suggestion of man's further evolution could be considered a reflection of sf writers' lack of faith in our ability to overcome our problems.

In order to get beyond the fact that human beings are

492

emotional, stubborn, and brutal animals, some writers seek a type of evolution where we become pure spirit. Arthur C. Clarke's story, Childhood's End (1953), is a case in point. In Clarke's novel, the human race is on the point of nuclear disaster when alien spaceships arrive and intervene. As a consequence of their intervention, the human race is able to evolve into one mass, spiritual mind possessing incredible powers.

Other writers believe that the human race must evolve further if the future is not to be "an endless treadmill of eternal recurrence."[66] Still others suggest our further evolution, not because they see no way out of our present predicament, but because they believe that in the future man's brain will have more survival value than his muscle. This attitude is embodied in Theodore Sturgeon's novel, More Than Human (1953). In this story, five individuals with strange gifts discover that together they make up Homo Gestalt, the next step in man's evolution. In this novel, as well as in Childhood's End, the author depicts a psychical evolution where each person loses his individual identity to become part of a greater, more powerful whole.

Sf writers also suggest the possibility that science and technology may aid, or even speed up, humanity's development. Paul A. Carter explains that "superman . . . has arrived in such instances not by the insensate trial and error of biological evolution, but by a creative--or demonic--fusion of engineering with medicine."[67] He points to "The Six Million Dollar Man" and "The Bionic Woman" as examples, but one is also reminded of Mary Shelley's Frankenstein.

Conclusion

Sf writers use science fiction as social diagnosis and warning in order to show readers that there is a limit to how far present trends can go before they cause a major crisis. For example, sf writers remind readers that the Earth can only sustain a certain number of people--soon it will reach that limit. They also warn readers that our economic system is on the point of getting out of control because it is beginning to require that we buy and consume things we neither want or need. Some sf writers explore male/female relations, suggesting that a possible limit in this area may be androgyny. Other sf writers believe that

493

there is a point at which religion will be discarded because it will have reached its limit of usefulness. They claim that as technological advances multiply and human beings move out into space, a vision of humanity as the creator will take religion's place.

Sf writers also address the question of limit when they speculate on the fate of mankind. Dystopian writers say that we should not strive for the "ideal" society; the perfect society is undesirable because it requires total conformity. Sf writers who suggest that humanity will continually move forward believe there is no limit to how far the human race can go. This optimistic view is contrasted to the more pessimistic view that history is cyclical. Although the latter idea also suggests that there is no limit, it offers no hope for the future. Sf writers who depict post-catastrophe worlds speculate on what may happen when we reach our limit and go beyond it. Still other writers describe the future evolution of the human race, thus proposing that we have reached our limit in the form or state that we know of as homo sapiens.

Furthermore, sf writers raise numerous probing questions which challenge readers to look at their society critically and try to determine what is best for it. For example, if what we are striving for is utopia, when we get there will we be happy? Or if utopia is not what we want, what should we be striving for? Can we really continue to go forward, or is that an optimistic dream of people who have too much faith in science and humanity? Or, if history is cyclical, should we bother to act at all? And finally, since the human race is not indestructible, ought we not to be concerned about the consequences of our actions?

At this point, one may well ask, are sf writers' attitudes toward the future justified? One may argue, for example, that post-industrial America is no closer to being a repressively conformist dystopia than any other society in history, that our machines are a long way from controlling us, and that we have the ingenuity to overcome our social problems and prevent our own destruction. On the other hand, the dangers of over-population, decadent capitalism, advertising, and nuclear war are well-known. Furthermore, concerned individuals are already beginning to warn against using the computer in government.

Many people claim that post-industrialism will result in more leisure time, allowing people to engage in self-actualizing activities. However, some sf writers may be only too right when they warn that the pace of life is speeding up instead of slowing down, and that people feel more alienated than they have in the past. Furthermore, a leisure-loving, wealthy, post-industrial America may take away the "challenge-and-response" necessary for society's survival.

One cannot shrug off the possibility that the visions of the future suggested by sf writers might become reality. We are often too close to our society to notice its subtle changing. Fortunately, the likelihood of this occurring is minimal because society will always have reactionaries, that is, people or groups unusually concerned about the direction in which their society is moving, who will refuse to be silent about dangerous trends. Sf, as an instrument of social diagnosis and warning, performs a reactionary function for its readers by addressing the question of limit. As we have seen, many sf writers create pictures of future worlds in which the idea of limit plays a vital role. Thus, these writers help to illustrate this idea and emphasize that its possible ramifications must be considered seriously in today's world in order to give sf readers, at least, a better perspective of the future.

End Notes

[1]Alan E. Nourse, "Man's Adaptation to Change," in
Science Fiction Today and Tomorrow, ed. Reginald Bretnor
(New York: Harper and Row, 1974), p. 80.

[2]Ibid., p. 118.

[3]Paul A. Carter, The Creation of Tomorrow (New York:
Columbia University Press, 1977), p. 270.

[4]Frank Herbert, "Introduction," in Saving Worlds,
eds. Roger Elwood and Virginia Kidd, 1973, quoted in The
Creation of Tomorrow, p. 271.

[5]Frank Herbert, "Science Fiction and a World in
Crisis," in Science Fiction Today and Tomorrow, p. 80.

[6]Nourse, p. 132.

[7]David Ketterer, "The Apocalyptic Imagination,
Science Fiction, and American Literature," in Science
Fiction, ed. Mark Rose (Englewood Cliffs, N.J.:
Prentice-Hall, Inc., 1956), p. 154-5.

[8]Kingsley Amis, New Maps of Hell (New York: Harcourt
Brace and World, 1960), p. 87.

[9]Theodore Sturgeon, "Future Writers in a Future
World," in The Craft of Science Fiction, ed. Reginald
Bretnor (New York: Harper and Row, 1976), p. 92.

[10]Ursula K. Le Guin, "Introduction," to The Left Hand
of Darkness (New York: Ace Books, 1969).

[11]Patricia Warrick, "Images of the Man-Machine
Intelligence Relationship," in Many Futures, Many Worlds,
ed. Thomas D. Clareson (Kent, Ohio: The Kent State
University Press, 1977), p. 220.

[12]Carolyn Rhodes, "Tyranny by Computer," in Many
Futures, Many Worlds, p. 66.

[13]Throughout this discussion, the publication date
will appear after the title of the work. However, in cases
where the publication date is unknown, a question mark will

appear instead.

[14]Roger Zelazny, "For a Breath I Tarry," in _Survival Printout_, eds. Leonard Allison, Leonard Jenkin, and Robert Perrault (New York: Vintage Books, 1973), p. 76.

[15]Warrick, p. 212.

[16]Mordecai Roshwald, _Level 7_ (New York: The New American Library, 1959), p. 112.

[17]Ibid., p. 48.

[18]Stanislaw Lem, "Robots in Science Fiction," trans. Franz Rittensteiner, in _SF: The Other Side of Realism_, ed. Thomas D. Clareson (Bowling Green, Ohio: Bowling Green University Popular Press, 1971), p. 320.

[19]Isaac Asimov, _I, Robot_ (Garden City, N.Y.: Doubleday and Company, Inc., 1950).

[20]Asimov, "Reason," in _I, Robot_, p. 63-4.

[21]Asimov, "The Evitable Conflict," in _I, Robot_, p. 218.

[22]Lem, p. 320.

[23]Ibid., p. 314.

[24]Warrick, p. 212.

[25]Carter, p. 273.

[26]Ibid., p. 221.

[27]Nourse, p. 119.

[28]_Extrapolation_: 10, May 1969, quoted in _The Creation of Tomorrow_, p. 269-270.

[29]Amis, p. 122.

[30]Bruce Franklin, "Foreword to J. G. Ballard's "The Subliminal Man,"" in _SF: The Other Side of Realism_, p. 201.

[31]Maskowitz, p. 90.

[32]Ibid.

[33]Beverly Friend, "Virgin Territory: Bonds and Boundaries of Women in Science Fiction," in Many Futures, Many Worlds, p. 149.

[34]Le Guin, "Introduction."

[35]Le Guin, The Left Hand of Darkness, p. 93-4.

[36]Maskowitz, p. 10.

[37]King, p. 241.

[38]Amis, p. 83.

[39]Ketterer, p. 154.

[40]Carter, p. 219.

[41]King, p. 257.

[42]Ibid., p. 250-1.

[43]Yevgeny Zamiatin, We, trans. Mirra Ginsburg (New York: The Viking Press, 1972), p. 41.

[44]Ibid., p. 157.

[45]Kurt Vonnegut, Player Piano (New York: Delacourte Press, 1952), p. 212.

[46]Ibid., p. 243.

[47]Ray Bradbury, Fahrenheit 451 (New York: Ballantine Books, 1952), p. 49-50.

[48]Ibid., p. 53.

[49]Ibid., p. 51.

[50]Ibid., p. 101.

[51]Warrick, p. 206.

[52]Herbert, p. 78.

[53]Maskowitz, p. 68.

[54]Carter, p. 213.

[55]Amis, p. 78.

[56]Carter, p. 254.

[57]Sturgeon, p. 99.

[58]Amis, p. 85n.

[59]Carter, p. 213.

[60]Ibid.

[61]Charles Elkins, "Asimov's Foundation Novels: Historical Materialism Distorted into Cyclical Psychohistory," in Isaac Asimov, ed. Joseph D. Olander (New York: Taplinger Publishing Co., 1977) The summary of the Foundation trilogy was derived from this essay.

[62]Carter, p. 253.

[63]Ibid., p. 225-6.

[64]Ibid., p. 199.

[65]Robert Scholes and Eric S. Rabkin, Science Fiction: History, Science, and Vision (London: Oxford University Press, 1977), p. 174.

[66]King, p. 253.

[67]Carter, p. 165.

Bibliography

Aldiss, Brian W. Billion Year Spree. New York: Schocken
    Books, 1973.

Allen, L. David. The Ballantine Teacher's Guide to Science
    Fiction. New York: Ballantine Books, 1975.

Amis, Kingsley. New Maps of Hell. New York: Harcourt
    Brace and World, 1960.

Asimov, Isaac. I, Robot. Garden City, N.J.: Doubleday and
    Company, Inc., 1950.

Bradbury, Ray. "The City." In Ray Bradbury, pp. 173-178.
    New York: Alfred A. Knopf, 1980.

Bradbury, Ray. Fahrenheit 451. New York: Ballantine
    Books, 1950.

Bradbury, Ray. "The Fire Balloons." In Ray Bradbury, pp.
    179-192. New York: Alfred A. Knopf, 1980.

Bradbury, Ray. "There Will Come Soft Rains." In Ray
    Bradbury, pp. 76-80. New York: Alfred A. Knopf,
    1980.

Bretnor, Reginald, ed. The Craft of Science Fiction. New
    York: Harper and Row, 1976.

Bretnor, Reginald, ed. Science Fiction Today and Tomorrow.
    New York: Harper and Row, 1974.

Carter, Paul A. The Creation of Tomorrow. New York:
    Columbia University Press, 1977.

Clarke, Arthur C. Childhood's End. New York: Ballantine
    Books, 1953.

Clareson, Thomas D., ed. Many Futures, Many Worlds. Kent,
    Ohio: The Kent State University Press, 1977.

Clareson, Thomas D., ed. SF: The Other Side of Realism.
    Bowling Green, Ohio: Bowling Green University Popular
    Press, 1971.

Elkins, Charles. "Asimov's Foundation Novels: Historical Materialism Distorted into Cyclical Psychohistory." In *Isaac Asimov*, pp. 97-110. Edited by Joseph D. Olander. New York: Taplinger Publishing Co., 1977.

Ellison, Harlan. "I Have No Mouth and I Must Scream." In *The Road to Science Fiction #3: From Heinlein to Here*, pp. 428-446. Edited by James Gunn. New York: New American Library, 1979.

Godwin, Tom. "The Cold Equations." In *The Road to Science Fiction #3: From Heinlein to Here*, pp. 244-270. Edited by James Gunn. New York: New American Library, 1979.

Hillegas, Mark R. *The Future as Nightmare: H. G. Wells and the Anti-Utopians*. New York: Oxford University Press, 1967.

Le Guin, Ursula K. *The Left Hand of Darkness*. New York: Ace Books, 1969.

Maskowitz, Sam. *Strange Horizons*. New York: Charles Scribner's Sons, 1976.

Pohl, Frederik. "The Census Takers." In *Nightmare Age*, pp. 39-46. Edited by Frederik Pohl. New York: Ballantine Books, 1970.

Pohl, Frederik. "The Midas Plague." In *Nightmare Age*, pp. 195-258. Edited by Frederik Pohl. New York: Ballantine Books, 1970.

Rose, Mark. *Science Fiction*. Englewood Cliffs, N.J.: Prentice-Hall, Inc., 1956.

Roshwald, Mordecai. *Level 7*. New York: The New American Library, 1959.

Scholes, Robert, and Rabkin, Eric S. *Science Fiction: History, Science, Vision*. London: Oxford University Press, 1977.

Sturgeon, Theodore. *More Than Human*. New York: Ballantine Books, 1953.

Vonnegut, Kurt. <u>Player Piano</u>. New York: Delacorte Press, 1952.

Zamiatin, Yevgeny. <u>We</u>. Translated by Mirra Ginsburg, New York: The Viking Press, 1972.   (first published in 1924)

Zelazny, Roger.   "For a Breath I Tarry."   In <u>Survival Printout</u>, pp.   68-110.   Edited by Leonard Allison, Leonard Jenkin, and Robert Perrault   New   York: Vintage Books, 1973.